Advanced Materials for Pharmaceutical Wastewater Treatment

Effluents generated from the pharmaceutical industry contain organic and inorganic contaminants that create potential threats to human health and the environment. Pharmaceuticals cannot be effectively removed by conventional wastewater treatment plants owing to the complex composition, high concentration of organic contaminants, high salinity, and biological toxicity of pharmaceutical wastewater. This book provides an overview of the production and environmental impacts of pharmaceutical compounds and their advanced treatment methods, with a focus on advanced materials used for removing pharmaceutical contaminants from wastewater.

- Provides an overview of the current state of advanced research and applications of materials for pharmaceutical wastewater treatment
- Discusses various adsorbents, photocatalysts, and electrodes, with a special focus on carbon materials
- Covers advanced material synthesis and fabrication
- Features case studies and chapters that are fully application-oriented

This book is essential reading for researchers and practitioners in materials science and engineering, environmental science and engineering, chemical engineering, and water treatment who are seeking to develop and implement advanced technologies for waste minimization and mitigation.

P.V. Nidheesh is a principal scientist at CSIR-National Environmental Engineering Research Institute, Nagpur, India.

Aydin Hassani is an associate professor in the Department of Materials Science and Nanotechnology Engineering at Near East University, Turkish Republic of Northern Cyprus.

Emerging Materials and Technologies

Series Editor: Boris I. Kharissov

The *Emerging Materials and Technologies* series is devoted to highlighting publications centered on emerging advanced materials and novel technologies. Attention is paid to those newly discovered or applied materials with potential to solve pressing societal problems and improve quality of life, corresponding to environmental protection, medicine, communications, energy, transportation, advanced manufacturing, and related areas.

The series takes into account that, under present strong demands for energy, material, and cost savings, as well as heavy contamination problems and worldwide pandemic conditions, the area of emerging materials and related scalable technologies is a highly interdisciplinary field, with the need for researchers, professionals, and academics across the spectrum of engineering and technological disciplines. The main objective of this book series is to attract more attention to these materials and technologies and invite conversation among the international R&D community.

Smart Micro- and Nanomaterials for Pharmaceutical Applications
Edited by Ajit Behera, Arpan Kumar Nayak, Ranjan K. Mohapatra, and
Ali Ahmed Rabaan

Friction Stir-Spot Welding: Metallurgical, Mechanical and Tribological Properties
Edited by Jeyaprakash Natarajan and K. Anton Savio Lewise

Phase Change Materials for Thermal Energy Management and Storage:
Fundamentals and Applications
Edited by Hafiz Muhammad Ali

Nanofluids: Fundamentals, Applications, and Challenges
Shriram S. Sonawane and Parag P. Thakur

MXenes: From Research to Emerging Applications
Edited by Subhendu Chakroborty

Biodegradable Polymers, Blends and Biocomposites: Trends and Applications
Edited by A. Arun, Kunyu Zhang, Sudhakar Muniyasamy and Rathinam Raja

Bioinspired Materials and Metamaterials: A New Look at the Materials Science
Edward Bormashenko

Computational Studies: From Molecules to Materials
Edited by Ambrish Kumar Srivastava

For more information about this series, please visit: www.routledge.com/Emerging-Materials-and-Technologies/book-series/CRCEMT

Advanced Materials for Pharmaceutical Wastewater Treatment

Edited by
P.V. Nidheesh and Aydin Hassani

CRC Press
Taylor & Francis Group
Boca Raton   London   New York

CRC Press is an imprint of the
Taylor & Francis Group, an **informa** business

Contents

Preface

The availability of clean water is essential for the survival of living organisms. With the increase in industrialization, urbanization, and human activities, there is a growing demand for clean water. However, the reliability of water sources has decreased while water pollution has increased, making it a significant environmental concern in the 21st century. To address the issue of water scarcity, water reclamation is an important strategy. Wastewater treatment has proven to be a promising solution in providing sustainable clean water to meet the increasing demand. In recent years, numerous treatment technologies and materials have been studied and implemented for wastewater treatment. Significant progress has been made in this field, and a new book titled *Advanced Materials for Pharmaceutical Wastewater Treatment* aims to compile the latest advancements made in the development of advanced materials for the applications of wastewater treatment. The book brings together the ideas of leading researchers from around the world, offering a comprehensive overview of the principles and applications of advanced materials in pharmaceutical wastewater treatment.

Innovative and advanced materials hold immense potential for enhancing present water and wastewater treatment processes. Recently, there has been a growing focus on environmental remediation, and employing nanomaterials has emerged as a promising solution for refining wastewater treatment processes. It is crucial to gather, summarize, and present data on the synthesis methods, physiochemical properties, and performance mechanisms of different nanomaterials to eliminate pharmaceutical pollutants. This will be beneficial for researchers, academics, and industries operating in this domain. However, certain technical hurdles need to be resolved before nanomaterials can be employed on a massive scale for wastewater treatment. These include the risk of aggregation, separation difficulties, and potential negative impacts on the ecosystem and human health. Hence, this book features chapters on the use of advanced nanomaterials, adsorbents, and nanocomposites as potential catalysts for pharmaceutical wastewater remediation, along with their present status and challenges for large-scale application. This presents an excellent learning opportunity for readers engaged in the area of wastewater treatment.

This book provides a thorough analysis of advanced nanomaterials for pharmaceutical wastewater treatment, including their synthesis and application as efficient catalysts and adsorbents. The book delves into the contemporary role of nanomaterials in water treatment processes and outlines the challenges that must be overcome. It offers a comprehensive understanding of the properties and applications of the latest advanced nanomaterials for wastewater treatment processes. This resource will prove invaluable to students, researchers, academicians, and industry professionals seeking to address basic and complex issues associated with the application of nanomaterials to pharmaceutical wastewater treatment processes and technologies.

The field of water reclamation is poised for continued growth, particularly with the advancements being made in materials science. This edited book can be used as an excellent reference for senior undergraduates, postgraduates, researchers, engineers, and professors working in this field. Its goal is to provide useful information on the latest progress in applications of advanced materials for water reclamation, and it is hoped that this book will prove to be an invaluable resource for this targeted group.

Finally, we extend our utmost gratitude to all the contributors to the chapters for their timely submissions and unwavering patience in formatting and enhancing them. We are deeply obliged to the esteemed publishers of this book for including this invaluable topic in their book series and for granting us the opportunity to be a part of this book project and edit this volume.

P.V. Nidheesh
Aydin Hassani

Contributors

Abimbola Aleshinloye
Florida Southern College
Lakeland, Florida, USA

Nahid Amini-Seresht
Department of Applied Chemistry
Faculty of Chemistry
Kharazmi University
Tehran, Iran

Hoda Ansari
Polymeric Materials Research Laboratory
Chemistry Department
Faculty of Arts and Science
Eastern Mediterranean University
TR North Cyprus, Turkey

Nur Ramadhan Mohamad Azaludin
Faculty of Applied Sciences
Universiti Teknologi Mara Pahang
Pahang, Malaysia

Hamidi Abdul Aziz
School of Civil Engineering
Engineering Campus
Universiti Sains Malaysia
Pulau Pinang, Malaysia

Lívia N. Cavalcanti
Renewable Energies and Environmental
    Sustainability Research Group
Institute of Chemistry
Federal University of Rio Grande do
    Norte
Natal, Rio Grande do Norte, Brazil

Letícia G.A. Costa
Renewable Energies and Environmental
    Sustainability Research Group
Institute of Chemistry
Federal University of Rio Grande do Norte
Natal, Rio Grande do Norte, Brazil

Shaari Daud
Faculty of Applied Sciences
Universiti Teknologi Mara Pahang
Pahang, Malaysia

Elisama Vieira dos Santos
Renewable Energies and Environmental
    Sustainability Research Group
Institute of Chemistry
Federal University of Rio Grande do
    Norte
Natal, Rio Grande do Norte, Brazil

Paria Eghbali
Department of Analytical Chemistry
Faculty of Pharmacy
Girne American University
Kyrenia, TRNC, Turkey

Mustafa Gazi
Polymeric Materials Research
    Laboratory
Chemistry Department
Faculty of Arts and Science
Eastern Mediterranean University
TR North Cyprus, Turkey

Farshid Ghanbari
Research Center for Environmental
    Contaminants
Abadan University of Medical
    Sciences
Abadan, Iran

Letícia Milena Gomes da Silva
Renewable Energies and
    Environmental Sustainability
    Research Group
Institute of Chemistry
Federal University of Rio Grande do
    Norte
Natal, Rio Grande do Norte, Brazil

Amanda D. Gondim
Renewable Energies and Environmental
    Sustainability Research Group
Institute of Chemistry
Federal University of Rio Grande do Norte
Natal, Rio Grande do Norte, Brazil

Venkateshwaran Gopal
Department of Civil Engineering
Indian Institute of Technology
Hyderabad, India

Santiago José Guevara-Martínez
University of Guadalajara
Guadalajara, Mexico

Mohamed Hamdani
Faculty of Sciences
Chemical Department
Ibn Zohr University
Agadir, Morocco

Aydin Hassani
Department of Materials Science and
    Nanotechnology Engineering
Faculty of Engineering
Near East University
Nicosia, TRNC, Turkey

Zul Adlan Mohd Hir
Faculty of Applied Sciences
Universiti Teknologi Mara Pahang
Pahang, Malaysia

Nurul Izzati Izhar
Faculty of Applied Sciences
Universiti Teknologi Mara Pahang
Pahang, Malaysia

Vijayasri K.
Bhilai Mahila Mahavidyalaya
Bhilai, Chhattisgarh, India

Afzal Husain Khan
School of Civil Engineering
Engineering Campus
Universiti Sains Malaysia
Pulau Pinang, Malaysia

Kun-Yi Andrew Lin
Department of Environmental
    Engineering and Innovation
Development Center of Sustainable
    Agriculture
National Chung Hsing University
Taichung, Taiwan

Carlos A. Martínez-Huitle
Renewable Energies and
    Environmental Sustainability
    Research Group
Institute of Chemistry
Federal University of Rio Grande do
    Norte
Natal, Rio Grande do Norte, Brazil

Karina Nava-Andrade
University of Guadalajara
Guadalajara, Mexico

Ezinne Favour Ogulewe
Polymeric Materials Research
    Laboratory
Chemistry Department
Faculty of Arts and Science
Eastern Mediterranean University
TR North Cyprus, Turkey

Akeem Adeyemi Oladipo
Polymeric Materials Research
    Laboratory
Chemistry Department
Faculty of Arts and Science
Eastern Mediterranean University
TR North Cyprus, Turkey

Christian Onfray
Instituto Universitario de Investigación
  y Desarrollo Tecnológico
Universidad Tecnológica Metropolitana
Santiago, Chile

Hartini Ahmad Rafaie
Faculty of Applied Sciences
Centre of Foundation Studies
Universiti Teknologi Mara
Selangor, Malaysia

José Eudes L. Santos
Renewable Energies and Environmental
  Sustainability Research Group
Institute of Chemistry
Federal University of Rio Grande do Norte
Natal, Rio Grande do Norte, Brazil

Zahra Sayyar
Department of Chemical Engineering
University of Bonab
Bonab, Iran

Nurul Syarima Nadia Sazman
Faculty of Applied Sciences
Universiti Teknologi Mara Pahang
Pahang, Malaysia

Ambika Selvaraj
Department of Climate Change
Indian Institute of Technology
Hyderabad, India

Mohsen Sheydaei
Department of Applied Chemistry
Faculty of Chemistry
Kharazmi University
Tehran, Iran

Ajaya Kumar Singh
Govt. V.Y.T.PG Autonomous
  College
Durg, Chhattisgarh, India

Abdoulaye Thiam
Instituto Universitario de
  Investigación y Desarrollo
  Tecnológico
Universidad Tecnológica
  Metropolitana
Santiago, Chile

Fatima Ezzahra Titchou
Faculty of Sciences Ain Chock
Chemical Department
Hassan II University
Casablanca, Morocco

Ali Yaghoot-Nezhad
Department of Chemical
  Engineering
Abadan Faculty of Petroleum
  Engineering
Petroleum University of
  Technology
Abadan, Iran

Tabata Natasha Feijoó Zambrano
Renewable Energies and
  Environmental Sustainability
  Research Group
Institute of Chemistry
Federal University of Rio Grande do
  Norte
Natal, Rio Grande do Norte, Brazil

1 Pharmaceuticals
Types and Production

Zahra Sayyar and Aydin Hassani

1.1 INTRODUCTION

Medicines are a group of developing organic compounds that have helped improve our nature of life. The growth, manufacturing, and marketing of patented and generic drugs are the responsibility of the pharmaceutical industry [1]. In 2014, total drug incomes worldwide increased by 1 trillion United States dollars (USD) for the first time. An annual rate of 5.8% growth was observed in the market from 2017 to 2020. In 2017, the universal drug market income was 1143 billion USD and reached 1462 billion USD in 2021 [2]. Furthermore, pharmaceuticals are ingredients that interact with the body systems of living organisms through chemical processes. These mutual effects usually happen when the ingredient binds to control molecules and activates or inhibits usual body processes [3, 4]. These ingredients can be used to exert a valuable healing influence, preventing and treating diseases and reducing toxic side effects in the patient. The study of drugs and their effects on life processes is known as pharmacology. In addition, toxicology is a part of pharmacology that studies the adverse effects of chemicals on active organisms, from particular cells to the environment, as seen in Figure 1.1 [5].

There are two great areas for the arrangement of chemical ingredients. The first area is related to medical pharmacology and toxicology, which is classified into two groups, that is, pharmacokinetics and pharmacodynamics, to understand the actions of pharmaceuticals as chemicals on living systems, particularly in humans and domestic animals [6, 7]. The absorption, distribution, and elimination of drugs are in the field of pharmacokinetics, and pharmacodynamics deals with the effects of the chemical ingredients on individual animals, plants, or single-celled life forms. The next area is dedicated to the toxicology of the environment, which is related to the chemical ingredients' effects on all organisms [7, 8]. The type, matter, and physical features of drugs causing interactions with biological systems, as well as the development and production of novel medicines, are discussed in this chapter.

1.2 THE HISTORY OF PHARMACOLOGY

Prehistoric people treated their pains or diseases with materials derived from plants, animals, and minerals. Undoubtedly, they knew the beneficial or toxic effects of these materials; however, some of these materials were valueless or harmful [3, 9]. The study of pharmaceuticals is less than 150 years old, and after that, dependence

DOI: 10.1201/9781003340164-1

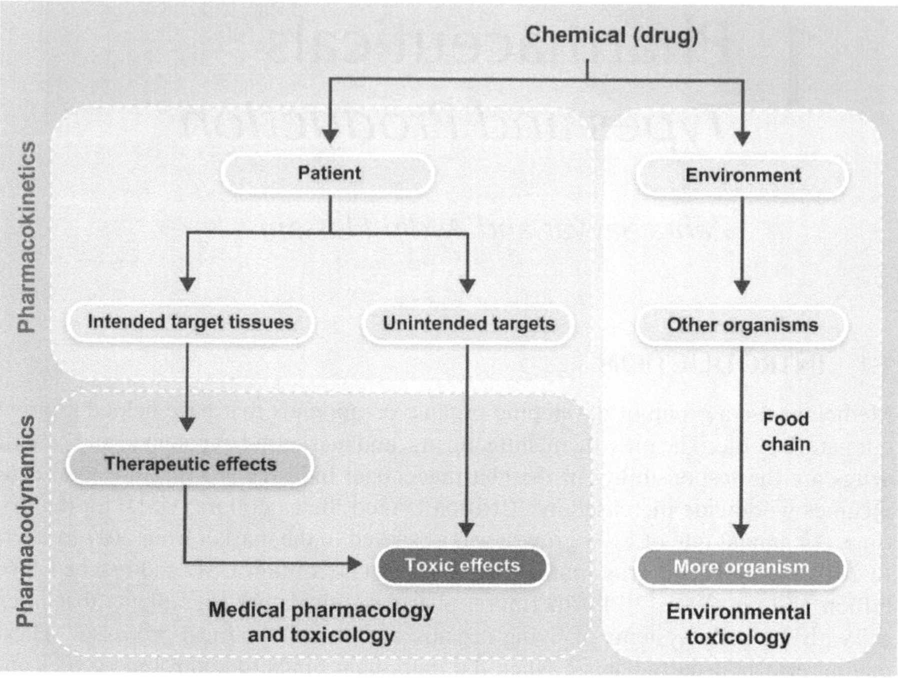

FIGURE 1.1   Main parts of study in the pharmaceutical field.

Source: Reproduced with modification from Ref. [5] with permission from McGraw-Hill Education.

on the observation and investigation of physiology was replaced with theory and clinical medicine [3, 7].

Thus, drug production science began in three phases and has grown as a forerunner in pharmacology. In the first phase, diseases and pains were treated using noxious plant and animal preparations. In the second phase, the experience enabled people to understand which substances were beneficial for relieving particular disease symptoms. The end phase was the scientific stage, in which advances in chemistry and physiology were observed and led to new knowledge of pharmacology [10, 11].

1.3   THE PHYSICAL NATURE OF DRUGS

Drugs can be defined as materials that cause a revolution in biological performance through their chemical actions. Mostly, agonist (activator) or antagonist (inhibitor) molecules of a drug interact with a receptor (target molecule) that shows a key part in the biological structure [11, 12]. Sometimes, different types of drugs as chemical antagonists can react with each other or water molecules [11]. Therefore, a suitable size, shape, electrical charge, and atomic composition are necessary for drug molecules to have the correct chemical interaction with their receptor [13]. Moreover, a useful medicine needs to have the required features to be transferred from the use place to the action place because a drug is used at a site far from the desired place of operation, for example, an oral tablet for headache relief. Medicine can have various forms, including solid, liquid, or gaseous at room temperature, which can affect their chemical actions. The size, shape, and charge of a drug molecule should display valuable features to avoid its binding to other receptors. Another parameter that affects the reaction of drugs is the size of the drug. However, the drug's molecular size varies from small to big. For instance, drugs with large molecular sizes such as proteins should often be used directly at the site of action because of their fast and high activation [11, 14].

Another important parameter is the drug's molecular shape. It needs to allow the connection to its recipient place through bonds. Hence, the shape of the drug molecule is supplementary to that of the receptor place like a key and a lock. Finally, the rational design of drugs is the most important part of drug production and displays the capability to predict the correct drug molecular structure based on available data about its biological recipient. However, there was no recipient to design such a drug with adequate detail. Drug design is a process of discovering newer active, potent, and safe molecules [15]. A small number of drugs that are now in use have been advanced via the molecular design on the basis of the three-dimensional (3D) structure information of the recipient place [16].

1.4 DRUG GROUPS AND NEW DRUG DEVELOPMENT

Nowadays, numerous drugs can be classified into approximately 70 groups. Pharmacodynamic functions and pharmacokinetic properties of various drugs in each group are similar. The significant properties of each group can be recognized by one or two prototype drugs [13]. A new drug needs the detection of its target such as the pathophysiologic development or disease type. The public sectors such as universities and research institutes are responsible for discovering this type of drug. However, to produce new drugs, industrial laboratories are usually used because of expensive chemical, pharmacological, and toxicological research to optimize a class of new drugs [17–19].

Initially, a novel drug target is often identified or a possible new pharmacological molecule is found. Second, the drug compound is extracted or synthesized from a natural source. Finally, medicines begin to be produced or developed. After that, interactions of the drug's molecules with its biological targets are investigated (Figure 1.2) [5].

The following approaches are considered to discover or develop a new drug:

1. Biologic activity of natural products, groups of earlier exposed chemical units, or huge collections of peptides, nucleic acids, and other organic molecules are screened (the activity and selectivity of the drug).
2. An identified active molecule is modified chemically.
3. A new drug target is recognized.
4. A new molecule is rationally designed on the basis of the knowledge of biological mechanisms and drug-recipient structure.

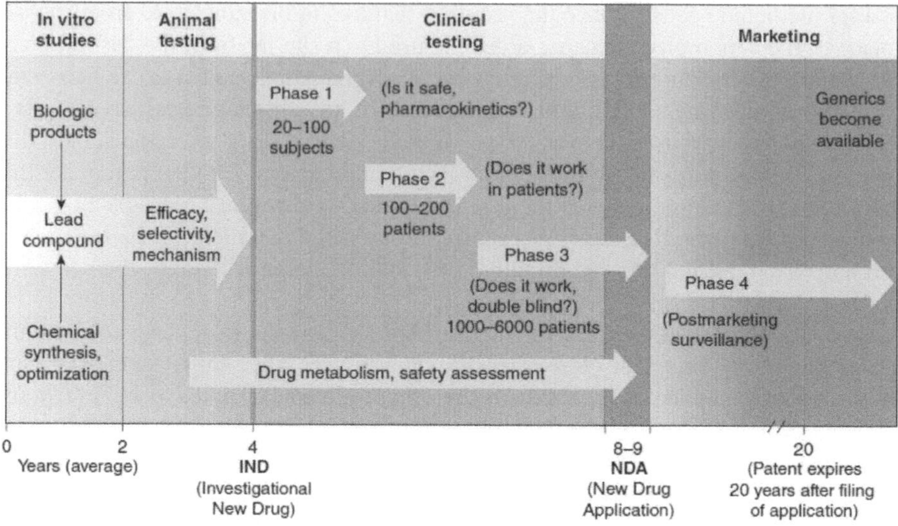

FIGURE 1.2   An analysis of necessary processes for a drug to be marketed.

Source: Reproduced from Ref. [5] with permission from McGraw-Hill Education.

Drug screening includes a variety of examinations at the molecular, cellular, tissue structure, and whole-animal levels to describe the pharmacologic profile. The type and the number of initial screening tests rely on the pharmacologic and beneficial aim [20–22].

1.5   PRECLINICAL SAFETY AND TOXICITY TESTING

Most of the chemical drugs with specific doses are poisonous. Before and during clinical analysis, the potential risks of drugs are investigated. Some preclinical toxicity analyses of drugs are presented in Table 1.1. Classifying potential human poisonousness, planning analyses to provide additional descriptions of the toxic mechanisms, and forecasting the greatest significant poisonousness to be observed in the clinical experiments are the main goals of preclinical poisonousness studies [5].

It is necessary to identify the limitations of preclinical analysis, for example:

1. Poisonousness analysis is time consuming and expensive because it takes several years to gather and examine information on toxicity before it is tested on humans.
2. Some animals are required to validate the preclinical information.
3. The extrapolation of poisonousness information from animals to humans is predictable for many but not for all cases.
4. For statistical reasons, unusual side effects are unlikely to be identified in the preclinical analysis.

TABLE 1.1
Safety Tests

Type of Test	Approach and Goals
Acute toxicity	Usually, two species are tested through two routes. The aim is to determine the no-effect dose and the maximum tolerated dose. In some cases, the acute dose that is lethal in about 50% of animals is also determined.
Subacute or subchronic toxicity	For clinical trials that are expected to last between two weeks and three months, three doses are tested on two species. The longer the expected duration of clinical use, the longer the subacute test. Biochemical and physiological effects are assessed.
Chronic toxicity	A rodent and at least one nonrodent species are tested for a minimum of six months. This test is required when the drug is intended for prolonged human use. It is usually conducted in parallel with clinical trials. The same endpoints as subacute toxicity tests are determined.
Effect on reproductive performance	Two species, typically one rodent and one rabbit, are tested to assess the effects on animal mating behavior, reproduction, parturition, progeny, birth defects, and postnatal development.
Carcinogenic potential	Two years, two species. This test is required when the drug is intended for prolonged human use. Gross and histologic pathology is determined.
Mutagenic potential	The drug's effects on genetic stability and mutations in bacteria (Ames test) or mammalian cells in culture, dominant lethal test, and clastogenicity in mice are tested.

Source: [5]

1.6 THE DRUG CYCLE

The drug cycle starts when a medication is ingested, applied, or injected. First, it moves into the bloodstream. The four key phases of this cycle are absorption, distribution, metabolism or biotransformation, and excretion [23].

1. **Absorption:** the process by which a substance transfers from its entered location into the bloodstream. Entirely and parenterally are two ways to administer the drug. The drug is administered directly into the gastrointestinal system orally, rectally, or through a tube entering, or through all other routes that do not contact the gastrointestinal system.
2. **Distribution:** after the drug has been absorbed into the bloodstream, distribution or the delivery of a drug to the target spot occurs. When a drug enters the bloodstream, it is rapidly distributed in all of the body. First, the brain, heart, kidneys, and lungs, as well-perfused organs, obtain the main stream of the drug. Then, skin, muscles, and fat, as the lesser perfused tissues, receive a small amount of the drug at a slower rate.
3. **Metabolism** or **biotransformation:** this takes place when the medications are chemically converted to a less active or inactive compound, which is

called a metabolite. The body does not typically need medication as an external material. Metabolism is essential to disrupt this external material into a compound that no longer impacts the body and can be successfully eliminated.

4. **Excretion:** the last and essential phase is extraction due to the lack of the accumulation and removal of the drugs or their transformed products from the body. If the accumulation of drugs happens, there is a possibility of causing other diseases [24–27].

1.7   CLASSIFICATION OF DRUGS

Natural-origin plants, animals, minerals, microbial-origin chemical synthesis, radioactive substances, and toxins are natural substances that are used to produce drugs by extracting active components from them, though most drugs are produced in laboratories. Therefore, researchers are examining natural resources such as plants, animals, marine animals, and microbes that are promising for finding new sources of medications [23].

Several drugs are prepared by combining natural with chemical products. For example, natural opium in the form of codeine is combined with acetaminophen to create hydrocodone. Today, the active ingredients of plants are extracted to produce drugs. For instance, digoxin (Lanoxin) is derived from the foxglove plant and it is used to treat heart failure. Also, procaine and aspirin are extracted from the coca plant and the bark of the white willow tree, respectively [28–30].

Furthermore, domesticated animals can also be utilized to produce various types of medication. For instance, sheep can provide lanolin, which is a skin medication extracted from their wool. Cows and pigs are excellent sources of hormones that can be used as replacements. Additionally, minerals such as calcium, iron, zinc, magnesium, copper, and selenium can be extracted from mineral materials and used as necessary supplement drugs. To treat certain diseases, minerals may be applied. For instance, gold, iodine, and magnesium sulfate are applied in the treatment of arthritis, goiter, and constipation, respectively [23, 31, 32].

Additionally, the biological activity of pharmaceuticals and the purpose of their specific uses, such as antibiotics to treat bacterial infections, analgesics to lessen pain, and antineoplastics to treat cancer, may be used to classify drugs [33–35]. Generally, medicines are divided into six groups in terms of the desirable effects of a drug on the body, which are presented in Table 1.2 [23]. Typically, a drug, such as a tablet or a liquid, contains one or more active pharmaceutical ingredients (APIs) and an inactive substance, including excipients, additives, mineral salts, and other organic chemicals such as sugars, flavors, pigments, and colors [36]. The main ways of consuming medicinal products are displayed in Table 1.2 along with general categories of the main groups of traditional medicinal forms. Essential time for onset of action or drug targeting issues is a parameter to select a suitable method of administration for a special bioactive. Likewise, properties of the bioactive material such as solubility and stability are parameters to choose drug delivery systems [32].

TABLE 1.2
Drug Categories

Category	Main Action	Example
Curative	Cures or treats a problem	• Penicillin to treat strep throat
Prophylactic	Prevents a problem	• Cefazolin (Ancef, Kefzol) to prevent infections from surgery • Vaccine to prevent measles, mumps, and rubella
Diagnostic	Helps diagnose a disease or condition	• Diatrizoate meglumine and diatrizoate sodium (Gastrografin) • Barium sulfate (Gastrografin and barium sulfate are used for computed tomography scans)
Palliative	Treats symptoms to make the patient more comfortable	• Morphine to relieve the pain of cancer • Oxygen to make breathing more comfortable
Replacement	Replaces a missing substance	• Levothyroxine • Natural thyroid to treat hypothyroidism
Destructive	Destroys tumors and/or microbes	• Carbimazole inhibit the production of thyroid hormone to treat hyperthyroidism

Source: [23]

1.7.1 PRINCIPLES OF DOSAGE FORM DESIGN

The essential keys in formulating drugs are efficacy, safety, and quality of health products. Quality by design (QbD) is used to design and improve the method in the pharmaceutical industry. This is the presentation of a logical method to advance a design, formulate drugs, and control processes [37].

Drugs are prescribed as formulated preparations or medicines, and they are not recommended as pure chemical materials. Therefore, these drugs can be in simple forms or the intricate delivery of drug organizations via the use of suitable additives or excipients in the formulations [27].

The main goal of the amount formula strategy can be to obtain an expectable healing reaction to a drug in the preparation that is manufactured on a grand scale with the quality of the repeatable product. Several properties are necessary to confirm product characteristics. Some of these properties are the chemical and physical stability of drugs, such as high protection against microbial pollution [27, 37]. Several dosage forms of medicinal material can be included for the suitable and effective action of illness. The design of pharmaceutical formulation for consumption is essential to maximize therapeutic response using a variety of delivery methods. Lists of the range of dosage forms to deliver drugs using different ways are shown in Figure 1.3 [27].

Therefore, it is essential that numerous parameters are noticed to formulate a suitable dosage form. They can be classified into three groups.

1. **Bio-pharmaceutical attentions:** including effective parameters on the drug absorption from diverse administration ways.

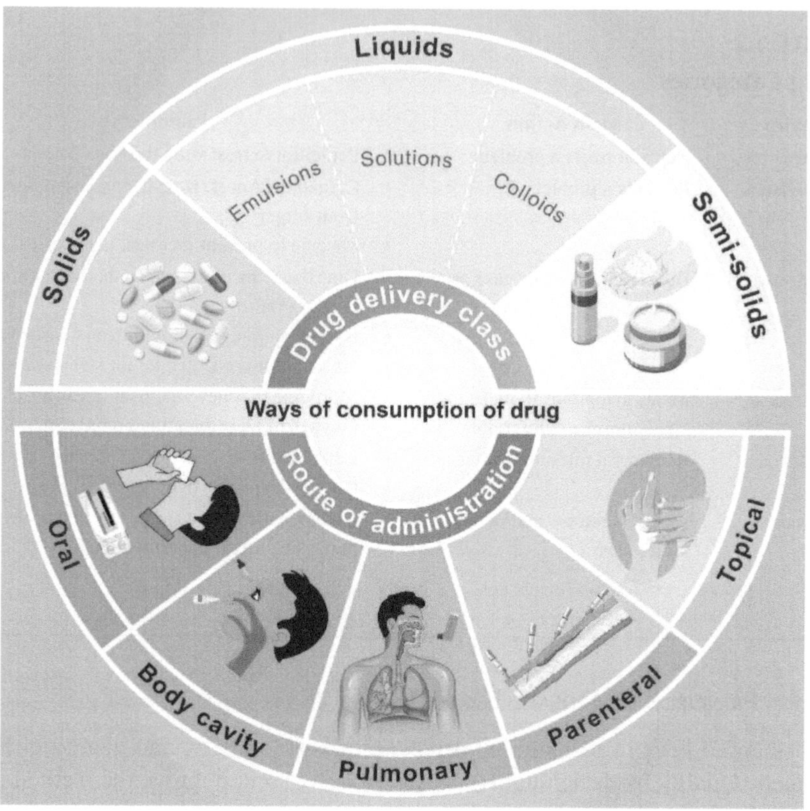

FIGURE 1.3 Methods of consumption of drug and drug delivery classes.

2. **Drug parameters:** they contain parameters such as the physical and chemical drug features.
3. **Healing attentions:** including attention to the clinical signs for treatment and patient parameters [27].

1.8 BIO-PHARMACEUTICAL FEATURES OF DOSAGE FORM DESIGN

The connection among the physico-chemical and biological sciences used for pharmaceuticals, amount formulas, and drug action can be found in bio-pharmaceutics.

The basics of this topic are vital in amount formula design, especially with respect to pharmaceutical absorption, as well as its distribution, metabolism, and excretion.

Overall, the drug in a solution may be absorbed through membranes and epithelia of the skin, gastrointestinal tract, and lungs [27].

There are two ways to absorb drugs, that is, passive diffusion and carrier-mediated transport mechanisms. In the first way, passive diffusion, drug molecules pass from high-concentration regions to low-concentration regions, and the concentration

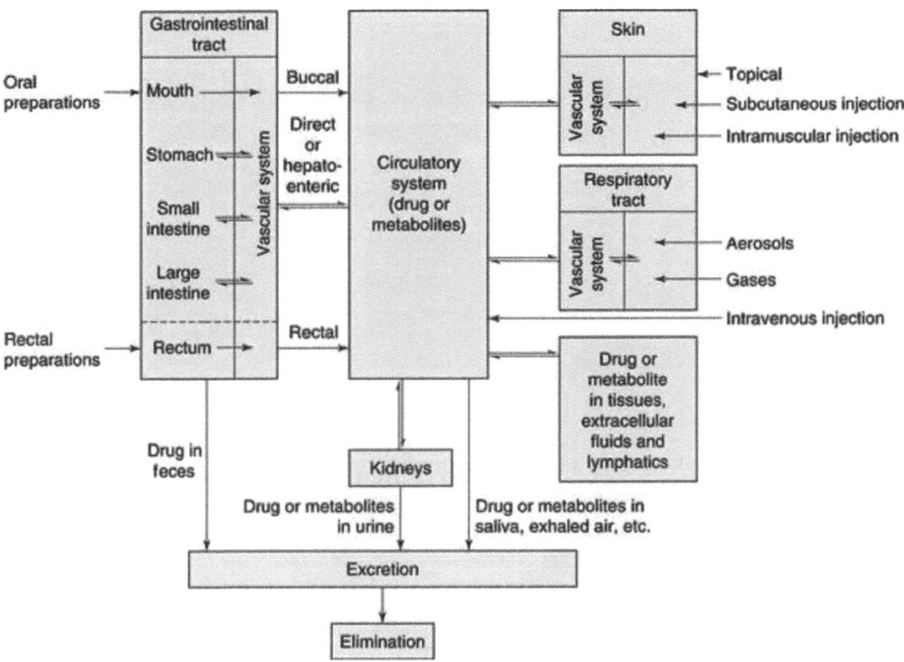

FIGURE 1.4 Pathways of a drug dosage formed by diverse routes.

Source: Reproduced from Ref. [27] with permission from Churchill Livingstone/Elsevier.

gradient across the cell barrier controls the absorption of numerous drugs [27, 38]. Current research on carrier-mediated transport mechanisms has used many data and facts to design and control new drug molecules. Numerous particular transport mechanisms have been hypothesized to account for the active and assisted vehicles. In a second way, the pharmaceuticals need to be transported in body fluids. Figure 1.4 shows the pathways of transportation of drugs throughout the body. The pattern of drug absorption changes significantly between each drug material and also between administration methods in diverse ways. Pharmaceutical formulations are designed in such a way that they prepare the drug in an appropriate formula for absorption by any chosen route of administration [27].

1.9 DRUG PARAMETERS TO DESIGN

The incorporation of physico-chemical, mechanical, and bio-pharmaceutical features of a selected drug plays a key role to improve and develop a bioavailable drug that has a favorable therapeutic effect. The selection of dosage form depends on the measurement of physico-chemical and mechanical features. Additionally, these measurements create an understanding of the ability to process and store drugs to confirm the desired quality of pharmaceutical production. The essential physico-chemical, mechanical, and bio-pharmaceutical features to design the formulation of drugs are displayed in Figure 1.5 [37].

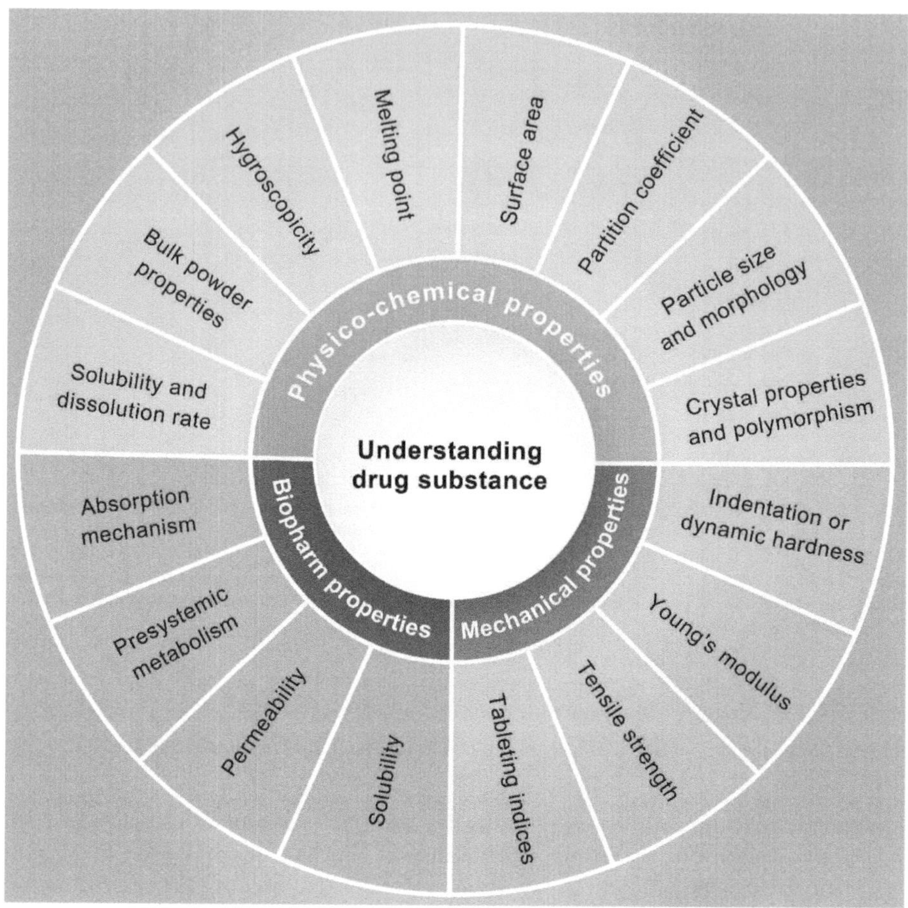

FIGURE 1.5 Properties of drug substances.

Source: Reproduced with modification from Ref. [37] with permission from John Wiley & Sons.

To obtain a stable and valuable drug, the physical and chemical features of any required dosage form of drug materials should be studied carefully. Some features such as dissolution, crystal size, polymorphic form, solid-state stability, and drug–additive interaction can have insightful impacts on the physiological accessibility and physical and chemical stability of the drug. This information and its combination with pharmacological and biochemical knowledge may be useful for designing a selected amount of drug to prepare the most suitable drug form and additives.

While a complete evaluation of properties is not essential for all types of formulation sorts, those features identified as significant in designing and processing the dosage form, potential stresses occurring during processes, and the techniques involved are shown in Table 1.3.

TABLE 1.3

Features of Drugs, Processing Stresses, and Manufacturing Procedures

Features	Processing Stresses	Manufacturing Procedures
Particle size	Pressure	Precipitation
Surface area	Mechanical	Filtration
Particle surface	Radiation	Emulsification
Chemistry	Exposure to liquids	Milling
Solubility	Exposure to gases, and liquid vapors	Mixing
Dissolution	Temperature	Drying
Partition coefficient		Granulation
Ionization constant		Compaction
Crystal properties		Autoclaving
Polymorphism		Crystallization
Stability		Handling
Organoleptic qualities		Storage
Molecular weight		Transport

Source: [27]

One of the important properties is particle size reduction, which increases the specific surface area (i.e., surface area per unit weight) of powders. Particle size, the distribution of particle size, and particle interaction with solid surfaces play an important role in drug dissolution and absorption rates, content uniformity in dosage forms, and stability. In many cases, the suitable physico-chemical characteristics for both drugs and additives are obtained by the reduction of particle size [27, 37].

1.10 PHARMACEUTICAL DEVELOPMENT

Overall, the growth of medicinal production differs from that of other industries with intensive investigation. In the industry of medicine, in particular, a constant requirement confirms the timely production and delivery of clinical supplies regardless of the current state of process efficiency or development. In addition, process growth, optimization, and scale-up have historically tended toward an iterative method. Clinical supply requests have been fulfilled using scaling of the model factory through phase I, phase II, and phase III. The medicinal industry devoted around $60 billion to R&D in 2007. Nowadays, tens of years are needed for drugs to reach the pharmacy from the laboratory. The distinctive growth activity timeline from detection to the introduction and a universal range of volunteers, and increasing the number of volunteers for developing clinical through phases I to III is displayed in Figure 1.6. Among many compounds that have been evaluated for possible healing efficacy, few of them overcome the safety, efficacy, and clinical hurdles that need to be confirmed [37, 39].

The cycle of a pharmaceutical from design to treatment and some chosen facts related to sustainability are exhibited in Figure 1.7 [34, 41].

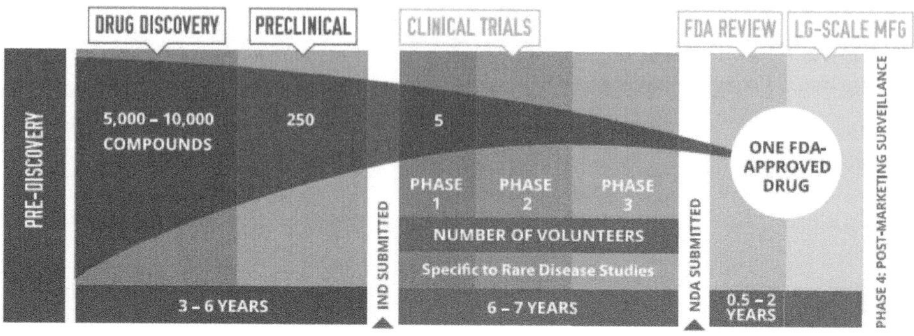

FIGURE 1.6   Drug research and development.

Source: Reproduced from Ref. [40] with permission from John Wiley & Sons.

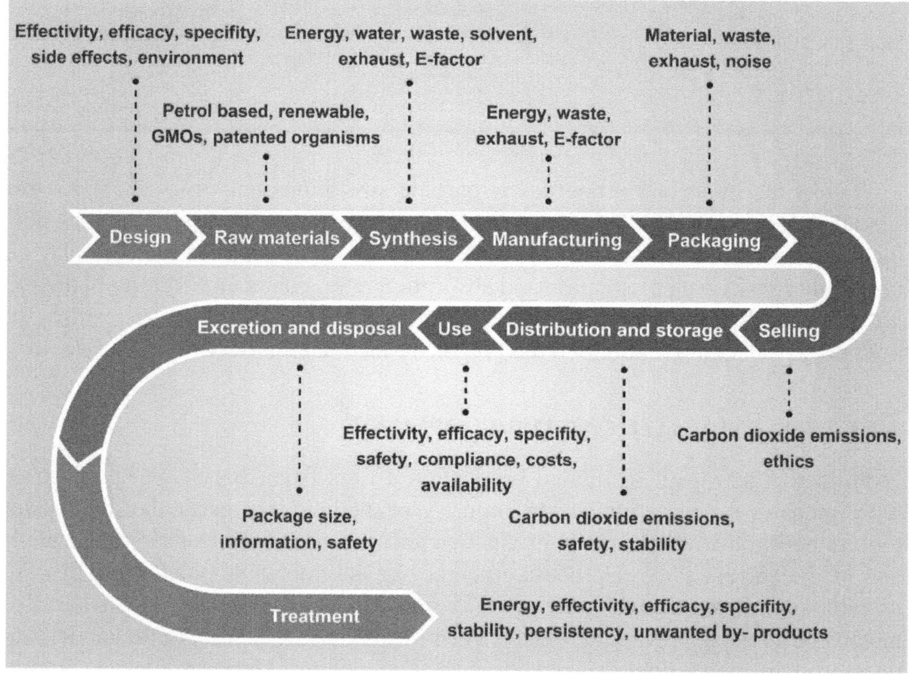

FIGURE 1.7   The life cycle of a pharmaceutical and some selected points that are relevant for sustainability.

Source: Reproduced with modification from Ref. [34] with permission from Springer.

1.11   PRESENT CHALLENGES AND CHANCES IN THE INDUSTRY OF MEDICINE

The existence of the pharmaceutical industry is based on the main challenge of enhancing the condition of human's lifetime using high product quality, strategies, decisions, and actions. One of the important challenges is modifications in the environment that have effects on people's health. Likewise, economic, social, and technical services are alternating commercial models of the industry. The forces challenging the industry are classified into four major types, including raised prices and dangers, revenue/price constraints, the globalization of activities, and the increasing complexity of medical knowledge. For each product from discovery to market, pharmaceutical companies naturally devote several thousand compounds to the detection of drugs, hundreds of compounds in clinical trials, and numerous failed medical tests during 9–15 years [39].

1.12   CHEMICAL ENGINEERING IN THE MEDICAL INDUSTRY

The challenges facing the industry of medicines generate numerous occasions for various vocations involving chemical engineering. Decreasing prices is directly related to technical principles of engineering that look for economies of scale and the use of effective knowledge.

The intricacies of pharmaceutical knowledge and the limitations of methods that must be appropriate for worldwide usage are intertwined with the use of chemical engineering equipment, technical revolution, and engineering analysis to solve cost problems and the active value of production. Chemical engineers can also apply modeling and physico-chemical property approximation, for example, thermodynamic solubility modeling and computational fluid dynamics (CFD). Therefore, they can maximize efficiency and minimize energy consumption and unwanted production related to suitable crops by novel technology. Also, chemical engineers may improve product value and manage strategic technologies [39].

1.12.1   SUSTAINABILITY ISSUES FOR THE PHARMACEUTICAL INDUSTRY

The United Nations have recognized sustainability since the 1980s as a way to examine issues and problems with the environment. In the industry of medicine, evaluating the sustainability of pharmaceutical manners may be an intricate assignment like multivariate optimization. A triple-bottom method for sustainability, with some sustainability occasions for the industry of medicine, is shown in Figure 1.8.

The design of drug production and methods that notably remove or decrease the usage and generation of dangerous or toxic materials or solvents and the avoidance or decline of environmental protection and health influences can be defined as a green pharmacy. Green pharmacy emphasizes the drug itself, which only contains aspects of the environment [42].

For instance, novel drugs may be defined as green pharmaceuticals because of the quality and quantity of waste created during their synthesis and the use of renewable raw materials; however, the drug may accumulate in the environment after elimination and make environmental or health problems [34]. After the publication of the basics of green chemistry, a primary effort to understand the principles of green engineering was made by publishing 12 suggested principles of green engineering. According to pharmaceutical processes, there are attempts to incorporate green chemistry and green engineering basics. In 2005, the American Chemical Society

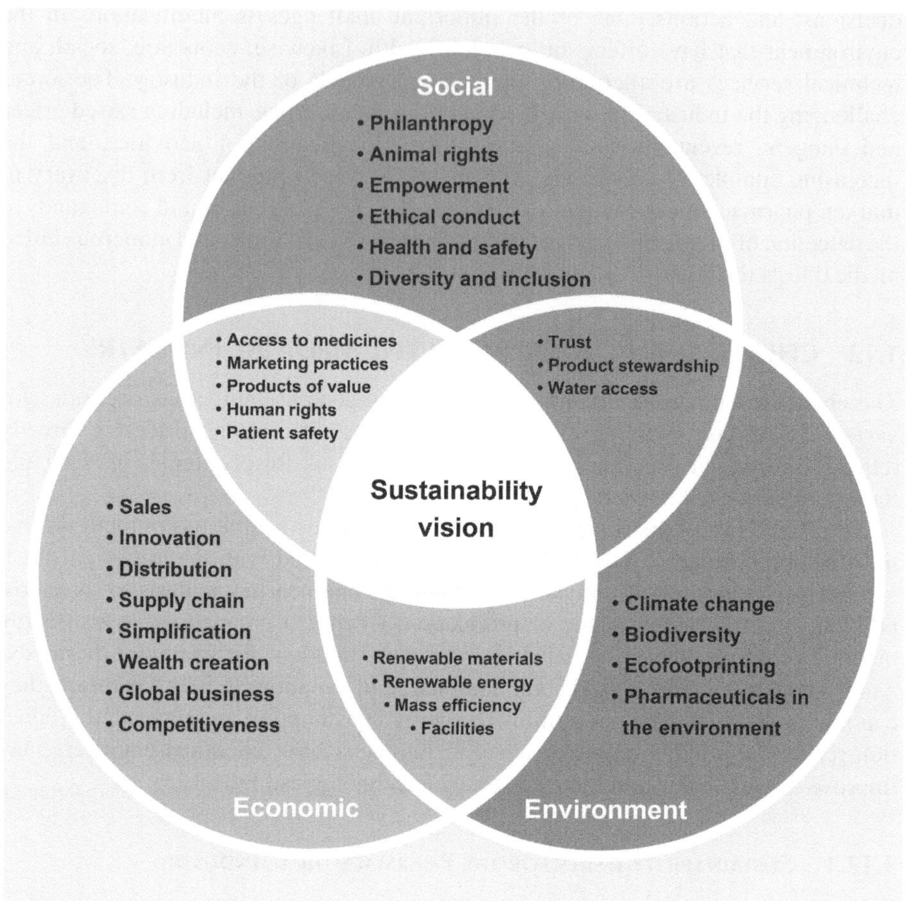

FIGURE 1.8   A triple-bottom method for sustainability in the pharmaceutical industry.

Source: Reproduced with modification from Ref. [40] with permission from John Wiley & Sons.

(ACS), the Green Chemistry Institute (GCI), and several main medical companies gathered together to organize the ACS GCI Pharmaceutical Roundtable (ACS GCI PR). The roundtable's strategic priorities included informing and influencing the industry's green processing research needs, classifying required originations, educating pharmaceutical leaders as well as others about the benefits of this method, and providing green processing knowledge to global pharmaceutical operations. Therefore, researchers designing sustainable pharmaceutical processes need to use these three generic basics to process growth approaches, for instance by using processes that are better, cheaper, faster, cleaner, and more sustainable by design such as optimizing the use of material and energy sources and minimizing environmental hazards and life cycle impacts [40].

1.12.2 CHEMICAL ENGINEERS DESIGNING SUSTAINABLE MEDICAL PROCESSES

Maximizing the usage of substances and energy sources is one of the key challenges for chemical engineers to design sustainable medical proceedings introduced by the ACS GCI PR. Therefore, it has selected process mass intensity (PMI) as a quantifier to improve efficiency in pharmaceutical syntheses. The members of the ACS GCI PR have applied this metric mass intensity to the common process (the total mass of material per product mass) to fairly compare the data of each company [43, 44]. Minimizing health and safety risks is necessary to ensure the sustainability of pharmaceuticals, and its importance has largely been combined into the work of evolving pharmaceutical processes. The challenge for chemical engineers will be to continue to develop chemical engineering in the pharmaceutical industry because it is used by the pharmaceutical industry to design more stable processes. The purpose of chemical engineering is to decrease both costs and environmental influence, that is, both resource consumption and waste production, while improving the community benefits [45, 46].

1.13 PRODUCTION OF PHARMACEUTICALS

The growth of fermentation knowledge has been mentioned as an accomplishment by chemical engineers that have had an insightful influence on process knowledge and medicine. For example, in the 1930s, after the detection of penicillin, deep-tank fermentation was used instead of surface culture by chemical engineers using the application of chemical engineering principles that developed in the 1930s and 1940s. Penicillin would remain in laboratory research for an indefinite period without using this knowledge. This technology was applied to the fermentation processes of all later antibiotics. Furthermore, continuous filtration and extraction were used to achieve large volumes of fermentation broth [47, 48].

1.13.1 BATCH OR CONTINUOUS PROCESSES

In the past, batch and semi-batch operations were commonly used in many processes due to their high efficiency and ease of start-up and shutdown [49]. Academic research in chemical engineering in the 1940s focused on continuous processing, which played a significant role in the isolation of penicillin. As a result, chemical engineers improved continuous processes in pharmaceutical industries and successfully implemented them in various process applications. Chemical engineers play a key role in developing and constructing safe and efficient process applications with minimal investment and operating costs [50]. The continuous process is crucial in explaining scale-up issues, such as reactor selection and design, separation methods, and crystallization processes. General reactors, including continuous stirred-tank reactors (CSTRs), packed beds, fluid beds, and trickle beds, are widely used in the chemical and medical industries [51]. However, improved protection, enhanced features, decreased energy prices, and more cycle efficiencies can be some advantages of organic synthesis reactions through a continuous mode. Therefore, these advantages cause smaller active reaction volumes and higher mass and heat transfer. Lately, pharmaceutical industries concentrate on continuous operations such as reactions, downstream extractions, solvent exchanges, and crystallizations. One of the greatest advantages of continuous processes is process safety within the pharmaceutical industry; this facilitates enhanced protection and transfer phenomena to reduce the inventory of reactive species. Moreover, continuous reactors are often used at higher temperatures compared to batch reactors, resulting in high conversion rates [40]. Additionally, increasing the heat and mass transmission ability of continuous processes can reduce the impurity of reactions because of operating conditions that can be controlled more consistently compared with batch reactors. As a result, minimizing reaction impurity cause an improvement in the quality of production and increase yield. Therefore, downstream workup and crystallization have been optimized to improve yield and control reaction conditions [52]. In many chemical reactions, substances need to be added over a long period. This leads to significant variation in the exposure time of the starting material, substrate, or product molecules to the reaction conditions, which can affect the quality of the final product and the operating conditions. These longer cycle times can have an impact on both the product quality and the selection of operating conditions.

The stability of a chemical intermediate has an inverse relationship with temperature and a direct relationship with time. In other words, it increases at lower temperatures and longer reaction times. Therefore, the stability of batch reactions happens at low temperatures and long reaction times, whereas it can be converted to high temperatures and short times in continuous processes [53]. The main steps of a continuous reaction process include 1) the feed solutions, 2) the reactor, and 3) the quench.

1.13.2 Comparison of CSTRs, PFRs, and Microreactors

As established previously, continuous processes have good potential for use in pharmaceutical industries, and their potential mainly depends on the type and size of the reactor. Continuous reactors include plug flow reactors (PFRs), microreactors, and continuous stirred-tank reactors (CSTRs) [40]. In this part, some universal comparisons among PFRs, microreactors, and CSTRs were investigated and are listed in Table 1.4. According to these results, PFRs and microreactors are commonly preferred over CSTRs. PFRs are preferable to microreactors because they have a high ability to transfer heat and mass for the desired reaction [40].

1.13.3 Role of Nanotechnology

Nanotechnology is one of the greatest methods to improve the quality and properties of drugs in the pharmaceutical industry. The distinctive properties of nanomaterials (NMs) have significant advantages in comparison with conventional drug preparations. When drug-loaded NMs pass via the digestive system, drug absorption gets better because the extraordinary dispersion of the drug can cause contact of the drug with mucous membranes. Table 1.5 presents some of the features and pharmaceutical applications of NMs [54].

TABLE 1.4

Characteristics of CSTRs, PFR, and Microreactors

Reactor Mode	Multiple CSTRs	PFR	Multiple Microreactors
Handling of solids	(++)	(−)	(−−)
Gas evolution	(++)	(−)	(−−)
Slow reaction kinetics	(++)	(−)	(−−)
Quickly achieve steady state	(−)	(++)	(++)
High conversions per volume	(−)	(++)	(++)
Narrow residence time distribution	(−)	(++)	(++)
Mitigates product reacting with starting material	(−)	(++)	(++)
Initial large heat sink	(+)	(−)	(−)
Low operational complexity	(−)	(++)	(−−)
Low level of equipment intensity	(+)	(++)	(−−)
Enhanced heat transfer	(−)	(+)	(++)
Enhanced mass transfer	(−)	(+)	(++)
Low cost	(+)	(+)	(−−)

Source: [40]

(+): Positive; (−): Negative; (++): Strong positive; (−−): Strong negative

TABLE 1.5

Characteristics and Applications of Some NMs in Pharmacies

Types of NMs	Characteristics	Pharmacy Applications	Ref.
Nanoparticles	Solid nanosized particles	For increased bioavailability, targeted delivery, and controlled release	[55]
Nanocapsules	Core-and-shell nanoparticles	Protecting entrapped agents	[56]
Liposomes	Lipid vesicles with entrapped compounds in inner phase or bilayers	For increased bioavailability, targeted delivery, and controlled release	[57]
Niosomes	Nonionic surfactant vesicles with similar properties to liposomes	For increased bioavailability, targeted delivery, and controlled release	[58]
Nanoemulsions	Nanoscale emulsions	As a way to make nanoparticles	[59]
Polymeric micelles	Micelles containing amphiphilic polymers	For solubilization, targeted delivery, and controlled release	[60]
Nanogels	Nanosized hydrogels consisting of cross-linked hydrophilic polymers	Loading various compounds for controlled release or targeting	[61]

Source: [54]

1.14 POLLUTION PREVENTION AND CONTROL

To control pollution, extremely poisonous and persistent components are replaced with degradable and less poisonous ones. Suggested pollution avoidance is to control the amounts of active components to minimize waste, recover used solvents, recirculate cooling water, return toxic material packaging, etc. [2, 62]. The removal of pharmaceuticals from water and wastewater through various methods such as oxidation processes, electrochemical processes, chlorination, membranes, microalga-based technology, adsorption, etc. have been investigated in a majority of previous studies [63–68].

REFERENCES

1. Taylor D. The pharmaceutical industry and the future of drug development. R Soc Chem. 2015;41:1–294.
2. González Peña OI, López Zavala MÁ, Cabral Ruelas H. Pharmaceuticals market, consumption trends and disease incidence are not driving the pharmaceutical research on water and wastewater. Int J Environ Res Public Health. 2021;18(5):2532.
3. Brenner GM, Stevens CW. Pharmacology. Elsevier Health Sciences; 2012.
4. Xu H, Zhang Y, Wang P, et al. A comprehensive review of integrative pharmacology-based investigation: A paradigm shift in traditional Chinese medicine. Acta Pharm Sin B. 2021;11(6):1379–1399.
5. Bertram G. Basic & Clinical Pharmacology. McGraw-Hill Education; 2018.
6. Spasov AA, Iezhitsa IN, Vassiliev PM, et al. Pharmacology of Drug Stereoisomers. Springer Nature; 2022. p. 76.
7. Katzung BG, Trevor AJ. Basic & Clinical Pharmacology. McGraw Hill Medical; 2012.
8. Ashelford S, Raynsford J, Taylor V. Pathophysiology and Pharmacology in Nursing. Sage Learning Matters; 2019.
9. Libertini G, Corbi G, Conti V, et al., Evolutionary Gerontology and Geriatrics: Why and How We Age. Springer Nature; 2021. p. 2.
10. Alamgir A, Medicinal, non-medicinal, biopesticides, color-and dye-yielding plants; secondary metabolites and drug principles; significance of medicinal plants; use of medicinal plants in the systems of traditional and Complementary and Alternative Medicines (CAMs). In: Therapeutic Use of Medicinal Plants and Their Extracts. Springer; 2017. Vol. 1, p. 61–104.
11. Katzung BG, Introduction: The nature of drugs & drug development & regulation, In: Katzung BG, Vanderah TW, editors. Basic & Clinical Pharmacology. McGraw Hill; 2021.
12. Anand U, Bandyopadhyay A, Jha NK, et al. Translational aspect in peptide drug discovery and development: An emerging therapeutic candidate. BioFactors. 2022;49:251–269.
13. Katzung BG, Masters SB, Trevor AJ. Basic & Clinical Pharmacology. McGraw-Hill; 2004.
14. Luke AM, Mathew S, Altawash MM, et al. Lasers: A review with their applications in oral medicine. J Lasers Med Sci. 2019;10(4):324.
15. Islam R, Maji S, Kabiraj S, et al. Role of in silico drug design in pharmaceutical sciences. Int J Appl Sci Eng Res. 2022;10(05):2318–2367.
16. Mills J, Dean PM. Three-dimensional hydrogen-bond geometry and probability information from a crystal survey. J Comput-Aided Mol Des. 1996;10(6):607–622.
17. Heinrich M, Jalil B, Abdel-Tawab M, et al. Best practice in the chemical characterisation of extracts used in pharmacological and toxicological research—The ConPhyMP—Guidelines. Fron Pharmaco. 2022;13:953205.

18. Deore AB, Dhumane JR, Wagh R, et al. The stages of drug discovery and development process. Asian J Pharm Res Dev. 2019;7(6):62–67.
19. Nayak RK, Avorn J, Kesselheim AS. Public sector financial support for late stage discovery of new drugs in the United States: Cohort study. BMJ. 2019;367.
20. Katzung BG. Development & regulation of drugs. Basic Clin Pharmacol. 2012;69.
21. Atanasov AG, Zotchev SB, Dirsch VM, et al. Natural products in drug discovery: Advances and opportunities. Nat Rev Drug Discov. 2021;20(3):200–216.
22. Karageorgis G, Foley DJ, Laraia L, et al. Principle and design of pseudo-natural products. Nat Chem. 2020;12(3):227–235.
23. Watkins CJ. Pharmacology Clear & Simple: A Guide to Drug Classifications and Dosage Xalculations. FA Davis; 2018.
24. Bertol E, Favretto D. Pharmacokinetics: Drug Absorption, Distribution, and Elimination, in Karch's Drug Abuse Handbook. CRC Press; 2023. p. 133–178.
25. Wang K, Mechanistic Model-based Drug Oral Absorption Analysis. Thesis; 2021.
26. Moroni I. Absorption and Pharmacokinetic Studies of Melatonin using Polymer-Coated Drug Delivery Systems. Macquarie University; 2022.
27. Aulton ME, Taylor K. Aulton's Pharmaceutics: The Design and Manufacture of Medicines. Elsevier Health Sciences; 2013.
28. Inoue M, Hayashi S. Blessings of medicinal plants—History and prospects. In: Medicinal Plants. Springer; 2021. p. 771–796.
29. Patocka J, Nepovimova E, Wu W, et al. Digoxin: Pharmacology and toxicology—A review. Environ Toxicol Pharmacol. Epub. 2020;79:103400.
30. La Mesa C, Corbo A, Gkouvi A, et al. Bio-active principles from the animal and plant kingdom: A review. Adv Res Org Inorg Chem(AROIC). 2020;1:1–12.
31. Nag A. Asymmetric Synthesis of Drugs and Natural Products. CRC Press; 2018.
32. Karsa DR, Stephenson RA. Chemical aspects of drug delivery systems. Royal Society of Chem. 1996;178.
33. Kümmerer K. Pharmaceuticals in the Environment: Sources, Fate, Effects and Risks. Springer Science & Business Media; 2008.
34. Kümmerer K. Pharmaceuticals in the environment. Annu Rev Environ resour. 2010;35:57–75.
35. Safdari R, Esmaeili M, Marashi Shooshtari SS, et al. Drug classification systems: Applications and characteristics. Health Manag Inform Sci. 2021;8(3):149–158.
36. Alamgir A. Therapeutic Use of Medicinal Plants and Their Extracts. Springer; 2017. p. 1.
37. Am Ende MT. Chemical Engineering in the Pharmaceutical Industry: Drug Product Design, Development, and Modeling. John Wiley & Sons; 2019.
38. Lennernas H, Nylander S, Ungell A-L. Jejunal permeability: A comparison between the ussing chamber technique and the single-pass perfusion in humans. Pharm Res. 1997;14(5):667.
39. Am Ende MT. Chemical Engineering in the Pharmaceutical Industry: Active Pharmaceutical Ingredients. John Wiley & Sons; 2019.
40. Am Ende MT. Chemical Engineering in the Pharmaceutical Industry: R&D to Manufacturing. John Wiley & Sons; 2011.
41. Levin M. Pharmaceutical Process Scale-Up. CRC Press; 2001.
42. Sheldon R. Introduction to green chemistry, organic synthesis and pharmaceuticals. Green Chem Pharm Ind. 2010;1–20.
43. Erythropel HC, Zimmerman JB, de Winter TM, et al. The green chemisTREE: 20 years after taking root with the 12 principles. Green Chem. 2018;20(9):1929–1961.
44. Aboagye EA, Chea JD, Yenkie KM. Systems level roadmap for solvent recovery and reuse in industries. Iscience. 2021;24(10):103114.

45. am Ende DJ, am Ende MT. Chemical engineering in the pharmaceutical industry: An introduction. In: Chemical Engineering in the Pharmaceutical Industry: Drug Product Design, Development, and Modeling. John Wiley & Sons; 2019. p. 1–17.

46. Plumb K. Continuous processing in the pharmaceutical industry: Changing the mind set. Chem Eng Res Des. 2005;83(6):730–738.

47. Boodhoo K, Flickinger M, Woodley J, et al. Bioprocess intensification: A route to efficient and sustainable biocatalytic transformations for the future. Chem Eng Process. 2022;108793.

48. Rosenberg N. Chemical engineering as a general purpose technology. Stud Sci Innov Process. 2009;303–328.

49. Bonvin D. Optimal operation of batch reactors—A personal view. J. Process Control. 1998;8(5–6):355–368.

50. Stanbury PF, Whitaker A, Hall SJ. Principles of Fermentation Technology. Elsevier; 2013.

51. Charpentier J-C. Modern chemical engineering in the framework of globalization, sustainability, and technical innovation. Ind Eng Chem Res. 2007;46(11):3465–3485.

52. Baxendale IR, Braatz RD, Hodnett BK, et al. Achieving continuous manufacturing: Technologies and approaches for synthesis, workup, and isolation of drug substance may 20–21, 2014 continuous manufacturing symposium. J Curr Chem Pharm Sci. 2015;104(3):781–791.

53. Tabatabaei M, Aghbashlo M, Dehhaghi M, et al. Reactor technologies for biodiesel production and processing: A review. Prog Energy Combust Sci. 2019;74:239–303.

54. Gad SC, Pharmaceutical Manufacturing Handbook: Production and Processes. John Wiley & Sons; 2008.

55. Jin Y. Nanotechnology in pharmaceutical manufacturing. In: Pharmaceutical Sciences Encyclopedia: Drug Discovery, Development, and Manufacturing. John Wiley & Sons; 2010. p. 1–32.

56. Fang JH, Lai YH, Chiu TL, et al. Magnetic core—Shell nanocapsules with dual-targeting capabilities and co-delivery of multiple drugs to treat brain gliomas. Adv Healthcare Mater. 2014;3(8):1250–1260.

57. Kesharwani R, Jaiswal P, Patel DK, et al. Lipid-Based Drug Delivery System (LBDDS): An emerging paradigm to enhance oral bioavailability of poorly soluble drugs. Biomed Mater Diagn Devices. 2022;1–16.

58. Chen S, Hanning S, Falconer J, et al. Recent advances in non-ionic surfactant vesicles (niosomes): Fabrication, characterization, pharmaceutical and cosmetic applications. Eur J Pharm Biopharm. 2019;144:18–39.

59. Tayeb HH, Sainsbury F. Nanoemulsions in drug delivery: Formulation to medical application. Nanomed. 2018;13(19):2507–2525.

60. Qiu N, Du X, Ji J, et al. A review of stimuli-responsive polymeric micelles for tumor-targeted delivery of curcumin. Drug Dev Ind Pharm. 2021;47(6):839–856.

61. Mohammadi M, Arabi L, Alibolandi M. Doxorubicin-loaded composite nanogels for cancer treatment. J. Control Release. 2020;328:171–191.

62. Cheremisinoff NP. Handbook of Air Pollution Prevention and Control. Elsevier; 2002.

63. Hassani A, Scaria J, Ghanbari F, et al. Sulfate radicals-based advanced oxidation processes for the degradation of pharmaceuticals and personal care products: A review on relevant activation mechanisms, performance, and perspectives. Environ Res. 2023;217:114789.

64. Yaghoot-Nezhad A, Wacławek S, Madihi-Bidgoli S, et al. Heterogeneous photocatalytic activation of electrogenerated chlorine for the production of reactive oxygen and

chlorine species: A new approach for bisphenol A degradation in saline wastewater. J Hazard Mater. 2023;445:130626.

65. Scaria J, Nidheesh PV. Pre-treatment of real pharmaceutical wastewater by heterogeneous Fenton and persulfate oxidation processes. Environ Res. 2023;217:114786.

66. Zhou T, Zhang Z, Liu H, et al. A review on microalgae-mediated biotechnology for removing pharmaceutical contaminants in aqueous environments: Occurrence, fate, and removal mechanism. J Hazard Mater. 2023;443:130213.

67. Alfonso-Muniozguren P, Serna-Galvis EA, Bussemaker M, et al. A review on pharmaceuticals removal from waters by single and combined biological, membrane filtration and ultrasound systems. Ultrason Sonochem. 2021;76:105656.

68. Yang Y, Ok YS, Kim K-H, et al. Occurrences and removal of Pharmaceuticals and Personal Care Products (PPCPs) in drinking water and water/sewage treatment plants: A review. Sci Total Environ. 2017;596–597:303–320.

2 Pharmaceutical Residues

Environmental Impacts, Guidelines, and Remediation in Various Environmental Matrices

Afzal Husain Khan

2.1 INTRODUCTION

Wastewater discharged from healthcare facilities has been a growing concern for environmental regulation agencies. These contaminants are hazardous to all environmental compartments due to the complex nature of effluents that contain pharmaceutical residues, pathogens, and other chemical compounds originating from research laboratories [1]. Usually, hospital wastewater (HWW) contains *microbes* like enteric pathogens (bacteria, viruses, and helminths), *hazardous chemicals* from hospitals to ensure proper disinfection, *drug residues* from patients, and *radioactive isotopes* generated from radiation therapy during cancer treatment [2].

While many effluent guidelines and standards exist for the disposal and treatment of municipal wastewater effluents, only limited countries have specific guidelines and standards for industrial effluents. Many countries do not treat hospital wastewater differently from domestic wastewater and dispose of it directly into the public sewer. These countries have passed legislation for wastewater disposal in public sewers. This legislation utilizes conventional parameters like biochemical oxygen demand (BOD), chemical oxygen demand (COD), pH, temperature, and total suspended solids (TSS), etc. However, some countries take hospital effluents as seriously as industrial wastewater in terms of the hazard potential and pre-treat hospital wastewater before its disposal in the drain. Therefore, it is essential to safeguard environmental systems, people, animals, and plants from such contaminants and their impacts [3]. Unlike urban wastewater, the physico-chemical and biological characteristics of HWW showed charged pollutants [4, 5]. However, the physico-chemical characteristics of HWW and urban wastewater are similar in some countries.

The highest poncentrations of pharmaceuticals detected in lower–middle-income countries indicate poor or inefficient wastewater treatment plants

DOI: 10.1201/9781003340164-2

(WWTPs) [6]. Some countries such as Switzerland, Brazil, the United States, and Laos show lower pharmaceutical concentrations due to restricted anthropogenic activities, regulated use of modern drugs, tertiary/advanced WWTPs, and lower contaminant discharge [6].

Generally, the average water demand in hospitals is around 200–1200 L bed^{-1}day^{-1} and discharges an equal amount as wastewater [7]. In addition, wastewater from other hospital units such as the kitchen, laundry, washrooms, and toilets are added to the overall effluent from the hospital. This HWW gets mixed up with the other streams of water and loaded for various purposes like irrigation and other domestic activities, thus making the situation more complex. This will increase the risk of disease transmission and regulatory compliance challenges, thereby reducing the effectiveness of wastewater treatment.

Several treatment technologies have been adopted for the treatment of pharmaceutical wastewater [8–10]. Some of the conventionally adopted methods such as sedimentation, coagulation, flocculation, and the activated sludge process (ASP) have been sufficiently effective only for conventional parameters such as electrical conductivity, COD, nitrogen, TSS, and BOD. Hence, the application of advanced technologies coupled with different treatment technologies for hospital wastewater treatment has been intensively investigated to overcome this constraint [11–14].

The present chapter presents a comprehensive review of pharmaceutical residue characteristics, occurrences, and impacts on the environment and human health. It is seen that the prioritization and risk assessment of pharmaceutical residues is quite rare in the extant literature. Hence, the present chapter sheds a new direction in this regard. In this context, the present work provides a prioritization indicator in tune with pharmaceutical residues. Finally, the present chapter highlights the treatment methods such as physico-chemical, biological, and chemical methods in connection with hybrid and advanced treatments.

2.2 OCCURRENCE OF ANTIBIOTIC AND OTHER PHARMACEUTICAL RESIDUES

Antibiotics are often discharged as drug residue, metabolites, or inactivating substances either through urine (50–80%) or feces (5–30%) [2,15]. This results in pharmaceutical residues, specifically antibiotics, in water bodies [16]. Therefore, it is evident that the concentration of pharmaceutical residues in HWW effluents is mainly a function of a) administered quantity, b) excretion, and c) laboratory waste [17], and their loads can vary by as much as 5 mg/L based on geographic location alone. In addition to antibiotics, cytotoxic drugs, which are commonly used in chemotherapy, are also sometimes found in the water. While found in scarce amounts, these drugs are usually genotoxic, mutagenic, and carcinogenic in nature. Moreover, they can interfere with DNA synthesis and disrupt cellular proliferation and endocrine function in several organisms [18, 19]

Despite the potential harmful effects of pharmaceutical residues, several countries lack discharge standards. As the occurrence, metabolism, and toxicity of

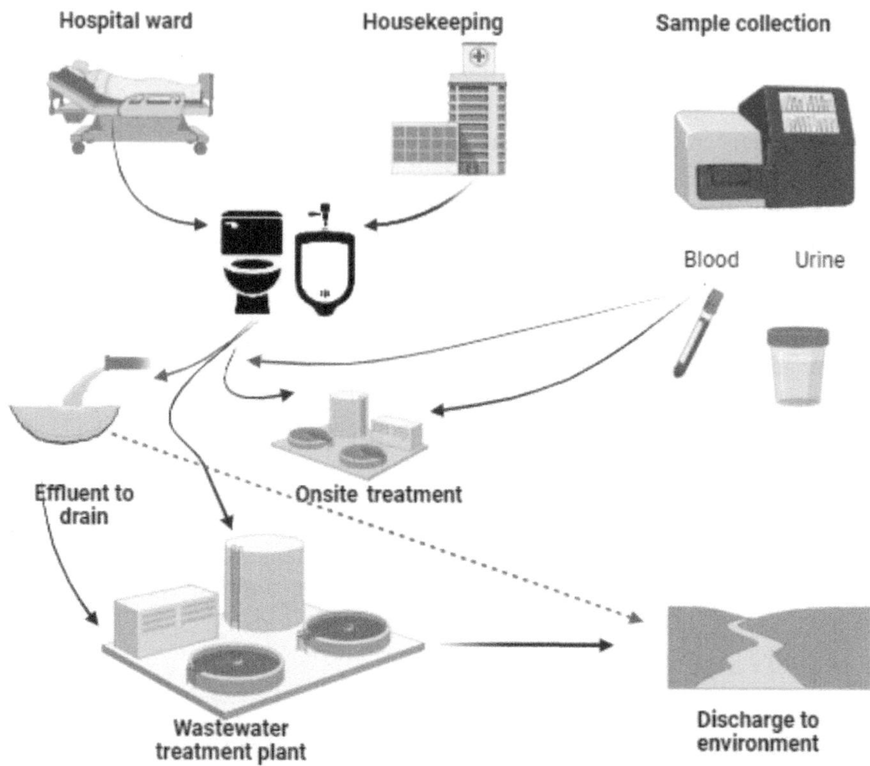

FIGURE 2.1   The pathway of pharmaceutical residues from HWW to the environment.

Source: (Courtesy of online drafting tool "Biorender" (https://biorender.com/).

pharmaceutical residues in the aquatic environment become the foci of current research globally, this chapter strives to emphasize the critical aspects for identifying pharmaceutical residues that could significantly impact the environment (Figure 2.1). In this regard, it is important to identify those hospital effluents that cause drastic harm when they find their way into water bodies. Lin and Tsai investigated hospital effluents in Taiwan and recorded the presence of non-steroidal anti-inflammatory drugs (NSAIDs), antibiotics, estrogen, lipids, and beta-blockers among other effluents [20]. Acetaminophen, erythromycin–H_2O, sulfamethoxazole, gemfibrozil, and NSAIDs were the dominant drug residues found in surface water in higher concentrations (acetaminophen, 417.5 g/L; and erythromycin, H_2O–7.84 g/L) [20].

In the United States of America, a nationwide survey of 95 pharmaceutical drugs in 139 streams between 1999 and 2000 confirmed the presence of 82 compounds in 80% of the streams. Sulfamethoxazole and erythromycin–H_2O were the most common drug residues found during the survey, representing a concerning discovery [21]. Similar studies in Germany revealed that more than 80%

of the 32 drugs commonly found in wastewater included antiphlogistic drugs, lipid regulators, psychiatric drugs, beta-blockers, and other metabolites that were not treated adequately [22]. A study on the river Elbe and its tributaries in Germany from 2004 found the concentration of NSAIDs, antibiotics, and beta-blockers to fall between 20 and 140 ng/L [23]. In contrast, psychoactive drugs and their metabolites were found as the dominant drug residues in the main rivers of Madrid and surrounding metropolitan areas [24]. A similar investigation in the Pearl River Delta in South China found the presence of eight antibiotics (chloramphenicol, fluoroquinolone, sulfonamide, and macrolide groups) with significant concentrations of ofloxacin, norfloxacin, roxithromycin, erythromycin–H_2O, and sulfamethoxazole [25].

Like antibiotics, biological treatment is ineffective against antihistamines such as cetirizine, acrivastine, fexofenadine, loratadine, desloratadine, and ebastine, and consequently, these antihistamines get discharged into water bodies [26]. Other active drug residues such as naproxen, ketoprofen, indomethacin, propyphenazone, ibuprofen, gemfibrozil, diclofenac, clofibric acid, and carbamazepine, belonging to NSAIDs, analgesics, antiepileptics, and lipid regulators, have been detected in water samples in Hanoi, Vietnam. This confirms that drug residues in hospital effluents pollute the environment significantly [27]. Therefore, it is understood from the literature that the removal of antibiotics and antihistamines as well as other pharmaceutical residues from wastewater is critical. For example, failure to remove antibiotic residues from the water can cause antibiotic resistance in humans, especially in developing countries like India [28]. Hence, most WWTPs employ several techniques such as biotransformation, air stripping, sorption, and photo transformation to convert pollutants into less harmful products [29].

Studies have been done in the past to compare predicted and measured concentrations of pharmaceutical residues in HWW effluents. In most of the studies, the consumption of pharmaceuticals has been considered the base factor. The main highlights of these studies are presented in Table 2.1.

TABLE 2.1

Past Research on the Prediction of Pharmaceutical Residues in Hospital Wastewater

Key Points	Reference
In this study, a model is presented to discharge pharmaceutical residues, including diclofenac and carbamazepine, in the wastewater from the military hospital of Berlin based on consumption data and pharmacokinetic data. Measured values have measured the predicted values over one week.	[30]
In this study, the same model is presented for metamizole.	[31]
Based on consumption, the authors evaluated the cytotoxic pharmaceuticals in the sewage of the Oncology Department of Vienna University Hospital in Austria. Thus, the consumption was compared with the measured concentration.	[32]

(Continued)

TABLE 2.1 (*Continued*)

Past Research on the Prediction of Pharmaceutical Residues in Hospital Wastewater

Key Points	Reference
The researchers investigated the concentration of common chemotherapy drugs and X-ray contrast media in a Swiss hospital with a high bed strength of 415. The composite sampling was done for 24 hours, and values were measured and then correlated with the predicted values. This comparison could conclude the amount of pharmaceuticals discharged into the sewer.	[33]
This study evaluated the pharmaceutical load in an Australian hospital with a bed strength of 190 in the associated WWTP. A total of 59 pharmaceuticals and personal care products were measured. The measured values were compared with the pharmacy data to evaluate if they can be used to predict the concentration without monitoring these compounds.	[34]
The authors studied three hospitals and three pharmaceuticals and evaluated the average measured concentrations. The values were predicted based on daily or annual consumption data, whichever is available. The predicted values were then compared with measured values for 14 days.	[35]
In this research, the authors reported evaluating predicted environmental concentration and predicted environmental load for many compounds in Switzerland. The ecotoxicological potential was evaluated for the 100 most commonly consumed pharmaceuticals in two hospitals. The ecotoxicological potential was assessed in raw hospital wastewater after dilution with urban wastewater and after treatment by the activated sludge process.	[36]
This study compared the predicted and measured loads of 30 pharmaceuticals, including cytotoxics and iodinated contrast media, in the wastewater discharged from a Swiss hospital having bed strength of 346.	[37]
This study attempted to evaluate the load of pharmaceutical residues in the WWTPs from hospitals and also to come up with a list of priority pharmaceuticals based on effect threshold and margin of exposure.	[38]
In this research, nine antibiotics were considered, and the monthly consumption variability was assessed in a hospital in Lausanne. The concentrations were predicted for the hospital and urban wastewater.	[39]
In a study under the European PILLS Project, the variation in the consumption of pharmaceuticals was studied in psychiatric and radiology departments. The annual consumption of six pharmaceuticals in different hospitals has also been discussed.	[40]
Six pharmaceuticals with predicted and measured concentrations were evaluated in a psychiatric hospital, a nursing home, and a general hospital in Germany.	[41]
The authors characterized the hospital effluent of the Geneva University hospitals to assess the predicted and measured values for 15 pharmaceuticals. The values were measured in the composite samples collected over 24 hours. The predicted values were based on consumption, and the daily load was also estimated.	[42]
In this study, the predicted and measured values of 38 pharmaceuticals were evaluated in hospital effluent. The values were measured in summer as well as winter. The predicted values were based on annual consumption.	[43]

2.3 IMPACT OF PHARMACEUTICAL RESIDUES ON THE ENVIRONMENT AND HUMAN HEALTH

The substantial growth in drug consumption due to the increasing population has been the driving force of water pollution from pharmaceutical residues [44]. Common residues found in surface, ground, and drinking water in several countries include NSAIDs, antibiotics, cytostatic drugs, synthetic hormones, beta-blockers, lipid regulators, psychiatric drugs, and antihistamines, primarily due to their inferior degradation during sewage treatment [45, 46]. These residues degrade water quality, increase microbial resistance, reduce surface water ability to self-purify, and negatively affect humans and aquatic life. Furthermore, some teratogenic, mutagenic, and genotoxic drugs can cause ecotoxicity in different organisms. Moreover, many drugs have the potential to accumulate in the body and persist in the environment.

On the other hand, the existing sewage treatment plants (STPs) are proving to be ineffective at handling the rising pollution loads. Consequently, many effluents find their way into the ecosystem. HWW, being a pharmaceutical cocktail having different chemicals and heavy metals, worsens the aquatic pollution problem since the effect of its constituent chemicals can potentially add up, causing significant harm to the overall ecosystem [1]. If the environmental conditions are not conducive to their degradation, the high degree of bioaccumulation and biomagnification of these pollutants poses both short- and long-term physical, chemical, and biological risks to living organisms exposed to the polluted environment [47]. For example, some ecotoxicological tests have shown that phenols and pharmaceutical residues cause acute or chronic toxicity, altered behavior, and stunted growth in lower organisms as well as reduced reproduction capability [48].

Unfortunately, this emerging concern regarding pharmaceutical compounds (antibiotics), chemical residues, radioelements, antibiotic-resistant strains, and pathogens is generally not covered by most regulatory mechanisms [49]. Hence, there is very little understanding of the adverse impact of these compounds on human health and the environment [49]. Not much research has been conducted on how WWTPs can reduce the presence of these compounds. It is important to realize that releasing large amounts of wastewater containing these substances could have disastrous effects on both the aquatic system and the human population [50]. Therefore, the proper disposal of hospital sludge is a principal focus zone to decrease public health concerns.

2.4 OVERVIEW OF GLOBAL HWW GENERATION

Hospitals usually consume large amounts of water for the upkeep of patients and facilities such as wards, kitchens, operation theaters, laboratories, and hemodialysis units. On average, hospitals consume 400–1200 liters/bed/day and discharge approximately 1000 liters/person/day of effluents [51–53]. However, the drug consumption is shown to vary across healthcare and hospital facilities globally [54]. For example, a study of drug consumption in Germany shows the mean annual individual use of 1000 grams of drugs consumed per bed and a variance of 1263 kg/year in general hospitals [55].

As this number grows steadily to meet the healthcare needs of local, national, and international patients, it is important to consider its effect on water consumption trends. Hospital effluents containing bactericides and disinfectants hinder biological processes necessary for wastewater treatment. Furthermore, a potential serious health risk and severe epidemics could arise due to hospital effluents being discharged untreated to the environment [56]. Therefore, the management of hospital effluents and wastewater treatment are critical issues that necessitate the on-site treatment of these effluents to prevent the contamination of water bodies.

Hospital effluents are treated before their disposal in the municipal sewage system to minimize toxins. Although the hospital effluent is highly diluted before entering the municipal sewage treatment plant, it can still disrupt the biological balance in the aquatic environment. Therefore, it is imperative to adequately treat hospital effluents before discharging them into the municipal sewage system to protect the aquatic system contamination [57].

2.5 GUIDELINES AND GLOBAL TRENDS IN HWW TREATMENT

Across the globe, various reference standards and guidelines are being used for the treatment of domestic wastewater effluents before they are discharged to the environment. The hospital effluent characteristics are not very consistent as compared to domestic effluents. Interestingly, domestic treatment technologies are applied to hospital effluents. A probable reason for such a state of affairs may be that most countries evaluate hospital-generated wastewater on similar lines as domestic wastewater [58]. Hence, such HWW is being directly discharged into the sewer system without pre-treatment. Very few countries regard hospital effluents as industrial waste and pre-treat these before discharging them into the municipal sewer network [59].

Even though there is very little global awareness about the need to treat hospital effluents before these are appropriately discharged, the few countries that are putting in some effort in this area are adopting different strategies. In countries like Australia, Egypt, Japan, Iran, South Africa, India, and Thailand, the norm is to release such effluents into the urban sewer system, where these are subjected to co-treatment along with the other effluents (already present) in the sewage treatment plant. At the other end of the spectrum are developing countries in Africa (Algeria, Congo, Ethiopia) and Asia (Bangladesh, Nepal, Pakistan, Taiwan, and Vietnam), where hospital effluents are discharged into drainage systems and water bodies without prior treatment. Specifically, hospitals in Pakistan and Taiwan lack adequate on-site wastewater treatment facilities and discharge their wastewater directly into nearby rivers with minimal treatment [60]. Meanwhile, only 48% of the government hospitals in Iran have proper wastewater treatment systems [61]. In Indonesia, nearly 65% of the hospitals discharge their wastewater into water bodies or using infiltration wells. Many hospital wastewater treatment plants have detected high concentrations of lead, phenol, ammonia, phosphates, and chlorine [62]. A study in Kunming city, China, showed that 36 of 45 hospitals had wastewater disinfection equipment. Similarly, a study of 50 hospitals in Wuhan City revealed that 46 had wastewater treatment facilities, but only 23 could meet the national discharge standards for effluent quality [63]. While most hospitals in strife-torn Iraq have their own treatment plants, these are unable to comply with the national standards, especially regarding removing nutrients and pathogens [64]. In such a depressing scenario, it is slightly encouraging to note that China, Indonesia, and the Republic of Korea have more stringent norms for wastewater quality due to provisions for on-site specific treatment of hospital wastewater.

This discussion leads us to conclude that there is still a wide gap between most wastewater treatments and the guidelines concerning the remediation of various environmental matrices. Even today, different treatment mechanisms, like specific treatment, co-treatment, and direct disposal into the environment, are being adopted by different countries.

When hospital wastewater is released into the surface water bodies, there is a great risk of surface water bodies being contaminated by antibiotic-resistant bacteria. The consequences are more intense in countries where wastewater treatment facilities are unavailable. A cost-effective alternative would be to provide on-site facilities for treating the hospital wastewater before it is discharged into municipal sewers. Other options include upgrading the existing WWTPs and streamlining the operation and maintenance of wastewater treatment processes.

As already mentioned, there cannot be a single universal treatment process applied across the board in all cases. No wonder there are instances where a combination of processes is being used. Still, it would be possible to effectively remove harmful bacteria and pharmaceutical residues by adopting better disinfection processes and membrane filtration.

TABLE 2.2

Guidelines in Some Countries for Pharmaceutical Residues

S. No.	Legislation	Year	Pharmaceuticals Regulated	Reference
1	**Australia:** Australian Guidelines For Water Recycling: Managing Health and Environmental Risks (Phase 2) Augmentation of Drinking Water Supplies	2008	Partially	[65]
2	**Belgium:** Directives 2004/27/EC and 2001/182/EC		No	[66]
3	**Canada:** Health Canada Environmental Impact Initiative	2009	Partially	[67]
4	**Denmark:** Directives 2004/27/EC and 2001/182/EC		Yes	[66]
5	**Finland:** Directives 2004/27/EC and 2001/182/EC		Yes	[66]
6	**Germany:** Federal Ministry for the Environment, Nature Conservation, and Nuclear Safety Wastewater Ordinance	2004	Yes	[68]
7	**Italy:** DRP No. 227/2011 on simplification of environmental law	2011	Yes	[66]

(Continued)

TABLE 2.2 (*Continued*)
Guidelines in Some Countries for Pharmaceutical Residues

S. No.	Legislation	Year	Pharmaceuticals Regulated	Reference
8	**Netherlands:** Directives 2004/27/EC and 2001/182/EC		Yes	[69]
9	**Portugal:** Directives 2004/27/EC and 2001/182/EC		Yes	[66]
10	**Spain:** Directive 57/2005, the Annexes to Law 10/1993, Discharges Industrial Liquids to the Comprehensive Sanitation System	2005	Yes	[70]
11	**Switzerland:** Swiss federal water protection legislation	2016	Yes	[71]
12	**United Kingdom:** National Guidance for Healthcare Wastewater Discharges	2011	Yes	[72]
13	**United States:** Removing the number of CFR regulations pertaining to its water programs and clarifying the legal status of these regulations for both the regulated community and the public	2016	Yes	[73]
14	**Vietnam:** National Technical Regulation on Health Care Wastewater	2010	Partially	[74]

2.6 HANDLING OF WWTP SLUDGE

The handling of WWTP sludge and the innovation of treatment technology have become constraints for water treatment industry members (European Directive 91/271/EEC of 20/05/91). Consequently, more than a 37% increase in global sewage sludge production was estimated, with over 103 million tons worldwide compared with 75 million tons in the year 2013, with the European Union alone accounting for around 15 million tons [75–77].

Around 55 and 80% of the active components in pharmaceuticals are excreted as either unaltered compounds, combinations of metabolites, or conjugated with an inactivating component [17]. Patients' urine and feces containing these excreted compounds are subsequently routed via municipal sewage systems, including those that are not eliminated in WWTPs being released into local surface or groundwater. Personal care items are frequently unloaded into the environment without being treated. As a result of unplanned water reuse, they may eventually be absorbed or digested by aquatic life forms or sediments, with the potential to be bioaccumulated into edible components of plants [78].

In summary, the key differences between STP/WWTP sludge with pharmaceutical wastewater mixing lie in their composition and potential contaminants. WWTP sludge that has been influenced by pharmaceutical wastewater is more likely to contain pharmaceutical residues and a wider range of chemical contaminants, making it more complex and potentially requiring specialized treatment and disposal practices to address the unique challenges associated with pharmaceutical wastewater.

2.7 TREATMENT TECHNOLOGIES

To conserve resources and cut costs, wastewater should be given tertiary treatment before being reused for such tasks as flushing and cleaning. The hospitals' rooftops are the ideal location for rainwater gathering systems. Hospital effluents should be treated before being released into the municipal sewage system to prevent contamination of the aquatic ecosystem. Hospital effluent sludge requires the same cautious handling as sludge from other municipal waste treatment facilities. There are still several challenges and limitations in implementing new regulations. Developing an effective policy framework to address these problems is essential.

Due to the diverse characteristics of wastewater, there is no specific treatment method that offers absolute screening against pharmaceuticals. Furthermore, conventional wastewater treatment plants are unable to remove pharmaceuticals completely. Hence, this approach cannot be relied on to prevent pharmaceuticals loading to the environment.

The presence of pharmaceuticals in the environment is solely dependent on wastewater treatment, and non-point sources influence raw water supply. Conventional treatment methods such as coagulation/flocculation using coagulants such as ferric or alum, followed by sedimentation and filtration, are incapable of eliminating pharmaceuticals. However, some pharmaceuticals can be removed during hospital disinfection [79].

Pharmaceuticals and other trace organic substances may be transformed by the oxidation processes used in chlorine, ClO_2, UV/chlorine, and ozone disinfection [80–82]. Recent studies reported that using UV–LED/chlorine treatment was inefficient for removing acetaminophen, ciprofloxacin, iopromide, and propranolol [83–86]. However, a study reported that the UV/chlorine process might increase or decrease the toxicity of pharmaceutical compounds [87]. In contrast, many pharmaceuticals may be transformed by more than 90% by the use of X-ray contrast media, ultraviolet light, and ozonation, which are all potent oxidants [88–90]. However, energy intake is recognized as a vital component in treatment selection [91].

However, it is impossible to determine if the removal of pharmaceuticals is better in a membrane bioreactor (MBR) as biodegradation has an impact on many factors other than reactor configuration. The conventional bioreactors may be connected to nano-filtration (NF) or reverse osmosis (RO) membranes to achieve removal efficiencies of more than 90%.

The most common pharmaceuticals effluents were acetaminophen 49–225 µg/L, caffeine 26–46 µg/L, naproxen 5.9–13 µg/L, ibuprofen 5.2–22 µg/L, and diclofenac 1.3–2.5 µg/L, with minimum–maximum concentrations [92]. For effluents, it was acetaminophen 0.054–7.8 µg/L, caffeine 0.084–4.8 µg/L, and diclofenac 0.28–3.3 µg/L, with minimum–maximum concentrations. Acetaminophen, sulfadiazine, cortisone, testosterone, metoprolol, and propranolol showed the maximum removal rates (100%). The lowest removal efficiencies were observed for carbamazepine (2.7%) and diclofenac (13.2%) [93].

TABLE 2.3

Treatment Technologies to Remove Pharmaceutical Residues from Wastewater

Treatment Category	Technology	Overall Efficiency (%)	Reference
Conventional	Coagulation–flocculation	~77%	[95]
	Membrane bioreactor	>80%	[76]
	Physico-chemical treatment	> 98	[52]
Biological	Anaerobic digestion	~75%	[96]
	Constructed wetlands		[97]
	Upflow anaerobic sludge blanket (UASB) reactor	84%	[98]
Advanced treatment	Advanced oxidation processes (AOPs): UV/H$_2$O$_2$/O$_3$	100%	[11]
	Adsorption	99%	[99]
Membrane technologies	Nanofiltration	~99%	[100]
Hybrid technologies	Disinfection with UV/chlorine and ozone	>99%	[80–82]
Hybrid technologies	UV–LED/chlorine	>98%	[83–86]
Hybrid technologies	X-ray contrast media, ultraviolet light, and ozonation	>90%	[88–90]
Hybrid technologies	UV/chlorine process	>92%	[87, 91]
Hybrid technologies	Integrated bioreactor	96%	[92]

In a comparative study for the removal 79 pharmaceuticals, largely parent compounds and their transition products, were examined using catalytic wet air oxidation, Fenton, and photo-Fenton technologies [94]. The complete removal of several pharmaceutical groups was successfully achieved by catalytic wet air oxidation, except for analgesics/anti-inflammatories and antibiotics (85%). Out of catalytic wet air oxidation, Fenton, and photo-Fenton technologies, Fenton oxidation is the most efficient technique for the complete elimination of pharmaceuticals (99.8%).

2.8 FUTURE PROSPECTS

Over the past decade, numerous governments and international organizations have developed regulation frameworks to combat the widespread presence of drugs in the natural environment. However, no universally accepted discharge limit guideline exists for pharmaceuticals in the effluents from hospitals, industries, municipalities, and WWTPs. Also, the treatment of wastewater should be handled entirely at the point of origin, as recommended by the Swedish Precautions and Risk Reduction Strategies [101]. Future research on pharmaceuticals should not target parent compounds alone, which may not represent a real hazard to human health, since there is a mixture of different pharmaceuticals present in the wastewater. Thus, due to the failure of the current WWTPs to remove pharmaceuticals, mainly due to the low concentration, many studies have focused on the application of advanced water treatment processes. However, sustainable measures that can tackle the problem should be prioritized rather than relying on the development and implementation of advanced treatment processes for the removal of developing pollutants from wastewater, which can be expensive and not sustainable.

2.9 CONCLUSION

The development of future pollutants and their effects could be mitigated with political, societal, and technical efforts. In summary, it can be emphasized that without critical thinking and a rethinking of the economy, growth, and environmental protection, pollution will not be mitigated. Pharmaceuticals in the environment, even in low quantities, pose health concerns, but there are no laws or regulations limiting pharmaceutical effluents from point sources like hospitals. Strict regulations can lessen the amount of pharmaceuticals that end up in water supplies.

The pre-treatment of pharmaceutical effluents at point sources is one example of how present and future research could lead to more effective treatment methods. Proper disposal has a significant negative effect on the environment, but this problem can be alleviated with the help of takeback programs, guidelines, and thorough consumer teaching.

Political, social, and technical measures could limit future pollution and its impacts. In conclusion, pollution cannot be reduced without critical thinking and reconsidering the economy, growth, and environmental preservation. These improvements in politics, technology, science, human behavior, habits, and public awareness of a more sustainable lifestyle may help reduce emerging contaminants (ECs) in the environment. In the end, tighter directives and increased environmental and health awareness would improve water quality worldwide.

ACKNOWLEDGMENT

The authors of the current work wish to thank Universiti Sains Malaysia, Pulau Pinang, for their facilities and lab support.

REFERENCES

[1] Emmanuel E, Perrodin Y, Keck G, et al. Ecotoxicological risk assessment of hospital wastewater: A proposed framework for raw effluents discharging into urban sewer network. J Hazard Mater. 2005;117:1–11.

[2] Carraro E, Bonetta S, Bertino C, et al. Hospital effluents management: Chemical, physical, microbiological risks and legislation in different countries. J Environ Manage. 2016;168:185–199.

[3] Khan AH, Aziz HA, Khan NA, et al. Impact, disease outbreak and the eco-hazards associated with pharmaceutical residues: A critical review. Int J Environ Sci Technol. 2022;19:677–688.

[4] Khan AH, Khan NA, Zubair M, et al. Sustainable green nanoadsorbents for remediation of pharmaceuticals from water and wastewater: A critical review. Environ Res. 2022;204:112243.

[5] Khan AH, Rudayni HA, Chaudhary AA, et al. Modern use of modified sequencing batch reactor in wastewater treatment. Ecol Quest. 2022;33:1–23.

[6] Wilkinson WJ, Boxall BA, Dana K, et al. Pharmaceutical pollution of the world's rivers. Proc Natl Acad Sci. 2022;119:e2113947119.

[7] Khan NA, Ahmed S, Farooqi IH, et al. Pathways and legislation towards hospital wastewater discharged around globe. Ecol Quest. 2021;32:91–100.

[8] Khan NA, Ahmed S, Farooqi IH, et al. Occurrence, sources and conventional treatment techniques for various antibiotics present in hospital wastewaters: A critical review. In: TrAC—Trends Anal. Chem. Elsevier BV; 2020. p. 115921.

[9] Singh S, Naik TSSK, Shehata N, et al. Novel insights into graphene oxide-based adsorbents for remediation of hazardous pollutants from aqueous solutions: A comprehensive review. J Mol Liq. 2023;369:120821.

[10] Ghosh S, Falyouna O, Onyeaka H, et al. Recent progress on the remediation of metronidazole antibiotic as emerging contaminant from water environments using sustainable adsorbents: A review. J Water Process Eng. 2023;51:103405.

[11] Khan AH, Khan NA, Ahmed S, et al. Application of advanced oxidation processes followed by different treatment technologies for hospital wastewater treatment. J Clean Prod. 2020;269:122411.

[12] Khan NA, Khan AH, Tiwari P, et al. New insights into the integrated application of Fenton-based oxidation processes for the treatment of pharmaceutical wastewater. J Water Process Eng. 2021;44:102440.

[13] Hasan MN, Altaf MM, Khan NA, et al. Recent technologies for nutrient removal and recovery from wastewaters: A review. Chemosphere. 2021;277:130328.

[14] Samaraweera H, Khan AH. Recent trends in graphene-based materials for pharmaceuticals wastewater treatment. In: Treat Pharm Wastewater. Elsevier; 2023. p. 53–68.

[15] Dimian AC, Bildea CS. Bioethanol manufacturing. Technology. 2008;28:429–460.

[16] Kotwani A, Holloway K. Trends in antibiotic use among outpatients in New Delhi, India. BMC Infect Dis. 2011;11.

[17] Verlicchi P, Galletti A, Petrovic M, et al. Hospital effluents as a source of emerging pollutants: An overview of micropollutants and sustainable treatment options. J Hydrol. 2010;389:416–428.

[18] Isidori M, Lavorgna M, Russo C, et al. Chemical and toxicological characterisation of anticancer drugs in hospital and municipal wastewaters from Slovenia and Spain *, ** Barcel o can. 2016;219.

[19] Chen H, Zhang M. Effects of advanced treatment systems on the removal of antibiotic resistance genes in wastewater treatment plants from Hangzhou, China. Environ Sci Technol. 2013;47:8157–8163.

[20] Lin AY-C, Tsai Y-T. Occurrence of pharmaceuticals in Taiwan's surface waters: Impact of waste streams from hospitals and pharmaceutical production facilities. Sci Total Environ. 2009;407:3793–3802.

[21] Kolpin DW, Furlong ET, Meyer MT, et al. Pharmaceuticals, hormones, and other organic wastewater contaminants in U.S. streams, 1999–2000: A national reconnaissance. Environ Sci Technol. 2002;36:1202–1211.

[22] Ternes TA. Occurance of drugs in german sewage treatment plants and rivers. Water Res. 1998;32:3245–3260.

[23] Wiegel S, Aulinger A, Brockmeyer R, et al. Pharmaceuticals in the river Elbe and its tributaries. Chemosphere. 2004;57:107–126.

[24] González Alonso S, Catalá M, Maroto RR, et al. Pollution by psychoactive pharmaceuticals in the Rivers of Madrid metropolitan area (Spain). Environ Int. 2010;36:195–201.

[25] Xu W, Zhang G, Zou S, et al. A preliminary investigation on the occurrence and distribution of antibiotics in the Yellow River and its Tributaries, China. Water Environ Res. 2009;81:248–254.

[26] Kosonen J, Kronberg L. The occurrence of antihistamines in sewage waters and in recipient rivers. Environ Sci Pollut Res. 2009;16:555–564.

[27] Tran NH, Urase T, Ta TT. A preliminary study on the occurrence of pharmaceutically active compounds in hospital wastewater and surface water in Hanoi, Vietnam. Clean—Soil, Air, Water. 2014;42:267–275.

[28] Kümmerer K. The presence of pharmaceuticals in the environment due to human use—Present knowledge and future challenges. J Environ Manage [Internet]. 2009;90:2354–2366. Available from: www.sciencedirect.com/science/article/pii/S030147970900022X.

[29] Zhang Y, Geißen SU, Gal C. Carbamazepine and diclofenac: Removal in wastewater treatment plants and occurrence in water bodies. Chemosphere. 2008;73:1151–1161.

[30] Heberer TFD. Contribution of effluents from hospitals and private households to the total loads of diclofenac and carbamazepine in municipal sewage effluents—Modeling versus measurements. J Hazard Mater. 2005;122:211–218.

[31] Feldmann DF, Zuehlke SHT. Occurrence, fate and assessment of polar metamizole (dipyrone) residues in hospital and municipal wastewater. Chemosphere. 2008;71:754–1764.

[32] Mahnik S, Lenz K, Weissenbacher N, Mader RFM. Fate of 5-fluorouracil, doxorubicin, epirubicin, and daunorubicin in hospital wastewater and their elimination by activated sludge and treatment in a membrane-bio-reactor system. Chemosphere. 2007;66:30–37.

[33] Weissbrodt D, Kovalova L, Ort C, et al. Mass flows of x-ray contrast media and cytostatics in hospital wastewater. Environ Sci Technol. 2009;43:4810–4817.

[34] Ort C, Lawrence MG, Reungoat J, et al. Determining the fraction of pharmaceutical residues in wastewater originating from a hospital. Water Res. 2010;44:605–615.

[35] Mullot JU, Karolak S, Fontova A, et al. Modeling of hospital wastewater pollution by pharmaceuticals: First results of Mediflux study carried out in three French hospitals. Water Sci Technol. 2010;62:2912–2919.

[36] Escher BI, Baumgartner R, Koller M, et al. Environmental toxicology and risk assessment of pharmaceuticals from hospital wastewater. Water Res. 2011;45:75–92.

[37] McArdell CS, Kovalova L, Siegrist H. Input and Elimination of Pharmaceuticals and Disinfectants from Hospital Wastewater—Final Report. Rep. eawag; 2011. p. 95.

[38] Le Corre KS, Ort C, Kateley D, et al. Consumption-based approach for assessing the contribution of hospitals towards the load of pharmaceutical residues in municipal wastewater. Environ Int. 2012;45:99–111.

[39] Coutu S, Rossi L, Barry DA, et al. Temporal variability of antibiotics fluxes in wastewater and contribution from hospitals. PLoS One. 2013;8:45–65.

[40] Helwig K, Hunter C, Mcnaughtan M, et al. Ranking prescribed pharmaceuticals in terms of environmental risk: Inclusion of hospital data and the importance of regular review. Environ Toxicol Chem. 2016;35:1043–1050.

[41] Herrmann M, Olsson O, Fiehn R, et al. The significance of different health institutions and their respective contributions of active pharmaceutical ingredients to wastewater. Environ Int. 2015;85:61–76.

[42] Daouk S, Chèvre N, Vernaz N, et al. Dynamics of active pharmaceutical ingredients loads in a Swiss university hospital wastewaters and prediction of the related environmental risk for the aquatic ecosystems. Sci Total Environ. 2016;547:244–253.

[43] Verlicchi P, Zambello E. Predicted and measured concentrations of pharmaceuticals in hospital effluents: Examination of the strengths and weaknesses of the two approaches through the analysis of a case study. Sci Total Environ. 2016;565:82–94.

[44] Khan AH, Abdul Aziz H, Palaniandy P, et al. Pharmaceutical residues in the ecosystem: Antibiotic resistance, health impacts, and removal techniques. Chemosphere. 2023;339:139647.

[45] Kummerer K. Drugs in the environment: Emission of drugs, diagnostic aids and disinfectants into wastewater by hospitals in relation to other sources—A review. Chemosphere. 2001;45:957–969.

[46] Lakshmi Prasanna B, Padmini VL, Navle K, et al. Analysis of drugs in aquatic environment. J Chem Pharm Res. 2015;7:71–79.

[47] Emmanuel E, Quisqueya U, Perrodin Y, et al. Effects of Hospital Wastewater on Aquatic Ecosystem. Congr Interam Ing Sanit y Ambient Cancún; 2002.

[48] Pinto P, Garcı PA. Ecotoxicity and environmental risk assessment of pharmaceuticals and personal care products in aquatic environments and wastewater treatment plants. Ecotoxicol. 2014;1517–1533.

[49] Peterson E, Kaur P. Antibiotic resistance mechanisms in bacteria: Relationships between resistance determinants of antibiotic producers, environmental bacteria, and clinical pathogens. Front Microbiol. 2018;9.

[50] Chaturvedi P, Shukla P, Giri BS, et al. Prevalence and hazardous impact of pharmaceutical and personal care products and antibiotics in environment: A review on emerging contaminants. Environ Res. 2021;194:110664.

[51] Al-ajlouni K, Shakhatreh S, Ibraheem NA, et al. Evaluation of wastewater discharge from hospitals in amman -JORDAN. Int J Basic Appl Sci IJBAS-IJENS. 2009;13:44–50.

[52] Gautam AK, Kumar S, Sabumon PC. Preliminary study of physico-chemical treatment options for hospital wastewater. J Environ Manage. 2007;83:298–306.

[53] Singh R, Mishra SS, Mishra A.1 Hospital solid waste and waste water analysis in Lucknow. Environ Sci Med. 2013; 6(1).

[54] Masood AS, Ali MS, Manzar MS, et al. Current situation of pharmaceutical wastewater around the globe. In: Treat Pharm Wastewater. Ist. Elsevier; 2023. p. 19–52.

[55] Lee Y, Kovalova L, McArdell CS, et al. Prediction of micropollutant elimination during ozonation of a hospital wastewater effluent. Water Res. 2014;64:134–148.

[56] Bogler A, Packman A, Furman A, et al. Rethinking wastewater risks and monitoring in light of the COVID-19 pandemic. Nat Sustain. 2020;3:981–990.

[57] Brandmayr C, Kerber H, Winker M, et al. Impact assessment of emission management strategies of the pharmaceuticals Metformin and Metoprolol to the aquatic environment using Bayesian networks. Sci Total Environ. 2015;532:605–616.

[58] Verlicchi P, Al Aukidy M, Galletti A, et al. Hospital effluent: Investigation of the concentrations and distribution of pharmaceuticals and environmental risk assessment. Sci Total Environ. 2012;430:109–118.

[59] Khan AH, Aziz HA, Khan NA, et al. Effect of seasonal variation on the occurrences of high-risk pharmaceutical in drain-laden surface water: A risk analysis of Yamuna River. Sci Total Environ. 2021;794:148484.

[60] Lin AY-C, Panchangam SC, Ciou P-S. High levels of perfluorochemicals in Taiwan's wastewater treatment plants and downstream rivers pose great risk to local aquatic ecosystems. Chemosphere. 2010;80:1167–1174.

[61] Amouei A, Asgharnia H, Fallah H, et al. Characteristics of effluent wastewater in hospitals of Babol University of medical sciences, Babol, Iran. Heal Scope. 2015;4:4–7.

[62] Aydin S, Aydin ME, Ulvi A, et al. Antibiotics in hospital effluents: Occurrence, contribution to urban wastewater, removal in a wastewater treatment plant, and environmental risk assessment. Environ Sci Pollut Res. 2019;26:544–558.

[63] He J-J, Zhao S-S, Zhang H, et al. Emergency management of medical wastewater in hospitals specializing in infectious diseases: A case study of huoshenshan hospital, Wuhan, China. Int J Environ Res Public Health. 2022;19(1):381.

[64] Ghawi AH. Study on the development of household wastewater treatment unit. J Ecol Eng. 2018;19:63–71.

[65] NWQMS. Australian guidelines for water recycling: Managing health and environmental risks (phase 2). Augmentation of Drinking Water Supplies. Natural Resource Ministerial Management Council (NRMMC), Environment Protection and Heritage Council (EPHC), Australian; 2008.

[66] OECD. Pharmaceutical Residues in Freshwater Hazards and Policy Responses. OECD Publ.; 2019.

[67] Gagnon E. Pharmaceutical disposal programs for the public: A Canadian perspective. Health Canada Environmental Impact Initiative; 2009. Available from: https://cdn.ymaws.com/www.productstewardship.us/resource/resmgr/imported/Takeback%20%282%29.pdf.

[68] UBA. Pharmaceuticals in the Environment—Avoidance, Reduction and Monitoring; 2014.

[69] EEA. Pharmaceuticals in the environment: Results of an EEA workshop. Copenhagen, European Environment Agency (EEA Technical Report No. 1); 2010.

[70] EU. Registration, evaluation, authorization and restriction of chemicals. Establishing a European Chemicals Agency, Amending Directive 1999/45/EC and repealing Council Regulation (EEC) No 793/93 and Commission Regulation (EC) No 1488/94 as Well as Council Dir; 2007.

[71] SR. Swiss Regulation of DETEC to Verify the Elimination Effect of Measures of Trace Organic Matter in Sewage Treatment Plants; 2016 [cited 2022 Mar 06]. Available from: www.admin.ch/opc/de/official-compilation/2016%0A/4049.pdf (in German). Fedlex.

[72] Water UK. National Guidance for Healthcare Wastewater Discharges; 2014. Available from: https://www.water.org.uk/national-guidance-healthcare-waste-water-discharges.

[73] Snyder SA, Adham S, Redding AM, et al. Role of membranes and activated carbon in the removal of endocrine disruptors and pharmaceuticals. Desalination. 2007;202:156–181.

[74] Carraro E, Bonetta S, Bonetta S. Hospital wastewater: Existing regulations and current trends in management. Handb Environ Chem. 2018;60:1–16.

[75] Eurostat. Sewage sludge production and disposal. Eur Comm. 2021.

[76] Radjenović J, Petrović M, Barceló D. Fate and distribution of pharmaceuticals in wastewater and sewage sludge of the Conventional Activated Sludge (CAS) and advanced Membrane Bioreactor (MBR) treatment. Water Res. 2009;43:831–841.

[77] Amedeo Vaccani ZS. International market developments in the sewage sludge treatment industry. A. Vaccani Partn. 2017;26.

[78] de Santiago-Martín A, Meffe R, Teijón G, et al. Pharmaceuticals and trace metals in the surface water used for crop irrigation: Risk to health or natural attenuation? Sci Total Environ. 2020;705:135825.

[79] Wang J, Shen J, Ye D, et al. Disinfection technology of hospital wastes and wastewater: Suggestions for disinfection strategy during coronavirus disease 2019 (COVID-19) pandemic in China. Environ Pollut. 2020;262:114665.

[80] Jia X-H, Feng L, Liu Y-Z, et al. Degradation behaviors and genetic toxicity variations of pyrazolone pharmaceuticals during chlorine dioxide disinfection process. Chem Eng J. 2018;345:156–164.

[81] Zhang A, Li Y, Song Y, et al. Characterization of pharmaceuticals and personal care products as N-nitrosodimethylamine precursors during disinfection processes using free chlorine and chlorine dioxide. J Hazard Mater. 2014;276:499–509.

[82] Lim S, Shi JL, von Gunten U, et al. Ozonation of organic compounds in water and wastewater: A critical review. Water Res. 2022;213:118053.

[83] Cha Y, Kim T-K, Lee J, et al. Degradation of iopromide during the UV-LED/chlorine reaction: Effect of wavelength, radical contribution, transformation products, and toxicity. J Hazard Mater. 2022;437:129371.

[84] Xiong R, Lu Z, Tang Q, et al. UV-LED/chlorine degradation of propranolol in water: Degradation pathway and product toxicity. Chemosphere. 2020;248:125957.

[85] Kim T-K, Kim T, Park H, et al. Degradation of ciprofloxacin and inactivation of ciprofloxacin resistant E. faecium during UV-LED (275 nm)/chlorine process. Chem Eng J. 2020;394:124803.

[86] Li B, Ma X, Li Q, et al. Factor affecting the role of radicals contribution at different wavelengths, degradation pathways and toxicity during UV-LED/chlorine process. Chem Eng J. 2020;392:124552.

[87] Yeom Y, Han J, Zhang X, et al. A review on the degradation efficiency, DBP formation, and toxicity variation in the UV/chlorine treatment of micropollutants. Chem Eng J. 2021;424:130053.

[88] Antoniou MG, Hey G, Rodríguez Vega S, et al. Required ozone doses for removing pharmaceuticals from wastewater effluents. Sci Total Environ. 2013;456–457:42–49.

[89] Okazaki S, Hirata A, Shogomori Y, et al. Radical reactions induced by ketoprofen in phospholipid membranes under ultraviolet light irradiation. J Photochem Photobiol B Biol. 2021;214:112090.

[90] Li J, Jiang J, Pang S, et al. Transformation of X-ray contrast media by conventional and advanced oxidation processes during water treatment: Efficiency, oxidation intermediates, and formation of iodinated byproducts. Water Res. 2020;185:116234.

[91] Meijide J, Lama G, Pazos M, et al. Ultraviolet-based heterogeneous advanced oxidation processes as technologies to remove pharmaceuticals from wastewater: An overview. J Environ Chem Eng. 2022;10:107630.

[92] Khan NA, Khan AH, Ahmed S, et al. Efficient removal of ibuprofen and ofloxacin pharmaceuticals using biofilm reactors for hospital wastewater treatment. Chemosphere. 2022;298:134243.

[93] Silva S, Cardoso VV, Duarte L, et al. Characterization of five portuguese wastewater treatment plants: Removal efficiency of pharmaceutical active compounds through conventional treatment processes and environmental risk. Appl Sci. 2021;11(16):7388.

[94] Segura Y, Cruz del Álamo A, Munoz M, et al. A comparative study among catalytic wet air oxidation, Fenton, and Photo-Fenton technologies for the on-site treatment of hospital wastewater. J Environ Manage. 2021;290:112624.

[95] Kooijman G, de Kreuk MK, Houtman C, et al. Perspectives of coagulation/flocculation for the removal of pharmaceuticals from domestic wastewater: A critical view at experimental procedures. J Water Process Eng. 2020;34:101161.

[96] Arikan OA. Degradation and metabolization of chlortetracycline during the anaerobic digestion of manure from medicated calves. J Hazard Mater. 2008;158:485–490.

[97] Zhang H, Wang XC, Zheng Y, et al. Removal of pharmaceutical active compounds in wastewater by constructed wetlands: Performance and mechanisms. J Environ Manage. 2023;325:116478.

[98] Alvarino T, Suárez S, Garrido M, et al. A UASB reactor coupled to a hybrid aerobic MBR as innovative plant configuration to enhance the removal of organic micropollutants. Chemosphere. 2016;144:452–458.

[99] Lima DR, Gomes AA, Lima EC, et al. Evaluation of efficiency and selectivity in the sorption process assisted by chemometric approaches: Removal of emerging contaminants from water. Spectrochim Acta Part A Mol Biomol Spectrosc. 2019;218:366–373.

[100] Yang G, Bao D, Zhang D, et al. Removal of antibiotics from water with an all-carbon 3D nanofiltration membrane. Nanoscale Res Lett. 2018;13:146.

[101] Schaaf N, Karlsson J, Borgendahl J, et al. Water and pharmaceuticals—A shared responsibility. Working Paper; 2016.

3 Metal Oxide Nanoparticle Adsorbents for Pharmaceutical Removal

Fatima Ezzahra Titchou and Mohamed Hamdani

3.1 INTRODUCTION

Pharmaceutical industries play an important role in producing synthetic chemicals and medications to help treat, prevent, or alleviate a wide range of illnesses and diseases. These synthetic chemicals are used for both human and veterinary applications, including the treatment of chronic diseases, infections, and other medical conditions. Generally, pharmaceutical ingredients are an essential part of modern healthcare. Approximately 3,000 different substances are used as pharmaceutical ingredients, including antibiotics, painkillers, antidepressants, anticoagulants, steroids, anticonvulsants, contraceptives, and chemotherapeutic agents [1]. The manufacturing of pharmaceuticals often results in the release of hazardous substances into the environment, leading to serious environmental pollution. This pollution can impact water sources, soil fertility, and air quality, significantly affecting the health and wellbeing of both humans and wildlife [2]. The presence of these products and their metabolites in urban wastewaters, resulting from human consumption, can pose a threat to the aquatic environment and can be toxic to its organisms. In addition, these residues may also enter the food chain through bioaccumulation and biomagnification processes [2, 3]. As such, it is critical that all pharmaceutical manufacturing processes take into consideration their potential impacts on the environment. Proper management techniques must be employed to ensure that any resulting waste is safely disposed of with minimal negative consequences.

Adsorption is the process by which a substance is bound to a surface or another material, usually through physical and/or chemical bonds [4]. It can be used to remove pollutants from a solution, to purify and isolate materials, or to remove gases from air. Adsorption plays an important role in many industries, such as pharmaceuticals, water treatment, and gas separation. Also, it has been used in various fields of science, including biochemistry, nanotechnology, and materials science [5]. This process can be used to treat both industrial wastewater and drinking water sources to ensure safe consumption [4, 6]. Metal oxide nanoparticles (MONPs) are increasingly being used as adsorbents for the removal of pharmaceutical products from water.

DOI: 10.1201/9781003340164-3

Metal oxides present a class of the most important and widely characterized solid catalysts. Depending on their synthesis methods, metal oxides may come in various shapes and sizes [7]. They can be used for several applications including the environmental remediation of pollutants, sensing and energy production, conversion, and storage. Generally, their application depends highly to their functional properties, including optical, electrical, chemical, and catalytic performances. Their proprieties are related to their crystal structure, composition, morphology, intrinsic defects, and doping [8]. Among the newly developed technologies, nanotechnology has gained much attention and achieved various applications. It has been applied for synthesizing MONPs, providing a new set of materials that exhibit distinct physical and chemical properties [9]. MONPs range in size from 1 to 100 nm (1 nm is equal to 10^{-9} m) and have various distinct features such as high reactivity, selectivity, lower process energy consumption, longer lifetime, and recyclability [9]. Many examples of these MONPs can be found, such as titania (TiO_2), iron oxides (Fe_2O_3 and Fe_3O_4), zinc oxide (ZnO), and ceria (CeO_2) [9]. These nanoparticles possess a high surface area that is capable of absorbing a wide range of pharmaceuticals, including antibiotics, hormones, and other drugs that represent a rapidly growing threat to ecosystems worldwide [6, 10, 11].

Considering MONPs, adsorption process is based on the interactions between these materials and the target pollutants in water [12]. It involves both physical and chemical interactions that allow for the efficient removal of pharmaceuticals from water sources. This chapter discusses the potential of MONPs as adsorbents for the removal of pharmaceutical compounds from wastewater and the challenges associated with their use.

3.2   METAL OXIDE NANOPARTICLES

From both a scientific and technological viewpoint, MONPs are an important class of materials. They can be prepared from joining two metal portions with oxy (–O–) or hydroxyl bridges (–OH–) to produce metal oxide- or metal hydroxide-based nanoparticles [13]. Their applications depend on their characteristics, which are mainly related to their synthesizing method and operational parameters [14]. Metal oxide nanoparticles can be synthesized using three distinct approaches: physical, chemical, and biological. Different methods, including gas phase synthesis, hydrothermal synthesis, chemical deposition, sol–gel technology, electrospinning, green technology employing natural extracts, intracellular synthesis with bacteria, electrochemical synthesis, and other combined methods are used for the preparation of different MONPs with different morphologies such as triangles, stars, spheres, nanowires, nanocubes, and nanorods (Figure 3.1) [15–17].

The most frequent synthesis methods of MONPs include: *i. Sol–gel method*, which involves the conversion of a metal-containing precursor solution (sol) into a solid gel network through hydrolysis and condensation reactions. The gel is then dried and calcined to obtain MONPs. Advantages of this method include low-temperature processing, control over particle size, and the ability to incorporate dopants. However, it can be a time-consuming process, sometimes needs expensive organometallic

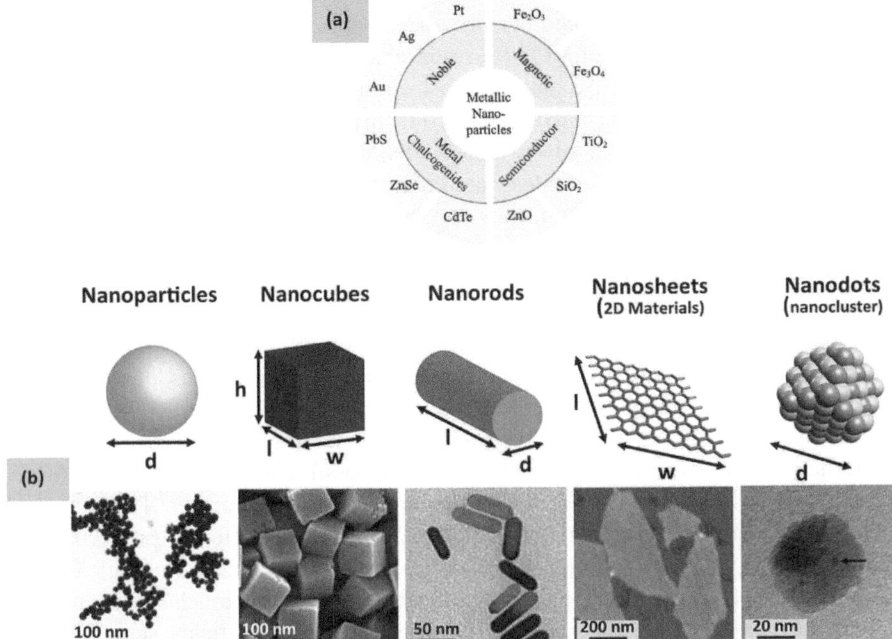

FIGURE 3.1   (a) Classification and (b) schematic representation of metallic nanoparticles.
Source: Reprinted from [6] and [18], Copyright (2022), with permission from Elsevier.

precursors, and may require multiple steps [19, 20]. *ii. Co-precipitation method,* where metal salts are simultaneously precipitated by the addition of a precipitating agent. This method is relatively simple, cost effective, and allows for the synthesis of nanoparticles with a narrow size distribution. However, it may suffer from many disadvantages, mainly the possible precipitation of impurities with the product, the reproducibility, time requirements, and limitations when treating reactants with significantly different precipitation rates [21]. *iii. Hydrothermal/solvothermal method,* which involves the reaction of metal salts or precursors at high pressure and temperature in an aqueous or organic solvent. This method allows for the production of well-defined MONPs with controlled size and morphology. It is versatile and applicable to a wide range of materials. On the downside, it requires specialized equipment, and optimizing the reaction conditions can be challenging [19]. *iv. Micro emulsion method,* in which metal salts are dissolved in a micro emulsion system consisting of water, oil, and surfactant. Nanoparticles are formed by chemical reduction or precipitation within the micro emulsion droplets. This method offers excellent control over particle size, narrow size distribution, and the ability to produce nanoparticles with specific compositions. However, it can be complex and requires precise control of the micro emulsion parameters [19]. It is important to recognize that each fabrication method has its own advantages and disadvantages. As scientists strive to meet specific application requirements, they are actively exploring hybrid techniques that combine the strengths of different approaches, aiming to create MONPs with customized properties and enhanced performances [14].

Besides the morphology of MONPs, researchers have highlighted the importance of optimizing their reactivity, stability, cost efficiency, and environmental sustainability [22]. These materials can be prepared from various kinds of metals and metal oxides, including iron oxide, silver, zinc oxide, and titanium dioxide, and they can be classified into pure metal oxides, rare earth elements, dopant-incorporated, and nanocomposites [23]. In addition, they can be coated with other materials such as surfactants, phytochemicals, biopolymers, antiviral therapeutics, or antibiotics to improve their functional performances. Their composition can be determined using X-ray diffraction (XRD) methods [24]. Typically, the size, shape, and aggregation state of these nanoparticles can be determined using microscopy methods, such as scanning electron microscopy (SEM), transmission electron microscopy (TEM), or atomic force microscopy (AFM). Moreover, their electrical characteristics (zeta-potential) are usually measured using laser-based electrophoresis instruments [24]. With the help of advanced technology, it is now possible to produce various mixed metal oxides by combining two or more metals in an oxide matrix. This approach offers unlimited flexibility for producing materials with different properties and characteristics, leading to relatively high performance in various technological applications [25]. As reported by Tee et al. [6], these nanoparticles are classified into four major categories: noble metal nanoparticles, magnetic nanoparticles, semiconductor nanoparticles, and metallic chalcogenide nanoparticles, as shown in Figure 3.1. Each category of metallic nanoparticles has its own distinct properties and uses. Noble metal nanoparticles such as gold and silver are used for medical imaging, catalysis, sensing, imaging, therapy, metamaterials, and light harvesting [26]. Magnetic nanoparticles such as iron oxides are used in magnetic resonance imaging as contrast agents and in drug delivery systems [22, 27]. Semiconductor nanostructures such as cadmium selenide (CdSe) quantum dots are used for photovoltaic applications and can be applied as catalysts, biological fluorescent labels, light-emitting diodes, solar cells, electrolytes for hydrogen production, photocatalysts, adsorption materials, fuel cells, electrocatalysts, and sensors [6, 22, 28].

3.3   METAL OXIDE NANOPARTICLES AS ADSORBENTS

The use of nanomaterials in adsorption process has been extensively studied due to its potential benefits over traditional techniques [29–31]. Adsorption can be broadly categorized into physical and chemical adsorption, where each category is determined by the type of interactions and the affinity of the target contaminant toward different functional groups. The mechanisms of pharmaceuticals adsorption by this class of materials may be due to: *i. Electrostatic interaction*, which is based on the adsorbent's surface charge and the ionization of the adsorbate [31]. *ii. Hydrogen interactions*, where hydrogen bond formation between the adsorbate and adsorbent's surface groups is a physical bond that occurs between a hydrogen atom (H) and an electronegative atom, such as fluorine (F), oxygen (O), or nitrogen (N) of another molecule (adsorbent or adsorbate) [32]. *iii. Hydrophobic interaction*, which may result from the tendency of non-polar molecules to aggregate together in an aqueous medium [33]. *iv. π interactions*, especially cation–π, anion–π, π–π, and π donor–acceptor interactions that occur between the electron-rich π system and a metal cation, anion, or another π system, respectively [31, 34]. *v. Ion exchange*, which is commonly observed in the case of clay materials, where adsorbate ions are exchanged with similar ones from the surface of the adsorbent [35]. *vi. Complexation*, which results in the formation of stable metal–ligand complexed with metals such as Ca^{2+}, Mg^{2+}, Cu^{2+}, Fe^{2+}, and Zn^{2+} [36].

To understand the mechanism of the adsorption of pharmaceuticals with MONPs, various models of adsorption isotherms have been proposed so far in establishing the adsorption equilibrium under constant temperature [37]. As reported by Singh et al. [29], an L curve pattern in isotherms is the most commonly obtained form by researchers for the adsorption of pharmaceutical products. Moreover, they affirmed that the most reported models illustrating the adsorption of pharmaceutical compounds and their metabolites are Freundlich, Langmuir, and Redlich–Peterson isotherms. The Freundlich model is based on the assumption of a heterogeneous/non-homogenous surface with non-uniform distribution of heat of sorption over the surface and suggests that the adsorption capacity depends on the equilibrium concentration of the pollutant. The Langmuir isotherm presents a model in which the adsorption occurs as a monolayer on a surface with a homogeneous energy distribution. Finally, the Redlich–Peterson isotherm describes the adsorption pathway with a Gaussian energy distribution onto a non-homogeneous surface based on the potential energy [4, 29].

Moreover, the adsorption of contaminants onto these materials depends on different parameters that affect the rate of the reaction. These parameters include the adsorbent–adsorbate affinity, initial concentration, contact time, adsorbent dosage, type of catalyst, pH, and temperature (Figure 3.2) [4, 38].

When considering catalysts for adsorption, it is important to discuss surface areas, reusability, and stability, which influence the adsorption capacity toward pollutants. The surface area of the adsorbent can be determined using Brunauer–Emmett–Teller (BET) analysis, which gives informations about the specific surface area, pore volume, and area, as well as the pore size distribution of the material and hence the diffusion of the adsorbate inside the adsorbent [39]. The surface area of a catalyst plays a crucial role in its adsorption capacity. A catalyst with a larger surface area provides more active sites for adsorption, resulting in enhanced contaminant removal [40]. The reusability of the adsorbate is an important aspect from an economic and environmental perspective. A catalyst that can be easily regenerated and reused without significant loss of its adsorption capacity is desirable. Generally, the reusability depends on factors such as the nature of the adsorbent and the strength of the adsorbate–catalyst interaction. If the adsorbent can be easily regenerated (e.g., by washing or desorption), the catalyst can be reused multiple times, making it cost effective and sustainable [41]. Considering the long-term performance of an adsorbent, catalyst stability presents a crucial parameter for efficient application. Stability refers to the ability of the catalyst to maintain its structural integrity and adsorption capacity under the operating conditions [42]. Stability π can be influenced by factors such as the chemical composition of the catalyst, its resistance to harsh environments, and its susceptibility to degradation or fouling. A stable catalyst ensures consistent and reliable adsorption performance over time [41, 43].

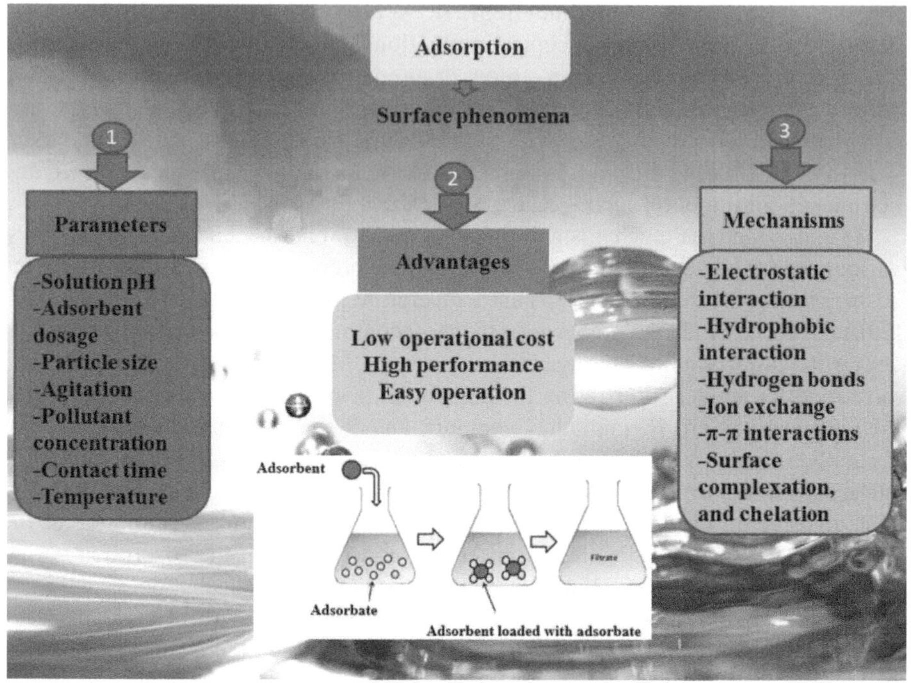

FIGURE 3.2   Adsorption phenomena of pollutants.

Source: Reprinted from [30]. Copyright (2022), with permission from Elsevier.

3.3.1 IRON OXIDES

Iron is one of the most used elements for water and wastewater remediation [44]. Its natural or synthetic materials are among the most cost-effective semiconductors studied in the literature for adsorption and photocatalytic applications, largely due to their wide band gaps [31]. Iron oxides can take many different forms, including goethite (α-FeOOH), lepidocrocite (γ-FeOOH), akaganeite (β-FeOOH), ferrihydrite ($Fe_5HO_8\cdot4H_2O$), hematite (Fe_2O_3), magnetite (Fe_3O_4), wustite ($Fe_{1-x}O$), maghemite (γ-Fe_2O_3), and hematite (α-Fe_2O_3) (Figure 3.3).

Among these phases, Fe_3O_4 and γ-Fe_2O_3 have easier separation and recovery after water treatment [45]. Generally, they can be prepared by various methods such as co-precipitation, thermal decomposition, sol–gel synthesis, and hydrothermal methods [45]. The characteristics of these materials and other iron oxide nanoparticles are discussed by Kharisov et al. [46].

From a general point of view, iron-based materials are the most cost-effective semiconducting materials. Various studies have shown their efficiency toward the removal of pharmaceutical pollutants from seawater [31]. Moreover, transition metal ferrites, both doped and undoped, and their composites are promising materials for catalytic purposes. They present attractive candidates for the treatment of pollutants. Based on their crystallinity, spinel (MFe_2O_4), hexagonal ($MFe_{12}O_{19}$), and garnet

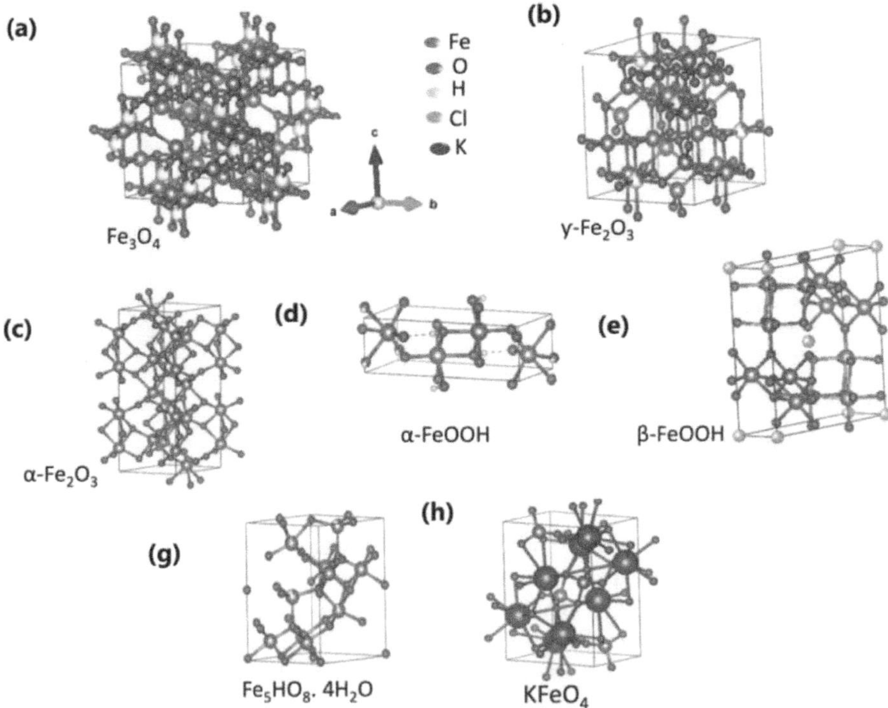

FIGURE 3.3   The crystallographic unit cell of (a) magnetite, (b) hematite, (c) maghemite, (d) goethite, (e) akaganeite, (f) ferrihydrite, and (g) potassium ferrate.

Source: Reprinted from [31], Copyright (2022), with permission from Elsevier.

($M_3Fe_5O_{12}$) forms, where M is one or more bivalent transition metals (Fe, Mn, Co, Cu, Ni, and Zn), represent the most reported ones [47]. Compositing or impregnating different materials allows for the development of hybrid materials characterized by dual functionality. Clay–iron oxide nanoparticles and activated carbon (AC)/graphene nanocomposite material–iron nanoparticles are also found to constitute a powerful class of materials for the treatment of pharmaceutical pollutants [48, 49].

3.3.2   ZINC OXIDES

ZnO NPs are inexpensive and safe semiconductors that have gained attention in wide range of applications. Zinc oxide is the second most abundant metal oxide found naturally on Earth after iron oxide, and it exists in three distinct phases (Figure 3.4) [50].

Zinc oxide nanoparticles are easily prepared materials that can be synthesized by sol–gel, microwave irradiation, colloid thermal synthesis, electromechanical synthesis, precipitation synthesis, micro emulsion systems, sono-chemical synthesis, precipitation pyrolysis, and thermal decomposition [45]. Besides their application in various industrial fields, such as catalysis, UV light-emitting devices, solar cells, paints,

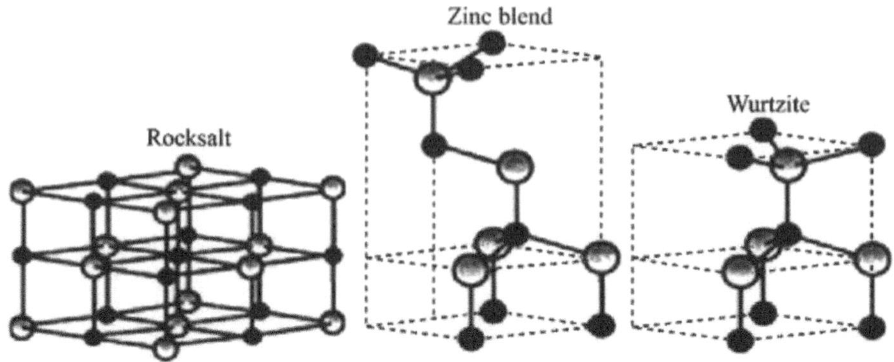

FIGURE 3.4   ZnO nanoparticles different phases.

Source: Reprinted from [51], Copyright (2022), with permission from Elsevier.

biomedicine, electronic devices, and cosmetics, in water bodies, zinc oxide nanoparticles have shown good adsorption ability for several contaminants in their organic and inorganic forms [52, 53]. These materials are mainly applied for the removal of metals and dyes, as reported by Samuel et al. [52]. Nevertheless, zinc oxide materials have a limited adsorption capacity due to their small specific surface area and higher colloidal stability in water, which results in difficult separation and recovery after use [54]. Zinc oxides can be synthesized in nanoforms of spheres, rods, flowers, disks, and walls [55]. As reported by Sharma et al. [56], zinc nanoflowers have a high surface area that provides more active sites for the adsorption of pollutants. Generally, the performance of zinc oxide nanoparticles can be improved by synthesizing new doped ZnO, ZnO-based nanocomposites with other metal oxides, carbon-based materials, and spinels [51]. ZnO nanoparticles synthesized through chemical approaches possess a significant tendency to aggregate due to their considerable surface-area-to-volume ratio. Therefore, researchers have aimed to develop green methods that may help in overcoming this limitation. Green ZnO nanoparticles synthesized from plants are characterized by their ease of use, time savings, stability, large-scale synthesis, and non-toxic by-products, as reported by Kavitha et al. [57]. The prepared activated carbon-based adsorbents modified with ZnO or $Zn(OH)_2$ nanoparticles have an improved surface area attributed to the increased number of oxygen-containing functional groups and the structural bonding of zinc to the surface of activated carbon. These enhancements contribute to the adsorption selectivity of the material [58]. In addition, metal–organic frameworks (MOFs) represent a new class of developed porous materials with metal clusters and organic ligands characterized by their high porosity and structural adjustment, flexibility, specific surface area, low density, and uniform size of pores [59].

3.3.3 Titanium Oxides

Generally, TiO_2 nanoparticles represent common nanophotocatalysts characterized by their nontoxicity, stability, and important photocatalytic activity [60]. They are

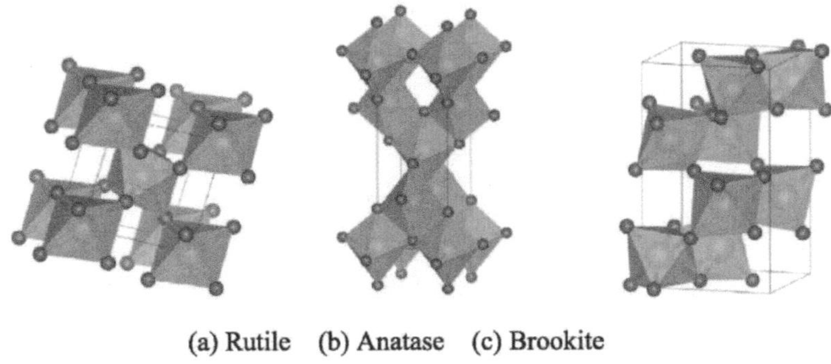

FIGURE 3.5   Different TiO_2 phases.

Source: Reprinted from [60], Copyright (2023), with permission from Elsevier.

synthesized through hydrothermal, solvothermal, sol–gel, chemical vapor deposition, electro spinning, ball milling, atomic layer deposition, sputtering, anodic oxidation, and microwave methods [45]. They are widely used as photocatalysts because of their low cost, plentiful availability, and good properties such as nontoxicity, eco-friendliness, biocompatibility, and high photostability [61]. They can be found in three different phases, namely rutile, brookite, and anatase, as presented in Figure 3.5, where anatase is known as the most promising phase for adsorbing pollutants [62].

However, due to their agglomeration in aqueous solutions, TiO_2 materials have poor pollutant adsorption abilities. Many studies have aimed to improve the surface area, porosity, and attraction toward the organic pollutants by preparing hybrid materials based on TiO_2 using carbonaceous materials such as activated carbon [63, 64]. For instance, in the case of ceftriaxone, the adsorption capacity of modified TiO_2 was much higher than that of pure TiO_2. Results showed that the maximum adsorption capacities of activated carbon-based TiO_2, activated carbon, and TiO_2 nanomaterials were determined to be up to 844.8, 728.3, and 272.4 mg g^{-1}, respectively [61]. Similarly, the modification of TiO_2-based junctions in carbon fibers increases the pores volume from 0.03 to 0.99 cm^3 g^{-1} [65].

3.3.4   SILICA OXIDES

Silicon dioxide, a major component of sand found naturally as quartz, is an amorphous material [66]. As in the case of the mentioned metal oxides, SiO_2 nanoparticles are also typical adsorbents used for photocatalysis applications. Many works have reported the importance of doping SiO_2 with metal atoms in order to increase the surface area of SiO_2 particles, thereby improving their adsorption ability [67]. The adsorption properties of SiO_2 nanoparticles are related to the porous structure and the chemical activity of each particle's surface. The surface characteristics depend on the concentration of OH groups, temperature, energy distribution of the silanols, and the presence of siloxane (Si–O–Si) bridges [68]. As presented in Figure 3.6, the SiO_2 surface may contain siloxane connections (Si–O–Si) and/or silanol (S–OH),

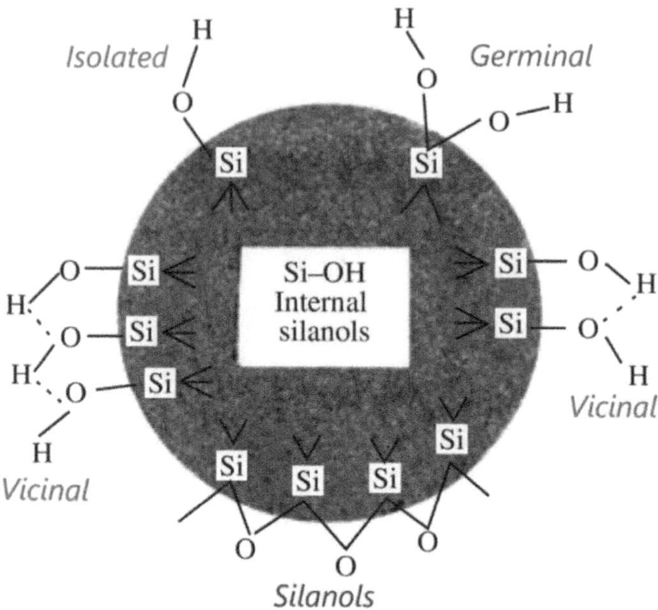

FIGURE 3.6 Chemistry of the SiO_2 surface.

Source: Reprinted from [69], Copyright (2018), with permission from Elsevier.

where this later can be classified as vicinal, germinal, and isolated ones [68, 69]. SiO_2-based nanoparticle materials may be prepared by sol–gel, water-in-oil (reverse) micro emulsion (micelle), flame spray pyrolysis, and blending processes, and also via applying $SiCl_4$ as the starting material [69]. Zr-doped SiO_2–Al_2O_3 mixed oxide [70], SiO_2–Al_2O_3 mixed oxides [71], cellulose nanocrystal/SiO_2 composites [72], polyamidoamine/silica [73], polyaniline/SiO_2 composite [74], and Fe_3O_4/SiO_2 [75] present some of the reported examples of SiO_2-based materials used for the adsorption of pharmaceuticals.

3.4   DESORPTION AND REGENERATION

After adsorption, various analytical techniques can be used to evaluate the adsorbent characteristics. Generally, instrumental techniques such as Fourier transform infrared spectroscopy (FTIR), SEM, energy-dispersive X-ray coupled with scanning electron microscopy (EDX-SEM), BET spectroscopy, and XRD spectroscopy represent the most used techniques to characterize the adsorbent after adsorption. This characterization aims generally to confirm the adsorption of pollutants. For instance, using FTIR, changes in the transmittance peaks such as the appearance, disappearance, or size reduction of some of them presents a confirmation of the involvement of the surface functional groups in the adsorption process. Also, a reduction in the surface area, pore size, pore volume, and a modification in the chemical composition may be observed [76]. Hence, the surface modification after adsorption should be taken into consideration for adsorbent reuse. An effective catalyst for adsorption should possess a high surface area to maximize the adsorption capacity, demonstrate reusability by maintaining adsorption efficiency over multiple cycles, and exhibit stability to withstand operating conditions without a significant degradation or loss of performance [43]. The cyclic application and use of regenerated nanomaterials are essential to consider these metal oxide nanomaterials for the treatment of pharmaceuticals from wastewater. The efficiency and the cost effectiveness of the support are related to its desorption–adsorption performance. Therefore, it is important to ensure that the regeneration process does not significantly degrade the adsorbent or reduce its adsorption capacity. Proper optimization and monitoring of desorption and regeneration processes are crucial for maintaining the long-term effectiveness and sustainability of adsorbents. Regeneration depends on desorption time, pH, recovery efficiency, and the eluent type. The most used eluents for the desorption of pharmaceuticals from nanomaterials are an acidic or alkaline solution, thiourea, HCl, and EDTA, which are selected according to the adsorption mechanism and the nature of adsorbent and adsorbates [29].

3.5 ADVANTAGES AND LIMITATIONS OF METAL OXIDE NANOPARTICLES OVER CONVENTIONAL ADSORBENTS

In terms of environmental remediation of pollutants, adsorption technology represents a versatile approach with several advantages such as easy and flexible operation, eco-friendly and low-cost methods, no generation of reaction intermediates, and easy reusability (adsorbent material) [77].

Metal oxide nanomaterials present an important class of materials because of their special characteristics and advantages. They are easily prepared through cost-effective and simple methods. The most reported studies in the literature demonstrated their long-term efficiencies in labs for a short period. In addition, their surface modification enhances the reactivity, photocatalytic activity, and selectivity toward contaminants [14].

As mentioned, MONPs show promising results in batch reactors, but their pilot-scale applications have different challenges, including their commercialization, shifting cost effectiveness, potential human and technical hurdles, and environmental risk. As a general point of view, the large-scale application of this technology is still challenging. It is also essential to consider engineering aspects [29]. Table 3.1 presents a comparison between metal oxide nanomaterials and conventional adsorbents for the removal of pharmaceuticals from wastewaters based on pharmaceutical uptake Q_t (mg g^{-1}), which represents the amount of adsorbed pollutant per gram of adsorbent (maximum adsorption capacity) [78].

During the adsorption process and when involving MONPs, there is a possibility of metal ion release into the solution [88]. This phenomenon can occur due to several reasons such as the surface reactivity, pH-dependent dissolution, surface ligand or coating detachment, and mechanical or environmental factors such as agitation and temperature. The release of metal ions during adsorption by MONPs can have implications for environmental and biological systems [89]. These released metal ions

TABLE 3.1

Adsorption of Pharmaceuticals Using MONPs and Conventional Adsorbents

Adsorbents	Pollutants	Conditions	Efficiency	References
Multi-wall carbon nanotubes/Al_2O_3	Ciprofloxacin (CIP)	$C_0 = 10-100$ mg L^{-1}, adsorbent dosage = 1.2 g L^{-1}, T = 45 °C, pH = 7.0, surface area = 441.2 m^2 g^{-1}	$Q_{max} = 83.47$ mg g^{-1}, contact time = 1 h	[79]
Biochar/Fe composites	Chlortetracycline	$C_0 = 25$ mg L^{-1}, adsorbent dosage = 0.8 g L^{-1}, T = 25 °C, pH = 4.0, surface area = 170.2 m^2 g^{-1}	$Q_{max} = 97.63$ mg g^{-1}, contact time = 24 h	[80]
Hydrous ferric oxide	Tetracycline (TC)	$C_0 = 40$ mg L^{-1}, adsorbent dosage = 1 g L^{-1}, T = 45 °C, pH = neutral, surface area = 226.796 m^2 g^{-1}	$Q_{max} = 99.49$ mg g^{-1}, contact time = 200 min	[81]
Fe_3O_4–chitosan nano-adsorbent	Metronidazole	$C_0 = 10$ mg L^{-1}, adsorbent dosage = 2 g L^{-1}, pH = 3.0, T = 25 °C, surface area = 112 m^2 g^{-1}	$Q_{max} = 97$ mg g^{-1}, contact time = 90 min	[82]
Iron-pillared montmorillonite	Levofloxacin	$C_0 = 20$ mg L^{-1}, adsorbent dosage = 0.5 g L^{-1}, pH = 6.8, T = 25 °C, surface area = 20 m^2 g^{-1}	$Q_{max} = 48.61$ mg g^{-1}, contact time = 48 h	[83]
Wood-chip biochar	Levofloxacin	$C_0 = 50$ and 150 mg L^{-1}, adsorbent dosage = 10 g L^{-1}, pH = 6.5, T = 30 °C, surface area = 312 m^2 g^{-1}	$Q_{max} = 7.72$ mg g^{-1}, contact time = 24 h	[84]
–TiO_2 –AC –AC–TiO_2 (ACT-4)	Ceftriaxone	$C_0 = 100$ mg L^{-1}, adsorbent dosage = 1 g L^{-1}, ambient temperature, pH = n.d., surface area of 1223, 1132, and 88 m^2 g^{-1} for ACT-4, AC, and TiO_2, respectively.	$Q_{max}TiO_2 = 272.4$ mg g^{-1} $Q_{max}AC = 728.3$ mg g^{-1} $Q_{max}ACT$-4 = 844.8 mg g^{-1}. Contact time = 24 h	[61]
Magnetic Fe_3O_4 nanoparticles	Ceftriaxone	$C_0 = 10$ mg L^{-1}, adsorbent dosage = 1.99 g L^{-1}, pH = 3.14, T = 25 °C, surface area = n.d.	$Q_{max} = 28.93$ mg g^{-1}, contact time = 90 min	[85]

TABLE 3.1 (*Continued*)

Adsorption of Pharmaceuticals Using MONPs and Conventional Adsorbents

Adsorbents	Pollutants	Conditions	Efficiency	References
Powdered AC	Carbamazepine (Cbz) and sildenafil citrate (Sil)	$C_0 = 5$ and 0.2 mg L^{-1} of the pharmaceuticals, adsorbent dosage = 100 mg L^{-1}, T = 25 °C, agitation = 90 rpm, pH = 6.15 and 5.97, surface area = 1328.3 m^2 g^{-1}	$Q_{max} = 220$ mg g^{-1} and 395 mg g^{-1} for Cbz and Sil, respectively, contact time = 32 h	[86]
Zr-MOF nanoparticles	Amoxicillin (AMX)	$C_0 = 100$ mg L^{-1}, adsorbent dosage = 0.5 g L^{-1}, pH = 7, T = 25.15 °C, surface area = 1084 m^2 g^{-1}	$Q_{max} = 2.3$ mg g^{-1}, contact time = 4 h	[87]

may contribute to water pollution, affect aquatic organisms, or have toxic effects, depending on their concentration and bioavailability [90]. To prevent this limitation, various strategies can be employed, such as modifying the surface of MONPs with stable coatings or functional groups, pH control, choosing MONPs that are known for their stability and low tendency for ions release, and applying post-treatment steps like rinsing or filtering that can help removing any residual metal ions or nanoparticles from the solution [64, 91].

3.6 REMOVAL OF PHARMACEUTICAL COMPOUNDS WITH DIFFERENT METAL OXIDE NANOPARTICLES

Synthesizing different metal oxide-based materials has gained increased interest in the literature, thereby promoting the application of this class of components for the treatment of wastewater, especially pharmaceutical effluents. In this section, different applications of synthesized metal oxide-based materials for the removal of pharmaceutical pollutants are reported.

Hybrid materials show better efficiency in removing pharmaceuticals compared to non-hybrid ones. For instance, the efficiency of the removal of CIP and TC was higher in the case of three-dimensional (3D) reduced graphene oxide/nano-Fe_3O_4 (3D-rGO/Fe_3O_4) compared to 3D-rGO, GO, rGO, and expanded graphite (EG). Results showed that the adsorption of 200 µmol L^{-1} of these pollutants followed a Langmuir isotherm, with a higher adsorption amount of 2.78 mmol g^{-1} for CIP and 4.76 mmol g^{-1} for TC using 0.33 g L^{-1} of synthesized 3D-rGO/Fe_3O_4 at pH 6 and for 72 h [92]. Moreover, authors confirmed the cost effectiveness of this material via adsorption–desorption analysis. This prepared adsorbent showed a higher removal efficiency and adsorption amount even after ten cycles, affirming its relatively stable successive adsorption–regeneration.

Similarly, Gu et al. [93] compared the adsorption of pharmaceuticals using a magnetic Fe/porous carbon hybrid material (MagFePC) and activated carbon. The removal of tetracycline, oxytetracycline, tetracycline hydrochloride, chlortetracycline, norfloxacin, levofloxacin, pefloxacin, sulfadiazine, and trimethoprim followed a Langmuir isotherm. For experiments using 0.05 g L^{-1} of the adsorbent and 80 mg L^{-1} of antibiotics at pH 5 and 25 °C for 24 h, results showed that the adsorption capacity of the magnetic Fe/porous carbon hybrid material was six times higher than that obtained in the case of activated carbon. Under the same operating conditions for tetracycline, the maximum adsorption capacity was 1110.59 mg g^{-1} in the case of MagFePC and 275.48 mg g^{-1} in the case of activated carbon.

Likewise, according to previous studies, ZnO-based materials showed good efficiency (Table 3.2). They are efficient even in complex matrices, especially for small-scale industries.

As reported by Dhiman et al. [107], in the case of ciprofloxacin, ofloxacin, diclofenac, and paracetamol mixtures, even if results indicate the presence of an antagonistic effect on the adsorption of each drug due to the presence of the second component, from adsorption capacity results, ZnO nanoparticles have the possibility for the simultaneous removal of similar drugs. The adsorption capacity of ZnO nanoparticles can be enhanced through synthesizing hybrid materials like the new composite ZnO-based nanoparticles applied for the adsorption of tetracycline from wastewater. Comparing a zeolitic imidazolate framework/ZnO nanocomposite (ZnO@ZIF-67) and zinc oxide, the hybrid composite had higher efficiency at pH 8 compared to the ZnO nanoparticles [59]. This new material showed higher adsorption capacity (Figure 3.7) due to its surface characteristics.

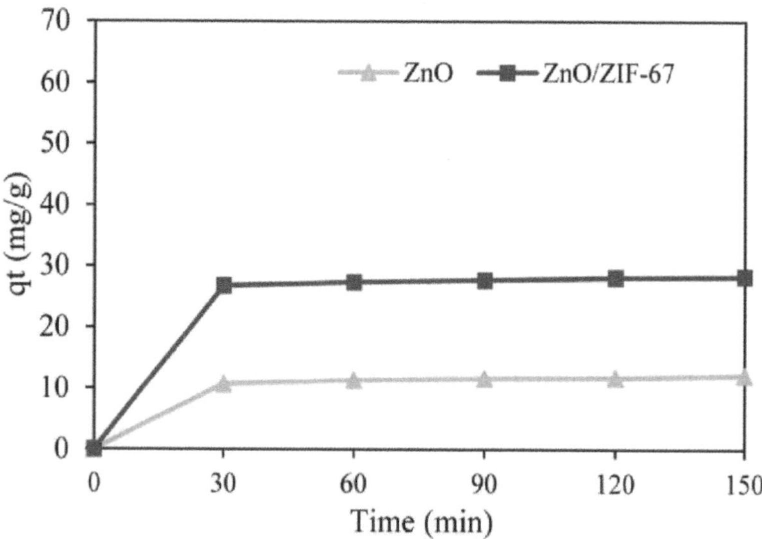

FIGURE 3.7   Removal of tetracycline by ZnO and ZnO@ZIF-67.

Source: Reprinted from [59], Copyright (2021), with permission from Elsevier.

TABLE 3.2

Applications of MONPs for the Removal of Pharmaceutical Compounds

Adsorbents	Pollutants	Conditions	Isotherms	Efficiency	References
Fe_3O_4/polyacrylonitrile composite electrospun nanofiber mat	TC	$C_0 = 22$ mg L^{-1}, adsorbent dosage = 0.5 g L^{-1}, pH = 6, T= 25 °C, surface area = 17.8 m^2 g^{-1}	Langmuir	$Q_{max} = 257.07$ mg g^{-1}, contact time = n.d.	[94]
Fe_2O_3 nanoparticle-assisted powdered AC	TC	$C_0 = 200$ mg L^{-1}, adsorbent dosage = 0.5 g L^{-1}, pH = 3, T= 20 °C, surface area = n.d.	Langmuir	$Q_{max} = 140$ mg g^{-1}, contact time = 300 min	[95]
Fe_3O_4 magnetic nanoparticles on graphene oxide	TC	$C_0 =$ from 12.5 mg L^{-1} to 75 mg L^{-1}, adsorbent dosage = 1.6 g L^{-1}, T = 25 °C, pH = n.d., surface area = n.d.	Freundlich	$Q_{max} = 95$ mg g^{-1}, contact time = 24 h	[96]
α-Fe_2O_3 nanoparticles	Cephalexin	$C_0 = 200$ mg L^{-1}, adsorbent dosage = 2 g L^{-1}, pH = 7.5, T = 55 °C, surface area = n.d.	Langmuir	$Q_{max} = 70$ mg g^{-1}, contact time = 24 h	[97]
$NiFe_2O_4$ nanoparticles	AMX	$C_0 = 200$ mg L^{-1}, adsorbent dosage = 1 g L^{-1}, T = 25 °C, pH = 7, surface area = 212 m^2 g^{-1}	Freundlich	$Q_{max} = 104$ mg g^{-1}, contact time = 24 h	[98]
ZnO nanoparticles	Ibuprofen	$C_0 = 100$ ppm, adsorbent dosage = 100 mg, T = room temperature, pH = n.d., surface area = n.d.	n.d.	$Q_{max} = 225$ mg g^{-1}, contact time = 1 h	[99]
Co/Cr-co-doped ZnO nanoparticle	TC	$C_0 = 1000$ mg L^{-1}, adsorbent dosage = 10 mg, pH = 7, T = room temperature, surface area = 75.38 m^2 g^{-1}	Langmuir	$Q_{max} = 874.46$ mg g^{-1}, contact time = 30 min	[100]
Montmorillonite coated with ZnO	Cephalexin	$C_0 = 25$ mg L^{-1}, adsorbent dosage = 4 g L^{-1}, pH = 2, T = 10 °C, surface area = n.d.	Langmuir	$Q_{max} = 17.09$ mg g^{-1}, contact time = 2 h	[101]
Pistachio shell powder coated with ZnO nanoparticles	TC, AMX, and CIP	$C_0 = 60$ mg L^{-1}, adsorbent dosage = 0.1 g L^{-1}, pH = 5, T = 25 °C, surface area = 4.24 m^2 g^{-1}	–AMX: Langmuir –TEC and CIP: Fruendlich	$Q_{max} = 92.450, 98.717$, and 132.24 mg g^{-1} for TC, AMX, and CIP, respectively, contact time = 2 h	[102]

(*Continued*)

TABLE 3.2 (*Continued*)

Applications of MONPs for the Removal of Pharmaceutical Compounds

Adsorbents	Pollutants	Conditions	Isotherms	Efficiency	References
g-C_3N_4-ZnO-BaTiO$_3$ nanocomposite	TC	C_0 = 60 mg L^{-1}, adsorbent dosage = 1 g L^{-1}, T = room temperature, pH = 4.59, surface area = n.d.	Langmuir	Q_{max} = 209.19 mg g^{-1}, contact time = 3 h	[103]
Pistachio shell coated with ZnO nanoparticles	TC	C_0 = 70 ppm, adsorbent dosage = 0.08 g mL^{-1}, T = 25 °C, pH = 4, surface area = n.d.	Freundlich	Q_{max} = 95.06 mg g^{-1}, contact time = 2 h	[104]
Montmorillonite coated with TiO$_2$	Cephalexin	C_0 = 25 mg L^{-1}, adsorbent dosage = 4 g L^{-1}, T = 10 °C, pH = 2, surface area = n.d.	Langmuir	Q_{max} = 49.26 mg g^{-1}, contact time = 2 h	[101]
AgO/MgO/FeO@Si$_3$N4 nanocomposite	TC	C_0 = 30 mg L^{-1}, adsorbent dosage = 30 mg, T = 30 °C, pH = 8, surface area = 8.34 m^2 g^{-1}	Freundlich	Q_{max} = 172.41 mg g^{-1}, contact time = 90 min	[105]
Silica–clay nanocomposites	CIP	C_0 = 50 mg L^{-1}, adsorbent dosage = 230 mg, pH = 7.5, T = 25 °C, surface area = n.d.	n.d.	n.d.	[106]

Synthesizing a porous TiO_2 nanocrystallite presented a challenge that has gained increased interest from researchers aiming to benefit from the porous structure of titanium oxide in order to remove pollutants from wastewaters. Therefore, as mentioned before, the aggregation of these nanoparticles affects their adsorption capacities. Gan et al. [108] studied the impact of post- treatment on the intrinsic characteristics of TiO_2 nanoparticles on their adsorption capacity. They evaluated the effect of hydrothermal, calcination, and combined hydrothermal–calcination methods to prepare mesoporous TiO_2 nanocrystallite aggregates from a titanium glycolate precursor. Results affirmed that the post-processing modes can affect the surface area, average crystallite size, and pore structure of the synthesized material. In term of maximum adsorption capacity, nanoparticles prepared through hydrothermal treatment exhibited superior adsorption capacity toward ciprofloxacin [108]. Moreover, activated carbon is also a commonly used material for synthesizing carbon-based support materials for the encapsulation of TiO_2. This method is known for its ability to reduce TiO_2 aggregation and improve its characteristics by forming hetero-junction structures [61, 109]. Multiple material-based TiO_2 composites such as TiO_2/SiO_2-decorated carbon nanostructured materials offer promising alternatives for enhancing the adsorption of pollutants [109]. Titanium oxide nanoparticles were applied for the removal of different pharmaceuticals such as ciprofloxacin [110], ceftriaxone sodium [61], and tetracycline [111]. Other applications are gathered in Table 3.2.

Pharmaceutical wastewater treatment using silica oxide-based nanoparticles is widely studied. Recently, a new polyamidoamine/silica nanohybrid material was synthesized in order to remove clofibric acid, diclofenac, ibuprofen, and ketoprofen as examples of pharmaceuticals. At optimized operating conditions of 25 min, 298 K, pH = 9, 1 mg L^{-1} of adsorbent dose, and 100 µg L^{-1} of adsorbate, a maximum adsorption efficiency of 94, 134, 124, and 112 mg g^{-1} was obtained for the mentioned pharmaceuticals, respectively. Results showed that the removal of these pollutants is endothermic and occurs spontaneously [73]. Likewise, a comparative study of the adsorption performance of a new material, $NiFe_2O_4$@SiO_2@aminopropyltrimethoxysilane, toward different pharmaceuticals was investigated by Chandrashekar Kollarahithlu et al. [112]. Its efficiency was evaluated in terms of its maximum adsorption capacity for ibuprofen (IBF), acetaminophen (ACE), and streptomycin (STR), as presented in Figure 3.8. It showed an excellent regeneration capacity for up to four cycles. Moreover, results affirmed that this material could remove 59, 58, and 49 mg g^{-1} of IBF, ACE, and STR at pH 7.0, 6.0, and 5.0, respectively.

3.7 CONCLUSIONS AND PERSPECTIVES

In the present chapter, different metal oxide nanoparticles were briefly reviewed and highlighted. Their ease and flexible preparation, besides their versatility and long-term applications, make them an important class of adsorbents. However, MONPs agglomeration in aqueous solutions impacts their pollutant adsorption ability. Therefore, as confirmed in different research works, the improvement of their

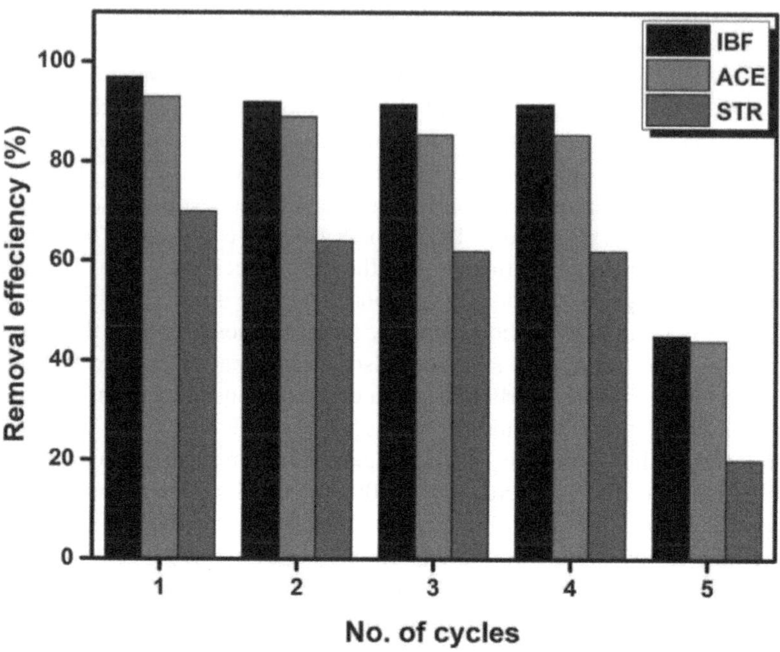

FIGURE 3.8   Regeneration capacity of $NiFe_2O_4@SiO_2@$aminopropyltrimethoxysilane.

Source: Reprinted from [112], Copyright (2021), with permission from Elsevier.

surface area, porosity, and therefore their attraction toward organic pollutants can be achieved by preparing new metal oxide hybrid nanomaterials. Their application for the removal of pharmaceuticals was taken into consideration in order to clarify their potential as efficient adsorbents. Different works aimed at the removal of pharmaceuticals have been reviewed and reported. They have demonstrated promising results in batch reactors. However, challenges remain for pilot-scale applications and real pharmaceutical wastewater treatment. These challenges include issues related to commercialization, fluctuating cost-effectiveness, potential human and technical obstacles, and environmental risks.

REFERENCES

[1] Beiras R. Chapter 6—Environmental risk assessment of pharmaceutical and personal care products in estuarine and coastal waters. In: Durán-Álvarez JC, Jiménez-Cisneros B, editors. Pharm Mar Coast Environ. Elsevier; 2021. p. 195–252.

[2] Kalia VC. 2—Pharmaceutical and personal care product contamination: A global scenario. In: Prasad MNV, Vithanage M, Kapley A, editors. Pharm Pers Care Prod Waste Manag Treat Technol. Butterworth-Heinemann; 2019. p. 27–61.

[3] Dey S, Bano F, Malik A. 1—Pharmaceuticals and Personal Care Product (PPCP) contamination—A global discharge inventory. In: Prasad MNV, Vithanage M, Kapley A, editors. Pharm Pers Care Prod Waste Manag Treat Technol. Butterworth-Heinemann; 2019. p. 1–26.

[4] Titchou FE, Zazou H, Afanga H, et al. Removal of Persistent Organic Pollutants (POPs) from water and wastewater by adsorption and electrocoagulation process. Groundw Sustain Dev. 2021;13:100575.

[5] Rao SN. Chapter 5—Adsorption. In: Ball V, editor. Interface Sci Technol. Elsevier; 2018. p. 251–331.

[6] Tee GT, Gok XY, Yong WF. Adsorption of pollutants in wastewater via biosorbents, nanoparticles and magnetic biosorbents: A review. Environ Res. 2022;212:113248.

[7] Velmurugan R, Incharoensakdi A. Chapter 18—Nanoparticles and organic matter: Process and impact. In: Tripathi DK, Ahmad P, Sharma S, et al., editors. Nanomater Plants Algae Microorg. Academic Press; 2018. p. 407–428.

[8] Danish MSS, Bhattacharya A, Stepanova D, et al. A systematic review of metal oxide applications for energy and environmental sustainability. Metals. 2020;10:1604.

[9] Murthy S, Effiong P, Fei CC. 11—Metal oxide nanoparticles in biomedical applications. In: Al-Douri Y, editor. Met Oxide Powder Technol. Elsevier; 2020. p. 233–251.

[10] Bose APH, McCallum ES, Avramović M, et al. Pharmaceutical pollution disrupts the behavior and predator-prey interactions of two widespread aquatic insects. iScience. 2022;25:105672.

[11] Souza H de O, Costa R dos S, Quadra GR, et al. Pharmaceutical pollution and sustainable development goals: Going the right way? Sustain Chem Pharm. 2021;21:100428.

[12] Kecili R, Hussain CM. Chapter 4—Mechanism of adsorption on nanomaterials. In: Hussain CM, editor. Nanomater Chromatogr. Elsevier; 2018. p. 89–115.

[13] Kumar JA, Krithiga T, Manigandan S, et al. A focus to green synthesis of metal/metal based oxide nanoparticles: Various mechanisms and applications towards ecological approach. J Clean Prod. 2021;324:129198.

[14] Danquah MK, Jeevanandam J. Chapter 5—Metal and metal oxide nanoparticles: Synthesis, properties, and applications as nanomedicines for diabetes treatment. In: Danquah MK, Jeevanandam J, editors. Emerg Nanomedicines Diabetes Mellit Theranostics. Elsevier; 2022. p. 111–142.

[15] Prasanna SRVS, Balaji K, Pandey S, et al. Chapter 4—Metal oxide based nanomaterials and their polymer nanocomposites. In: Karak N, editor. Nanomater Polym Nanocomposites. Elsevier; 2019. p. 123–144.

[16] Dharmalingam P, Palani G, Apsari R, et al. Synthesis of metal oxides/sulfides-based nanocomposites and their environmental applications: A review. Mater Today Sustain. 2022;20:100232.

[17] Nair GM, Sajini T, Mathew B. Advanced green approaches for metal and metal oxide nanoparticles synthesis and their environmental applications. Talanta Open. 2022;5:100080.

[18] Jabbar KQ, Barzinjy AA, Hamad SM. Iron oxide nanoparticles: Preparation methods, functions, adsorption and coagulation/flocculation in wastewater treatment. Environ Nanotechnol Monit Manag. 2022;17:100661.

[19] Modan EM, Plăiaşu AG. Advantages and disadvantages of chemical methods in the elaboration of nanomaterials. Ann "Dunarea Jos" Univ Galati Fascicle IX Metall Mater Sci. 2020;43:53–60.

[20] Athar T. Chapter 17—Smart precursors for smart nanoparticles. In: Ahmed W, Jackson MJ, editors. Emerg Nanotechnologies Manuf Second Ed. William Andrew Publishing; 2015. p. 444–538.

[21] Rane AV, Kanny K, Abitha VK, et al. Chapter 5—Methods for synthesis of nanoparticles and fabrication of nanocomposites. In: Mohan Bhagyaraj S, Oluwafemi OS, Kalarikkal N, et al., editors. Synth Inorg Nanomater. Woodhead Publishing; 2018. p. 121–139.

[22] Chavali MS, Nikolova MP. Metal oxide nanoparticles and their applications in nanotechnology. SN Appl Sci. 2019;1:607.

[23] Danquah MK, Jeevanandam J. Chapter 5—Metal and metal oxide nanoparticles: Synthesis, properties, and applications as nanomedicines for diabetes treatment. In: Danquah MK, Jeevanandam J, editors. Emerg Nanomedicines Diabetes Mellit Theranostics. Elsevier; 2022. p. 111–142.

[24] Alavi M, Kamarasu P, McClements DJ, et al. Metal and metal oxide-based antiviral nanoparticles: Properties, mechanisms of action, and applications. Adv Colloid Interface Sci. 2022;306:102726.

[25] Juma AO, Arbab EAA, Muiva CM, et al. Synthesis and characterization of CuO-NiO-ZnO mixed metal oxide nanocomposite. J Alloys Compd. 2017;723:866–872.

[26] Wang C, Sun F, He G, et al. Noble metal nanoparticles meet molecular cages: A tale of integration and synergy. Curr Opin Colloid Interface Sci. 2023;63:101660.

[27] Sun C, Lee JSH, Zhang M. Magnetic nanoparticles in MR imaging and drug delivery. Adv Drug Deliv Rev. 2008;60:1252–1265.

[28] Okpara EC, Olatunde OC, Wojuola OB, et al. Applications of transition metal oxides and chalcogenides and their composites in water treatment: A review. Environ Adv. 2023;11:100341.

[29] Singh S, Kumar V, Anil AG, et al. Adsorption and detoxification of pharmaceutical compounds from wastewater using nanomaterials: A review on mechanism, kinetics, valorization and circular economy. J Environ Manage. 2021;300:113569.

[30] Qumar U, Hassan JZ, Bhatti RA, et al. Photocatalysis vs adsorption by metal oxide nanoparticles. J Mater Sci Technol. 2022;131:122–166.

[31] Olusegun SJ, Souza TGF, Souza G de O, et al. Iron-based materials for the adsorption and photocatalytic degradation of pharmaceutical drugs: A comprehensive review of the mechanism pathway. J Water Process Eng. 2023;51:103457.

[32] Chen M, Coasne B, Guyer R, et al. Role of hydrogen bonding in hysteresis observed in sorption-induced swelling of soft nanoporous polymers. Nat Commun. 2018;9:3507.

[33] Meyer EE, Rosenberg KJ, Israelachvili J. Recent progress in understanding hydrophobic interactions. Proc Natl Acad Sci. 2006;103:15739–15746.

[34] Anslyn EV, Dougherty DA. Modern Physical Organic Chemistry. University Science Books; 2006.

[35] Couper JR, Penney WR, Fair JR, et al., editors. Chapter 15—Adsorption and ion exchange. In: Chem Process Equip Second Ed. Gulf Professional Publishing; 2005. p. 523–554.

[36] Khurana P, Pulicharla R, Kaur Brar S. Antibiotic-metal complexes in wastewaters: Fate and treatment trajectory. Environ Int. 2021;157:106863.

[37] Dragan ES, Humelnicu D, Dinu MV, et al. Kinetics, equilibrium modeling, and thermodynamics on removal of Cr(VI) ions from aqueous solution using novel composites with strong base anion exchanger microspheres embedded into chitosan/poly(vinyl amine) cryogels. Chem Eng J. 2017;330:675–691.

[38] Calisto V, Ferreira CIA, Oliveira JABP, et al. Adsorptive removal of pharmaceuticals from water by commercial and waste-based carbons. J Environ Manage. 2015;152:83–90.

[39] Pellenz L, de Oliveira CRS, da Silva Júnior AH, et al. A comprehensive guide for characterization of adsorbent materials. Sep Purif Technol. 2023;305:122435.

[40] Ozelcaglayan ED, Parker WJ. β-Cyclodextrin functionalized adsorbents for removal of organic micropollutants from water. Chemosphere. 2023;320:137964.

[41] Gkika DA, Mitropoulos AC, Kyzas GZ. Why reuse spent adsorbents? The latest challenges and limitations. Sci Total Environ. 2022;822:153612.

[42] Velempini T, Ahamed MEH, Pillay K. Heavy-metal spent adsorbents reuse in catalytic, energy and forensic applications-a new approach in reducing secondary pollution associated with adsorption. Results Chem. 2023;5:100901.

[43] Adesina Adegoke K, Samuel Agboola O, Ogunmodede J, et al. Metal-organic frameworks as adsorbents for sequestering organic pollutants from wastewater. Mater Chem Phys. 2020;253:123246.

[44] Ren B, Han C, Al Anazi AH, et al. Iron-based nanomaterials for the treatment of emerging environmental contaminants. Interact Nanomater Emerg Environ Contam. American Chemical Society; 2013. p. 135–146.

[45] Naseem T, Durrani T. The role of some important metal oxide nanoparticles for wastewater and antibacterial applications: A review. Environ Chem Ecotoxicol. 2021;3:59–75.

[46] Kharisov BI, Dias HVR, Kharissova OV, et al. Iron-containing nanomaterials: Synthesis, properties, and environmental applications. RSC Adv. 2012;2:9325–9358.

[47] Kharisov BI, Dias HVR, Kharissova OV. Mini-review: Ferrite nanoparticles in the catalysis. Arab J Chem. 2019;12:1234–1246.

[48] Fadillah G, Yudha SP, Sagadevan S, et al. Magnetic iron oxide/clay nanocomposites for adsorption and catalytic oxidation in water treatment applications. Open Chem. 2020;18:1148–1166.

[49] Brito CHV, Gloria DCS, Santos EB, et al. Porous activated carbon/graphene oxide composite for efficient adsorption of pharmaceutical contaminants. Chem Eng Res Des. 2023;191:387–400.

[50] Lakshmipriya T, Gopinath SCB. 1—Introduction to nanoparticles and analytical devices. In: Gopinath SCB, Gang F, editors. Nanoparticles Anal Med Devices. Elsevier; 2021. p. 1–29.

[51] Hessien M. Recent progress in zinc oxide nanomaterials and nanocomposites: From synthesis to applications. Ceram Int. 2022;48:22609–22628.

[52] Samuel Z, Ojemaye MO, Okoh OO, et al. Adsorption of simazine herbicide from aqueous solution by novel pyrene functionalized zinc oxide nanoparticles: Kinetics and isotherm studies. Mater Today Commun. 2023;34:105435.

[53] Al-Arjan WS. Zinc oxide nanoparticles and their application in adsorption of toxic dye from aqueous solution. Polymers. 2022;14:3086.

[54] Lei C, Pi M, Jiang C, et al. Synthesis of hierarchical porous Zinc Oxide (ZnO) microspheres with highly efficient adsorption of Congo red. J Colloid Interface Sci. 2017;490:242–251.

[55] Raj VJ, Ghosh R, Girigoswami A, et al. Application of zinc oxide nanoflowers in environmental and biomedical science. BBA Adv. 2022;2:100051.

[56] Sharma P, Hasan MR, Mehto NK, et al. 92 years of zinc oxide: Has been studied by the scientific community since the 1930s-An overview. Sens Int. 2022;3:100182.

[57] Kavitha A, Doss A, Praveen Pole RP, et al. A mini review on plant-mediated zinc oxide nanoparticles and their antibacterial potency. Biocatal Agric Biotechnol. 2023;48:102654.

[58] Altıntıg E, Yenigun M, Sarı A, et al. Facile synthesis of zinc oxide nanoparticles loaded activated carbon as an eco-friendly adsorbent for ultra-removal of malachite green from water. Environ Technol Innov. 2021;21:101305.

[59] Ahmadi SAR, Kalaee MR, Moradi O, et al. Synthesis of novel Zeolitic Imidazolate Framework (ZIF-67)—Zinc Oxide (ZnO) nanocomposite (ZnO@ZIF-67) and potential adsorption of pharmaceutical (tetracycline (TCC)) from water. J Mol Struct. 2022;1251:132013.

[60] Jia Z-M, Zhao Y-R, Shi J-N. Adsorption kinetics of the photocatalytic reaction of nano-TiO_2 cement-based materials: A review. Constr Build Mater. 2023;370:130462.

[61] Abdullah M, Iqbal J, Ur Rehman MS, et al. Removal of ceftriaxone sodium antibiotic from pharmaceutical wastewater using an activated carbon based TiO_2 composite: Adsorption and photocatalytic degradation evaluation. Chemosphere. 2023;317:137834.

[62] Singh P, Sen K. Selective adsorption of epigallocatechin gallate onto highly reusable gallium doped mesoporous TiO_2 nanoparticles adsorbent. J Indian Chem Soc. 2022;99:100576.

[63] Wei D, Li B, Huang H, et al. Biochar-based functional materials in the purification of agricultural wastewater: Fabrication, application and future research needs. Chemosphere. 2018;197:165–180.

[64] Manuja A, Kumar B, Kumar R, et al. Metal/metal oxide nanoparticles: Toxicity concerns associated with their physical state and remediation for biomedical applications. Toxicol Rep. 2021;8:1970–1978.

[65] Zhang Y, Xiong M, Sun A, et al. MIL-101(Fe) nanodot-induced improvement of adsorption and photocatalytic activity of carbon fiber/TiO_2-based weavable photocatalyst for removing pharmaceutical pollutants. J Clean Prod. 2021;290:125782.

[66] Reed ML, Fedder GK. 2—Photolithographic microfabrication. In: Fukuda T, Menz W, editors. Handb Sens Actuators. Elsevier Science B.V.; 1998. p. 13–61.

[67] Yu Y, Hu X, Li M, et al. Constructing mesoporous Zr-doped SiO_2 onto efficient Z-scheme TiO_2/g-C_3N_4 heterojunction for antibiotic degradation via adsorption-photocatalysis and mechanism insight. Environ Res. 2022;214:114189.

[68] Zhuravlev LT, Potapov VV. Density of silanol groups on the surface of silica precipitated from a hydrothermal solution. Russ J Phys Chem. 2006;80:1119–1128.

[69] Mallakpour S, Naghdi M. Polymer/SiO_2 nanocomposites: Production and applications. Prog Mater Sci. 2018;97:409–447.

[70] Ma Y, Liang R, Wu W, et al. Enhancing the activity of MoS_2/SiO_2-Al_2O_3 bifunctional catalysts for suspended-bed hydrocracking of heavy oils by doping with Zr atoms. Chin J Chem Eng. 2021;39:126–134.

[71] Di Felice L, Catherin N, Piccolo L, et al. Decalin ring opening over NiWS/SiO_2-Al_2O_3 catalysts in the presence of H_2S. Appl Catal Gen. 2016;512:43–51.

[72] Ruan C, Ma Y, Shi G, et al. Self-assembly cellulose nanocrystals/SiO_2 composite aerogel under freeze-drying: Adsorption towards dye contaminant. Appl Surf Sci. 2022;592:153280.

[73] Lotfi R, Hayati B, Rahimi S, et al. Synthesis and characterization of PAMAM/SiO_2 nanohybrid as a new promising adsorbent for pharmaceuticals. Microchem J. 2019;146:1150–1159.

[74] Abdel-Galil EA, Tourky AS, Kasem AE. Sorption of some radionuclides from nuclear waste effluents by polyaniline/SiO_2 composite: Characterization, thermal stability, and gamma irradiation studies. Appl Radiat Isot. 2020;156:109009.

[75] Cai W, Guo M, Weng X, et al. Adsorption of doxorubicin hydrochloride on glutaric anhydride functionalized Fe_3O_4@SiO_2 magnetic nanoparticles. Mater Sci Eng C. 2019;98:65–73.

[76] Ahmad T, Danish M. A review of avocado waste-derived adsorbents: Characterizations, adsorption characteristics, and surface mechanism. Chemosphere. 2022;296:134036.

[77] de Jesus RA, de Assis GC, de Oliveira RJ, et al. Environmental remediation potentialities of metal and metal oxide nanoparticles: Mechanistic biosynthesis, influencing factors, and application standpoint. Environ Technol Innov. 2021;24:101851.

[78] Titchou FE, Ait Akbour R, Assabbane A, et al. Removal of cationic dye from aqueous solution using Moroccan pozzolana as adsorbent: Isotherms, kinetic studies, and application on real textile wastewater treatment. Groundw Sustain Dev. 2020;11:100405.

[79] Balarak D, McKay G. Utilization of MWCNTs/Al$_2$O$_3$ as adsorbent for ciprofloxacin removal: Equilibrium, kinetics and thermodynamic studies. J Environ Sci Health Part A. 2021;56:324–333.

[80] Zhao N, Liu K, Yan B, et al. Chlortetracycline hydrochloride removal by different biochar/Fe composites: A comparative study. J Hazard Mater. 2021;403:123889.

[81] Zang J, Wu T, Song H, et al. Removal of tetracycline by hydrous ferric oxide: Adsorption kinetics, isotherms, and mechanism. Int J Environ Res Public Health. 2019;16:4580.

[82] Asgari E, Sheikhmohammadi A, Yeganeh J. Application of the Fe$_3$O$_4$-chitosan nano-adsorbent for the adsorption of metronidazole from wastewater: Optimization, kinetic, thermodynamic and equilibrium studies. Int J Biol Macromol. 2020;164:694–706.

[83] Liu Y, Dong C, Wei H, et al. Adsorption of levofloxacin onto an iron-pillared montmorillonite (clay mineral): Kinetics, equilibrium and mechanism. Appl Clay Sci. 2015;118:301–307.

[84] Yi S, Gao B, Sun Y, et al. Removal of levofloxacin from aqueous solution using rice-husk and wood-chip biochars. Chemosphere. 2016;150:694–701.

[85] Yegane Badi M, Azari A, Pasalari H, et al. Modification of activated carbon with magnetic Fe$_3$O$_4$ nanoparticle composite for removal of ceftriaxone from aquatic solutions. J Mol Liq. 2018;261:146–154.

[86] Delgado N, Capparelli A, Navarro A, et al. Pharmaceutical emerging pollutants removal from water using powdered activated carbon: Study of kinetics and adsorption equilibrium. J Environ Manage. 2019;236:301–308.

[87] Liu L, Cui W, Lu C, et al. Analyzing the adsorptive behavior of Amoxicillin on four Zr-MOFs nanoparticles: Functional groups dependence of adsorption performance and mechanisms. J Environ Manage. 2020;268:110630.

[88] Su Y, Qian F, Qian Z. Enhancing adsorption capacity and structural stability of Li$_{1.6}$ Mn$_{1.6}$ O$_4$ adsorbents by anion/cation co-doping. RSC Adv. 2022;12:2150–2159.

[89] Disha, Kumar M. Metal oxide nanomaterials for photocatalytic degradation of antibiotics. Mater Today Proc. 2023, https://doi.org/10.1016/j.matpr.2023.03.422.

[90] Kerin H, Nagaraj K, Kamalesu S. Review on aquatic toxicity of metal oxide nanoparticles. Mater Today Proc. 2023, https://doi.org/10.1016/j.matpr.2023.02.183.

[91] Bao L-R, Zhang J-Z, Tang W-P, et al. Synthesis and adsorption properties of metal oxide-coated lithium ion-sieve from salt lake brine. Desalination. 2023;546:116196.

[92] Shan D, Deng S, Jiang C, et al. Hydrophilic and strengthened 3D reduced graphene oxide/nano-Fe$_3$O$_4$ hybrid hydrogel for enhanced adsorption and catalytic oxidation of typical pharmaceuticals. Environ Sci Nano. 2018;5:1650–1660.

[93] Gu W, Huang X, Tian Y, et al. High-efficiency adsorption of tetracycline by cooperation of carbon and iron in a magnetic Fe/porous carbon hybrid with effective Fenton regeneration. Appl Surf Sci. 2021;538:147813.

[94] Liu Q, Zhong L-B, Zhao Q-B, et al. Synthesis of Fe$_3$O$_4$/polyacrylonitrile composite electrospun nanofiber mat for effective adsorption of tetracycline. ACS Appl Mater Interfaces. 2015;7:14573–14583.

[95] Zhou J, Ma F, Guo H. Adsorption behavior of tetracycline from aqueous solution on ferroferric oxide nanoparticles assisted powdered activated carbon. Chem Eng J. 2020;384:123290.

[96] Zhang Y, Chen B, Zhang L, et al. Controlled assembly of Fe$_3$O$_4$ magnetic nanoparticles on graphene oxide. Nanoscale. 2011;3:1446–1450.

[97] Nassar MY, Ahmed IS, Hendy HS. A facile one-pot hydrothermal synthesis of hematite (α-Fe$_2$O$_3$) nanostructures and cephalexin antibiotic sorptive removal from polluted aqueous media. J Mol Liq. 2018;271:844–856.

[98] Caetano P, Simões N, Pinto P, et al. Application of nickel ferrite nanoparticles in adsorption of amoxicillin antibiotic. J Braz Chem Soc. 2020;31:2452–2461.

[99] Ulfa M, Nisa D, Muhammad FP, et al. Investigating the hydrophilicity of zinc oxide nanoparticles using xylene and water for ibuprofen adsorption. J Chem Technol Metall. 2021;56:761–768.

[100] Li Z, Sun Y, Xing J, et al. One step synthesis of Co/Cr-codoped ZnO nanoparticle with superb adsorption properties for various anionic organic pollutants and its regeneration. J Hazard Mater. 2018;352:204–214.

[101] Khosravi R, Zarei A, Heidari M, et al. Application of ZnO and TiO_2 nanoparticles coated onto montmorillonite in the presence of H_2O_2 for efficient removal of cephalexin from aqueous solutions. Korean J Chem Eng. 2018;35:1000–1008.

[102] Mohammed AA, Al-Musawi TJ, Kareem SL, et al. Simultaneous adsorption of tetracycline, amoxicillin, and ciprofloxacin by pistachio shell powder coated with zinc oxide nanoparticles. Arab J Chem. 2020;13:4629–4643.

[103] Ciğeroğlu Z, Kazan-Kaya ES, El Messaoudi N, et al. Remediation of tetracycline from aqueous solution through adsorption on g-C_3N_4-ZnO-$BaTiO_3$ nanocomposite: Optimization, modeling, and theoretical calculation. J Mol Liq. 2023;369:120866.

[104] Mohammed AA, Kareem SL. Adsorption of tetracycline fom wastewater by using Pistachio shell coated with ZnO nanoparticles: Equilibrium, kinetic and isotherm studies. Alex Eng J. 2019;58:917–928.

[105] Sharma G, Bhogal S, Kumar A, et al. AgO/MgO/FeO@Si_3N_4 nanocomposite with robust adsorption capacity for tetracycline antibiotic removal from aqueous system. Adv Powder Technol. 2020;31:4310–4318.

[106] Levard C, Hamdi-Alaoui K, Baudin I, et al. Silica-clay nanocomposites for the removal of antibiotics in the water usage cycle. Environ Sci Pollut Res. 2021;28:7564–7573.

[107] Dhiman N, Sharma N. Removal of pharmaceutical drugs from binary mixtures by use of ZnO nanoparticles: (Competitive adsorption of drugs). Environ Technol Innov. 2019;15:100392.

[108] Gan Y, Wei Y, Xiong J, et al. Impact of post-processing modes of precursor on adsorption and photocatalytic capability of mesoporous TiO_2 nanocrystallite aggregates towards ciprofloxacin removal. Chem Eng J. 2018;349:1–16.

[109] Rasheed T, Adeel M, Nabeel F, et al. TiO_2/SiO_2 decorated carbon nanostructured materials as a multifunctional platform for emerging pollutants removal. Sci Total Environ. 2019;688:299–311.

[110] Zheng X, Xu S, Wang Y, et al. Enhanced degradation of ciprofloxacin by Graphitized Mesoporous Carbon (GMC)-TiO_2 nanocomposite: Strong synergy of adsorption-photocatalysis and antibiotics degradation mechanism. J Colloid Interface Sci. 2018;527:202–213.

[111] Feng X, Li X, Su B, et al. Two-step construction of WO_3@TiO_2/CS-biochar S-scheme heterojunction and its synergic adsorption/photocatalytic removal performance for organic dye and antibiotic. Diam Relat Mater. 2023;131:109560.

[112] Chandrashekar Kollarahithlu S, Balakrishnan RM. Adsorption of pharmaceuticals pollutants, ibuprofen, acetaminophen, and streptomycin from the aqueous phase using amine functionalized superparamagnetic silica nanocomposite. J Clean Prod. 2021;294:126155.

4 Magnetic Adsorbents for the Removal of Pharmaceutical Contaminants

Vijayasri K. and Ajaya Kumar Singh

4.1 INTRODUCTION

One of the elements that determines environmental safety and public health is the quality of natural waters [1]. However, aquatic ecosystems deteriorate as a result of drugs getting into the water. Over the past 20 years, medicines have been recognised as emerging pollutants in the aquatic environment because of their lengthy persistence, low biodegradability, negative effects on human and animal health, and harm to aquatic species [2]. These compounds are classified into several groups, such as blood-lipid regulators, steroidal hormones, antimicrobial agents, and anti-inflammatory agents [3]. Several primary origins of these pharmaceutical contaminants in water include hospital waste, personal hygiene products, and animal or human excretion through urine or faeces [4]. They may unintentionally find their way into the human food chain through other channels, like fertilisers [5]. Pharmaceuticals from various medical facilities, hospitals, and towns are ending up in bottom sediments, aquatic flora, and aquatic fauna, harming the aquatic ecosystem and making it less safe for the environment [6]. Pharmaceutical chemicals have been found to pose a serious environmental threat to all aquatic environmental matrices, including drinking water, groundwater, wastewater treatment plant effluents, and surface water (rivers, lakes, etc.) in numerous different nations [7].

Pharmaceuticals enter the reservoir, dissolve in the water, and then sink to the bottom, where they produce organic biomass as bottom sludge, which is constantly broken down by fungi and putrefactive bacteria [6]. During the breakdown process, organic molecules strongly consume the dissolved oxygen in water, releasing decomposition products. A reservoir's biological equilibrium is first disturbed, then its biological self-cleaning is inhibited, and ultimately the ecology of the pond or lake is altered when there is an abundance of organic waste present [8]. Because of this, the wastewater needs to be fully treated using a number of treatment methods. The water system contaminated by pharmaceutical contaminants is depicted in Figure 4.1.

DOI: 10.1201/9781003340164-4

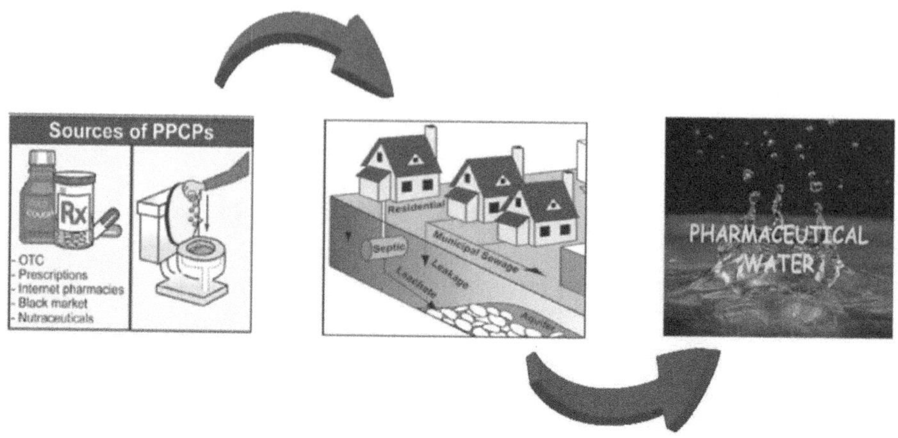

FIGURE 4.1   Contamination of Water system by pharmaceutical pollutants.

The dangerous compounds included in industrial wastewater are not entirely neu-tralised by conventional treatment techniques. Pharmaceuticals continue to accu-mulate and do not totally vanish from the environment because of their durability and several purification processes [8]. When these compounds are present in the environment in excess, they harm the microflora and fauna. Another problem is that resistant bacteria can develop in the environment even at low concentrations of phar-maceuticals [9]. Several researchers are focusing on creating better adsorbents in order to extract medicines.

Over the past ten years, a number of technologies, including biological proce-dures [10], advanced oxidation [11], photodegradation [12], ozonation [13], adsorp-tion [14], electrochemical degradation, and others [15], have been evaluated with regard to the removal of pharmaceutical pollutants. Most of these methods have disadvantages, including being expensive (membrane technology), energy intensive (oxidation, coagulation, electrochemical treatment), slow (photochemical and elec-trochemical treatment), and short lived (ozonation) [16]. They also produce a lot of sludge or by-products. Due to its low cost, simplicity of usage, speed, lack of sludge formation, and reusability, adsorption has garnered a lot of interest in the last few years for the removal of pharmaceuticals from wastewater. [17].

Several types of adsorbent materials have been used to recover pharmaceuticals from aquatic environments. "Green" biochar and activated charcoal were studied by X. Zhu et al. [18] for the adsorption of medications. Using "green" synthesis, Magesh et al. [19] produced a zinc oxide biosorbent covered with activated carbon to extract ciprofloxacin from aqueous solutions. 99.8% of the maximum amounts of debris was eliminated, according to the results [19]. With a 92% removal efficiency, ciprofloxacin was extracted from water in a study [20] using a biosorbent made from tamarind shells that had been treated with acid. Another "green" ceramic/organic xerogel adsorbent was developed by P. Arabkhani et al. [21] to remove antibiot-ics from water. Under optimised adsorption conditions, the removal efficiency was ca. 99, 99, 98, and 98% for amoxicillin, tetracycline, cephalexin, and penicillin G, respectively [21].

Magnetic oxide-based adsorbents are among the many types of adsorbents that are important for environmental restoration [22–24]. They often provide a sizable specific surface area and are readily recoverable. Every magnetic sorbent has enhanced mechanical strength, a high capacity for adsorbing both organic and inorganic contaminants, and strong magnetic properties. These properties enable the sorbent to be collected by the use of permanent or electric magnets.

The majority of this chapter will focus on magnetic oxide-based adsorbents that can be used to extract various pharmaceuticals from aquatic environments. The most common and dangerous pharmaceutical classes for the environment, like non-steroidal anti-inflammatory medicines and antibiotics, are examined in depth and given particular focus. The physicochemical properties of adsorbents, including adsorption modes, are described in detail.

4.2 PHARMACEUTICAL MICROPOLLUTANTS

Despite the fact that numerous nations have acknowledged that medications have the potential to contaminate surface and ground waterways, there are currently no appropriate regulations or legislative frameworks in place to manage pharmaceuticals. Because of this, households, hospitals, and pharmaceutical industries regularly leak pharmaceutical and personal care products (PPCPs) into the environment through their sewage discharges [25]. Table 4.1 provides an overview of the classification and essential characteristics of a few medications that are regularly studied in the literature. It demonstrates that the main pharmaceutical pollutants are hormones, lipid regulators, beta-blockers, antidepressants, and non-steroidal anti-inflammatory drugs (NSAIDs). These medications are present in water, contributing to pollution and posing a risk to human health and aquatic life.

4.3 MAGNETIC SUBSTANCES AND THEIR TYPES

Certain materials, such as iron, cobalt, and nickel components and their alloys, can be drawn to a magnet. Magnetic materials have been categorised based on their magnetizability. Consequently, ferromagnetic, paramagnetic, and diamagnetic materials are the three categories into which magnetic materials have been divided. The size of the particles in these materials largely determines their magnetic characteristics. Iron oxide nanoparticles are among the most well-known paramagnetic materials and one of the finest options for heavy metal separation due to their huge specific area and good environmental biocompatibility. When exposed to an external magnetic field, the crystals will exhibit two distinct behaviours: one where they align parallel to the field, and another where they act like a pure magnetic field. Utilising magnetised nanoparticles, this special feature has been successfully applied for the separation of contaminants [64].

Magnetised ceramic has demonstrated certain advantages over other magnetic materials in terms of ease of synthesis, resistance to corrosion and abrasion, and alloy and metallic base magnetised materials. The three most common magnetic

TABLE 4.1

The Classification and Main Characteristics of Some Pharmaceutical Contaminants in Wastewater

Pharmaceuticals	Compounds	Formula	Molecular Weight (g mol^{-1})	Solubility (mg L^{-1})	pK$_1$	pK$_2$	log K$_{ow}$[a]	References
Antibiotics	Sulphamethoxazole	$C_{10}H_{11}N_3O_3S$	253.28	610	5.60		0.97	[26]
	Sulphadiazine	$C_{10}H_{10}N_4O_2S$	250.28	77	6.28		−0.14	[27]
	Sulphadimethoxine	$C_{12}H_{14}N_4O_4S$	310.33	343	5.97		1.76	[28]
	Sulphamethazine	$C_{12}H_{14}N_4O_2S$	278.33	1500	7.42		0.05	[29]
	Tetracycline	$C_{22}H_{24}N_2O_8$	444.4	231	3.4	7.6/9.7	−1.37	[30]
	Amoxicillin	$C_{16}H_{19}N_3O_5S$	365.4	3430	3.2	11.7	0.87	[31]
	Penicillin V	$C_{16}H_{18}N_2O_4S$	334.4	210	2.74		1.83	[32]
	Ciprofloxacin	$C_{17}H_{18}FN_3O_3$	331.4	150	5.9	8.8	0.28	[33]
	Oxytetracycline	$C_{22}H_{24}N_2O_9$	460.4	313	3.27	9.5	−0.9	[34]
	Enrofloxacin	$C_{19}H_{22}FN_3O_3$	359.4	612	5.69	6.68	−2.8	[35]
	Chlortetracycline	$C_{22}H_{23}ClN_2O_8$	478.9	259	3.3	7.4/9.33	−3.60	[36]
	Trimethoprim	$C_{14}H_{18}N_4O_3$	290.32	400	7.12		0.9	[37]
	Tylosin	$C_{46}H_{77}NO_{17}$	916.1	5000	7.1		1.63	[38]
	Triclosan	$C_{12}H_7Cl_3O_2$	289.5	10000	7.9		4.76	[39]
	Lincomycin	$C_{18}H_{34}N_2O_6S$	406.53	1693	7.6		0.2	[40]
	Norfloxacin	$C_{16}H_{18}FN_3O_3$	319.33	280	6.22	8.51	0.46	[41]
	Fluconazole	$C_{13}H_{12}F_2N_6O$	306.27	1390	1.72		0.4	[42]
NSAIDs	Paracetamol	$C_8H_9NO_2$	151.16	14,000	9.38		0.46	[43]
	Naproxen	$C_{14}H_{14}O_3$	230.26	15.9	4.19		3.18	[44]
	Ketoprofen	$C_{16}H_{14}O_3$	254.28	51	4.45		3.12	[45]
	Diclofenac	$C_{14}H_{11}Cl_2NO_2$	296.1	2370	4.15		4.51	[46]
	Ibuprofen	$C_{13}H_{18}O_2$	206.28	21000	4.91		3.97	[47]
	Acetylsalicylic acid	$C_9H_8O_4$	180.16	4600	3.49		1.19	[48]

Antidepressants	Fluoxetine	$C_{17}H_{18}F_3NO$	309.33	1.7	9.8		4.6	[49]
	Oxazepam	$C_{15}H_{11}ClN_2O_2$	286.71	20	1.55	10.9	2.8	[50]
Hormones	17 α-Ethinyl estradiol	$C_{20}H_{24}O_2$	296.4	11300	10.47		3.9	[51]
	Mestranol	$C_{21}H_{26}O_2$	310.4	1132	17.59		4.61	[52]
	17-β Estradiol	$C_{18}H_{24}O_2$	272.4	3900	10.46		4.01	[53]
β -blockers	Tramadol	$C_{16}H_{25}NO_2$	263.37	1151	9.23	13.08	2.4	[54]
	Atenolol	$C_{14}H_{22}N_2O_3$	266.34	13300	9.6		0.16	[55]
	Metoprolol	$C_{15}H_{25}NO_3$	267.36	60	9.7		1.88	[56]
Lipid regulators	Clofibric acid	$C_{10}H_{11}ClO_3$	214.64	583	4		2.84	[57]
	Gemfibrozil	$C_{15}H_{22}O_3$	250.33	11	4.5		4.77	[58]
Diuretics	Furosemide	$C_{12}H_{10}ClN_2O_5S$	330.74	73.1	3.5	9.0	2.03	[59]
H₂ blockers	Ranitidine hydrochloride	$C_{13}H_{23}ClN_4O_3S$	350.86	79.5	2.7	8.2	0.99	[60]
Stimulants	Caffeine	$C_8H_{10}N_4O_2$	194.19	21600	14		–0.07	[61]
Anti-epileptics	Carbamazepine	$C_{15}H_{12}N_2O$	236.27	18	2.3	13.9	2.25	[62]

[a] K_{ow}: defined as the ratio of the concentration of a chemical in n-octanol and water at equilibrium, a relative indicator of the tendency of an organic compound to adsorb to soil and living organisms.

ceramics found in nature are haematite (α-Fe$_2$O$_3$), maghemite (γ-Fe$_2$O$_3$), and magnetite (Fe$_3$O$_4$). In practical contexts, these three substances have proven to be highly effective and technologically advanced, especially at the nanoscale [63].

4.3.1 Haematite (α-Fe$_2$O$_3$)

The first iron oxide is haematite, which is found in many rocks and soils. It is also known by the terms renal stone, iron ore, martite, and ferric acid. Haematite is actually the hydroxide part of a complicated ferrous oxide solution, with different amounts of water and hydroxyl groups that affect the crystal's chemical makeup and the magnetisation of the mine. Haematite is normally a black or grey, coarse-grained mineral, however when it develops as fine-grained crystals, this iron oxide turns red. Haematite is often a byproduct of other iron oxides such as maghemite and is particularly stable in the presence of oxygen [63].

4.3.2 Maghemite (γ-Fe$_2$O$_3$)

The spinel structure of maghemite, a ferrous material, is comparable to that of magnetite. As a byproduct of magnetite oxidation, iron oxide in the Earth's crust is spontaneously converted into different species of iron oxide under the impact of meteorological factors on the magnetite or by adding heat. This oxide tends to become haematite at high temperatures and lose its magnetic characteristics. Usually, the so-called sol–gel method is employed to produce haematite nanoparticles. The presence of gaps beneath its cation lattice and the regularly spaced tetrahedral crystals of trivalent iron are the only structural differences between this oxide and magnetite [64].

4.3.3 Magnetite (Fe$_3$O$_4$)

This mineral is called black magnetic magnetite. Other names for it include ferrite iron, magnetised iron rock, iron oxide, and magnet. Because of its significant magnetic characteristics, this material is one of the essential iron oxides. Magnetite features an inverse spinel structure and an alternating octahedral configuration in terms of its crystal structure. The synthesis method and magnetite's particle size have the strongest effects on its magnetic characteristics. The effective adsorbent and active reducing agent properties of this oxide's nanoparticles have been demonstrated in their ability to remove heavy metals from the environment. A strong magnetic field can be used to easily isolate the particles from their surroundings. Many benefits stem from their recoverability and reusability as well, which has spurred extensive research on the development and application of these nanoparticles.

The general formula for complex oxides of spinel, known as AB$_2$O$_4$, is as follows: A = Mg^{2+}, Fe^{2+}, Ni$^{2+,}$ Zn^{2+}, Mn^{2+}, Be^{2+}, Co^{2+}; B = Fe^{3+}, Al^{3+}, Cr^{3+}, Mn^{3+}, and infrequently Ti^{4+}, V^{3+} [63]. Haematite (γ-Fe$_2$O$_3$) is another member of the spinel group. Spinels are compounds that have a wide range of A and B cations and well-developed isomorphisms [64]. The spinel structures are classified as normal (M^{2+} cations are located in tetrahedral sites, while M^{3+} cations are in octahedra), inverted (M^{3+} cations are equally placed on both sites, while M^{2+} occupy only octahedral sites), and mixed spinels (both types of cations arbitrarily occupy tetrahedral and octahedral sites) based on the cation distribution between the crystallographic positions. The cation distribution can be described as $(A1_{-x}B_x)[A_x/2B1\text{-tetrahedral}]$, where x is the inversion degree. $MgAl_2O_4$, $ZnFe_2O_4$, $FeAl_2O_4$, $(Mn, Fe) Al_2O_4$, and so on all have inherent typical spinel structures. A hallmark of Fe_3O_4, $MgFe_2O_4$, $CoFe_2O_4$, $\gamma\text{-}Fe_2O_3$, and so on is the inverted spinel structure. Magnetic ferrites with the general formula MFe_2O_4, which are typically entirely or partially inverted, are significant from a practical standpoint. The explanation is that the crystal field theory states that the Fe^{3+} ion with five d-electrons has no preference for octahedral locations. As a result, the structure is inverted because bigger divalent ions primarily occupy octahedral locations, while Fe^{3+} ions are dispersed among octahedral and tetrahedral places. One of the most well researched spinel magnetic adsorbents is magnetite, Fe_3O_4.

Recently, there has been a lot of interest in micro- and nano-scaled magnetised materials as possible adsorbents for eliminating pharmaceutical pollutants from water and wastewater. In this way, the technology of magnetised materials has played a major role in solving environmental issues. The mechanism of magnetic adsorption consists of three steps:

(i) The film diffusion step, in which the water film transfers the mass of the adsorbent to the adsorbent surface.
(ii) The adsorbate diffuses through the pores.
(iii) At the equilibrium stage, the adsorbent's surface is occupied by physisorption or chemisorption.

4.3.4 Techniques for Developing Magnetised Absorbents

Synthesis techniques have a significant impact on the adsorption capacity of magnetic adsorbents [65]. In order to create spinal adsorbents, S.K. Paswan et al. [66] employed both "top-down" and "bottom-up" classification techniques. Because materials are made from smaller units or molecules, the "bottom-up" process allows for exquisite control over the size, shape, content, and surface features of the magnetic adsorbent [66]. Materials are synthesised by breaking down large structures into smaller particles or components in a process known as "top-down" synthesis [67]. "Bottom-up" approaches include co-precipitation, sol–gel, hydrothermal, solvothermal, microemulsion, and other methods [66–68]. "Top-down" methods include mechanical machining and pulsed laser ablation [67]. The desired characteristics of the magnetic adsorbent, the equipment that is available, the cost, and the synthesis technique's scalability will all influence the choice of synthesis method.

To synthesize magnetised particles, three different sorts of procedures have often been used: physical, chemical, and biological. Roughly 90% of magnetised adsorbents are produced by chemical methods. Techniques in chemistry include thermal, electrochemical, hydrothermal, and precipitation decomposition. Furthermore, gas phase coating, laser-induced pyrolysis, and power ball milling are examples of physical approaches. Biological approaches also include protein-mediated, bacteria-mediated, and aerosol-mediated procedures.

4.3.4.1   Chemical Techniques

Chemical procedures are simple yet efficient. The size, shape, and composition of the magnetised nanoparticles produced using these techniques are all influenced by the kind of salt employed, the Fe^{2+}-to-Fe^{3+} ratio, the pH level, and the ion strength. By adjusting these variables, one may create nanoparticles with the required characteristics and manage the synthesis process [69]. Numerous methods exist for the chemical synthesis of magnetised adsorbents. For example, thermal decomposition was used to produce Tris (acetylacetonato) iron (III) (Fe(acac)$_3$), hydrothermal methods were used to create $Cd_{0.8}Fe_{0.2}S$ magnetised nanoparticles, microemulsion-based techniques were used to create $NiMnO_3$ nanostructures, and sono-chemical techniques were used to create Zr-ferrite nanoparticles [70–73].

4.3.4.2   Physical Techniques

Physical procedures are accurate and, depending on the reaction occurring during the process, can occur in solid, gaseous, or liquid forms. However, these methods are not able to control the size of particles at the nanoscale. Moreover, these methods do not include any chemical interactions, and the nanoparticles are produced only by physical processes rather than any chemical ones. Other physical techniques are the gas phase deposition method for face-centred-cubic cobalt (FccCo) nanoparticles [74], the aerosol-based method for carbon-13 NMR synthesis [73], the electron beam lithography method for cylindrical nanoparticles [75], the laser-induced pyrolysis method for Fe-based nanoparticles [76], and the ball milling method for iron nanoparticle synthesis [77].

4.3.4.3   Biological Techniques

Because of its great efficacy and low cost, chemical-based synthesis is one of the most used biological procedures. For this reason, the most significant biological technique is the microbial culture method [78].

4.4   CHARACTERISATION OF MAGNETISED ADSORBENTS BY USING DIFFERENT ANALYTICAL METHODS

The physical and chemical properties of magnetised materials have been studied using a variety of methods, including atomic force microscopy (AFM), Fourier transform infrared spectroscopy (FTIR), zeta potential, scanning tunnelling microscopy (STM), scanning electron microscopy (SEM), field emission scanning electron microscopy (FESEM), transmission electron microscopy (TEM), value stream mapping (VSM), and nuclear magnetic resonance (NMR). In this regard, FTIR analysis is used to pinpoint functional groups, dynamic light scattering (DLS) analysis confirms the magnetised adsorbents' nanoscale or particle size distribution, and XRD analysis is used to pinpoint the material's phase and crystal structure. STM, TEM, and AFM analyses are used to determine the crystal accumulation, structure, size, surface heterogeneity, and particle dispersion, while NMR analysis is typically used to assess the purity and concentration of the substance. VSM analysis is used to measure characteristics and determine the magnetic behaviour of the material (ferromagnetic, paramagnetic, and diamagnetic).

4.4.1 TEM ANALYSIS

In a study by Min-Hung Liao et al. [79], the size and form of magnetic nanoparticles were investigated by TEM using a JEOL Model JEM1200EX at 80 kV. To create a sample for TEM analysis, a drop of ethanol solution containing the magnetic nanoparticles was put on a copper grid coated with Formvar and allowed to evaporate at room temperature. To increase particle dispersion on the copper grid, the dispersed solution was sonicated for one minute before the sample was removed. Using more than 60 particles from different grid positions, the mean diameter of the particles was determined for each sample. It was clear that the bare Fe_3O_4 particles were essentially very tiny and monodisperse, with a mean diameter of 13.2 nm. After binding to polyacrylic acid (PAA), the particles maintained their uniqueness and separation while sharing a similar mean diameter. This indicates that there was no discernible agglomeration or alteration in the particle sizes as a result of the binding procedure. This might be explained by the fact that the reaction only happened on the surface of the particles and that the average number of PAA molecules associated to each particle was just two.

4.4.2 XPS AND FTIR ANALYSIS

Santhosh et al. [80] looked at the full-scan X-ray photoelectron spectroscopy (XPS) spectra of woodchips and modified biochar sludge. The XPS spectra reveals that C, Fe, and O were the main components in both samples. The ratios of C, O, and Fe in MS-450 and MS-700 were, respectively, 46.48%, 22.72%, and 15.8% and 29.75%, 42.56%, and 15.28%. In MWC-450 and MWC 700, on the other hand, the proportions of C, O, and Fe were 36.1%, 40.32%, and 19.01% and 47.36%, 39%, and 18.91%, respectively. When there was enough Fe present, Fe_2O_3 nanoparticles were successfully added to charcoal materials, as indicated by the XPS data.

The FTIR spectra of the modified and unmodified biochar samples were analysed. The major peaks that matched the functional groups' vibration in the unmodified biochar materials were as follows: the values for –OH were 3456 cm^{-1}, aromatic C=C and C=O were 1595 and 1698 cm^{-1}, and C–O–C was 1048 cm^{-1}. A comparison of the spectra of the modified and untreated biochar materials demonstrates that FeO, iron oxide, is responsible for the notable absorption peak at 562 cm^{-1}. This indirectly confirmed the presence of Fe_3O_4 in the modified biochar materials.

4.4.3 SEM ANALYSIS

Morphology, surface characteristics, and particle structure are examined using SEM and FESEM studies, and the size and surface charge of the particles are assessed using zeta potential analysis [80].

4.4.4 XRD Analysis

According to XRD analysis, every magnetic adsorbent is crystalline. A magnetic adsorbent's large surface area, tiny nanoscale particle size, heterogeneous surface structure, and high porosity enable its use in the removal of pharmaceutical contaminants from water and wastewater. Liu et al. [81] used XRD to describe the crystallographic properties of the magnetic particles (MPs) that were generated. The poorly crystallised ferrihydrite was linked to the primary peaks in MPs^{-1} and MPs^{-2}, which corresponded to 2θ of 34° and 61°, respectively [82]. The major peak of the poorly crystallised iron oxide at 2θ of 34° was covered by the strengths of diffraction peaks for iron mud, which were reflective of impurities such quartz and albite. The reflection peaks at 2θ of 30.1°, 35.4°, 43.2°, 53.4°, 57.2°, and 62.7°, attributed to diffractions from the planes of (220), (311), (400), (422), (511), and (440) of magnetite Fe_3O_4 (JCPDS 76–1849) with a cubic spinel structure, were observed in the synthesised MPs-3w, MPs 3–5, and the magnetite control, indicating that Fe_3O_4 was present in MPs. In addition, the peaks observed in MPs-3w at 2θ = 20.9° and 29.3° were affiliated with gypsum (JCPDS 33–0311) [83], which was due to $CaSO_4$ precipitation at pH 9.5 because of high concentrations of Ca^{2+} and SO_4^{2-} in the acid wastewater.

4.5 DIFFERENT PHARMACEUTICAL ADSORPTION MECHANISMS

Creating the efficient use of magnetic adsorbents requires an understanding of the specific physicochemical processes that take place at the surface of the adsorbent and are required for the adsorption of adsorbates, or medicines. Electrostatic or non-electrostatic interactions may be responsible for the modalities of drug adsorption onto magnetic adsorbents, depending on the pH of the solution and the charge (or at least polarity) of the adsorbent and adsorbate. Electrostatic interactions occur when the adsorbate is protonated or dissociated in an aqueous solution in an experimental setting. Other types of interactions include hydrogen bonds, van der Waals forces, and others [84]. Figure 4.2 shows the adsorption mechanism between magnetic adsorbents and pharmaceutical pollutants. The following are the primary methods of pharmaceutical adsorption:

4.5.1 Electrostatic Interactions

The electrostatic interaction between the adsorbate and the adsorbent surface can be predicted using the charges of their oppositely charged functional groups [85]. The ionic charge is influenced by the point of zero charge (pH_{pzc}), dissociation constants, and solution pH. If the charge of the adsorbate molecules and the adsorbent surface are the same, electrostatic repulsion will cause a reduction in adsorption [86]. When charges are at odds with one another because of electrostatic attraction, adsorption increases. Therefore, the surface charges of the adsorbent and adsorbate are the main elements regulating adsorption efficacy [87].

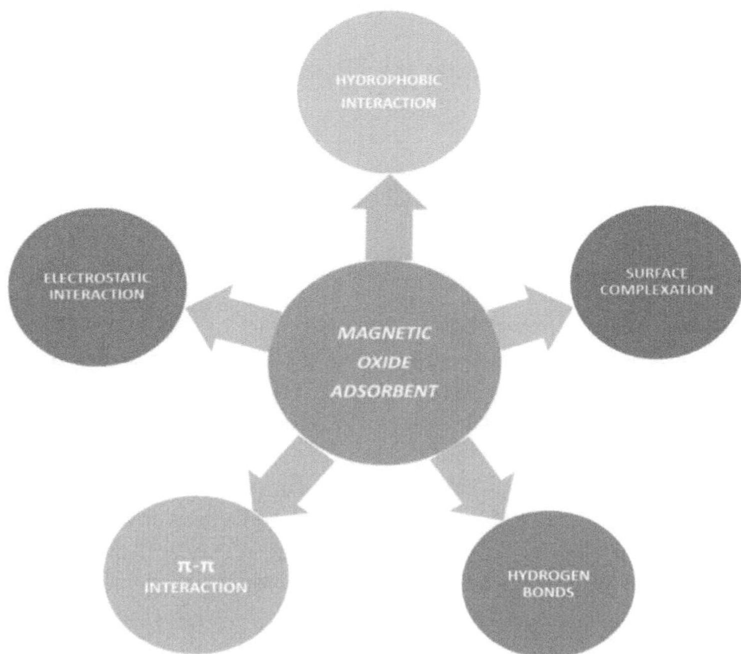

FIGURE 4.2 Types of interactions between Magnetic oxide adsorbents and drugs.

4.5.2 VAN DER WAALS FORCES

Van der Waals forces form between the molecules of the adsorbate and the adsorbent surface. These relatively weak intermolecular interactions are rapidly reduced by the distance between the molecules of the adsorbent and adsorbate. Before the adsorbate molecule can be desorbed, the surface force field of the magnetic nanocomposite keeps it there for a while [87]. The adsorbent surface has adsorption centres, or areas where adsorption is more intense, in accordance with a model of localised adsorption.

4.5.3 HYDROGEN BONDS

Hydrogen bonds are created when an atom with a high electronegativity from one molecule attracts another atom with an electrostatic force [87]. Magnetic adsorbents feature a large number of oxygen-containing functional groups, such as -OH, -COOH, etc., on their surface [88]. These pieces have the ability to form hydrogen bonds with nitrogen- or oxygen-containing therapeutic groups. The surface hydroxyl groups have the potential to bind several anionic species [89]. For example, the C=O and -NH functional groups of the magnetic adsorbent facilitate chemisorption by establishing hydrogen bonds with the -COOH group of the ciprofloxacin [90].

4.5.4 π–π Interactions

The aromatic groups of the adsorbent and the adsorbate interact through π–π interactions. The functional groups affixed to the adsorbent's benzene ring mostly affect the strength of the π–π bond [87]. Pharmaceutical functional groups possessing a strong electron-withdrawing ability can function as π acceptors, whereas OH groups on benzene rings, for instance, allow the adsorbent to function as a π donor [91].

4.5.5   Surface Complexation

This takes place when functional groups on the surface of magnetic adsorbents interact with drug molecules to form complexes [92]. The surface functional groups of adsorbents can contain hydroxyl, carboxyl, amino, or thiol groups [93]. During the complex-formation process, the medicinal molecules and iron ions from the magnetic adsorbent structure may create inner-sphere or outer-sphere complexes [94, 95]. Inner-sphere complexes are produced when a direct chemical bond is established between the adsorbent surface and the medicinal molecule known as the adsorbate. Outer-sphere complexes are formed when water molecules or hydroxyl groups cling to the surface of the magnetic adsorbent and the oppositely charged adsorbate (a pharmaceutical molecule). The concentration of medicines in the solution, the ionic strength of the solution, the pH of the solution, and the physicochemical characteristics of the adsorbent all affect the formation of surface complexes.

4.5.6   Hydrophobic Interactions

These takes place when non-polar molecules or hydrophobic drug groups interact with hydrophobic regions on the magnetic adsorbent surface. Non-polar molecules or groups interact in a hydrophobic manner because they seek to minimise their interaction with water. This kind of adsorption mode is affected by the concentration of the adsorbate in the solution, the surface area of the adsorbent, and the density of hydrophobic groups [96]. As an illustration, the structures of ofloxacin, ciprofloxacin, norfloxacin, propranolol, and clomipramine [97] all include both hydrophilic and hydrophobic groups. When ofloxacin, ciprofloxacin, and norfloxacin are utilised as adsorbents, or when carbon- or polysaccharide-based magnetic composites are used [86, 98], hydrophobic interactions are also seen.

4.5.7   Ion Exchange

This is observed when ions in the solution have the ability to swap places with other ions that are drawn to the functional groups on the surface of the adsorbent [87]. Magnetic adsorbents can be functionalised with ion exchange sulphate, carboxylate, or amino groups to draw and exchange ions from wastewater [99]. The selectivity of the ion exchange is dependent on the precise functional group and ionic charge of the magnetic adsorbent, as well as the concentration and charge of the ions in the solution [85]. The ion exchange process can be optimised by varying the temperature, adsorption contact duration, and solution pH. Since ionised antibiotics can either be deprotonated or protonated based on the pH of the solution, this process is particularly important for them [85, 100].

4.5.8 π–Cation Interactions

The mechanism of contact is adsorption, which can take place between positively charged species like metal ions or quaternary ammonium ions that are provided on the surface of magnetic adsorbents and aromatic groups of antibiotics (π-electron system of aromatic rings). The pharmaceutical structure, cationic forms, pH, ionic strength, and other factors all affect how strong the π–cation connection is [101]. An illustration of this can be found in the interaction between the aromatic π-electrons of an antibiotic molecule, such as tetracycline, and the Fe atom of the spinel lattice [102].

The key factor influencing the degree of medicine adsorption is the surface charge of magnetic adsorbents, which is regulated by the pH of the solution. Adsorption can be controlled by keeping an eye on the point of zero charge (pH_{pzc}) of the adsorbents. Through electrostatic attraction, the adsorbent may remove cationic adsorbates from the aqueous solution because the functional groups are deprotonated, and the surface is negatively charged at pH > pH_{pzc}. The adsorbent may efficiently remove anionic adsorbates because the protonation of functional groups causes the adsorbent surface to become positively charged at pH pH_{pzc}. Therefore, the adsorption capacity of magnetic nanoparticles can be changed by adjusting the pH value of the solution [103]. For example, as investigated in a study [104], the adsorption capacity of magnetic graphene oxide adsorbent increases with increasing pH value, and the maximum adsorption was observed at pH 10. The increase in pH led to the deprotonation of the -OH and -COOH groups of graphene sheets, making the surface charge more negative [104].

4.6 APPLICATION OF MAGNETIC ADSORBENTS FOR THE ADSORPTION OF PHARMACEUTICALS

Drugs such as hormones, antibiotics, anticonvulsants, and antipyretics pose a serious threat to the environment in surface and ground water. Even in trace concentrations, the aquatic life is gravely threatened by their contamination. This section discusses significant findings and the effectiveness of different magnetic adsorbents in eliminating pharmaceutical chemicals in light of recent research.

The primary features of magnetic nanoparticles that are utilised in the adsorption of medicines are high specific surface area, high chemical stability, high surface concentration of adsorption sites, adjustable particle size, magnetism, etc. Several investigations have demonstrated the effectiveness of employing magnetic oxide nanoparticles as adsorbents for the rapid and simple extraction of different medications from wastewater. Figure 4.3 shows applications of magnetic oxides used as adsorbents for the removal of various pharmaceuticals from aquatic systems.

For example, P. Yadav et al. [105] synthesised Fe_3O_4-functionalised MIL101(Fe) chitosan composite beads for the removal of tetracycline, doxycycline, and ciprofloxacin from an aqueous medium (removal efficiency was over 99%). A. Parashar

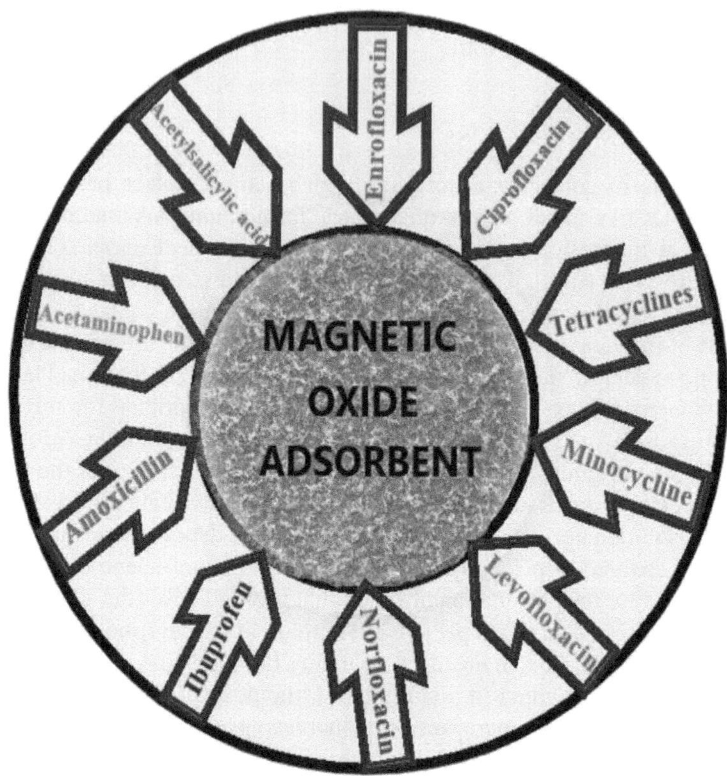

FIGURE 4.3   Applications of magnetic oxides used as adsorbents for the removal of various pharmaceuticals from aquatic system.

et al. [106] used magnetic iron oxide nanoparticles for the effective removal (94.12–99.52%) of oxcarbazepine from wastewater.

Of the more than ten classes of antibiotics, tetracyclines, fluoroquinolones, and aminoglycosides represent the greatest threat to the environment and public health. These are the most commonly used antibiotics in veterinary and human medicine. Thus, they are the main pollutants in surface water. The frequent use of many antibiotics contributes to the rapid establishment of pathogenic strains that are resistant to multiple drugs [107]. It is believed that antibiotic resistance in pathogenic flora is a serious medical problem that endangers people's health [107].

4.6.1 Adsorption of Ciprofloxacin

Ciprofloxacin is a member of the fluoroquinolone antibiotic class [108]. Ciprofloxacin exists in three distinct chemical forms in solution: zwitterionic [110], anionic [109], and cationic [109], which have deprotonated carboxylic acid groups at pH levels over 8.89 ± 0.11 and protonated amino groups at pH values below 5.90 ± 0.15. According to [111], the maximum concentration of ciprofloxacin found in surface waters in Western Europe is 13.6 g/L, while the average value is 0.008 g/L. Many investigations have looked for effective magnetic adsorbents for the removal of ciprofloxacin from aquatic environments. According to a review of the literature, the most often utilised magnetic nanocomposites are those based on Fe_3O_4 [86, 112–114], magnetic biochar [115], granular ferric hydroxide [116], magnetic chitosan-grafted graphene oxide nanocomposite [117], etc. Ciprofloxacin is extracted from wastewater using Fe_3O_4 nanoparticles as adsorbents [86, 113, 114]. The adsorption of ciprofloxacin on magnetic $Fe_3O_4(s)$ nanoparticles, which were produced as a secondary corrosion product of iron nanoparticles, was assessed by S. Rakshit et al. [118]. The adsorption of ciprofloxacin was found to have risen from 45% at pH = 3.44 to 80% at pH = 5.97. According to in situ ATR-FTIR studies, functional carboxyl groups are involved in the coordination of ciprofloxacin on the magnetite surface [118]. H. Rasoulzadeh et al. [86] removed ciprofloxacin from aqueous solutions using a chitosan nanocomposite (Fe-CS) supported by Fe_3O_4. Maximum adsorption capacity was 142 mg/g, and maximum removal efficiency was 68% [44]. In this, the charge of functional groups depends on the solution pH. At pH < 5.9 and pH > 8.9, the same signs of surface charge of Fe-CS and ciprofloxacin (CIFO) molecules caused electrostatic repulsion and reduced adsorption efficiency. However, at a pH within the range of 5.9–8.9, the electrostatic attraction was responsible for the adsorption process.

Another significant pathway of adsorption at neutral pH may include hydrophobic interactions between the adsorbent and the zwitterionic species of ciprofloxacin [86]. H. Mao et al. [119] investigated the efficacious removal of ciprofloxacin from an aqueous medium via magnetic Fe_3O_4-RM-NPs. The greatest rate of ciprofloxacin clearance was achieved at pH = 6. This can be explained by the fact that the secondary amine protonates on the piperazine group of ciprofloxacin, causing it to appear in the cationic form when the pH of the solution is below 5.9. The zwitterionic form of ciprofloxacin is present in solutions with pH values between 6.1 and 8.7. Ciprofloxacin is accessible in the anionic form with pH values greater than 8.89 [119].

4.6.2 Adsorption of Tetracyclines

The tetracycline family is one of the most widely used groups of broad-spectrum bacteriostatic medicines. Its chemical structure is based on an octahydronaphthacene core [120]. Methacycline, oxytetracycline, oxychlortetracycline, doxycycline, and tetracycline are examples of tetracyclines. Tetracycline hydrochloride bases or salts are used. Amphoteric compounds having dimethylamine and hydroxyl functional groups are called tetracyclines. They can be ionised or nonionised [121] depending on the pH: cationic at pH 3.3, zwitterionic at pH = 3.3–7.7, and anionic at pH > 7.7 [122]. Tetracycline's molecular structure has four dissociable functional groups and five dominant forms in water [123]. These forms exhibit a range of morphological and biological traits.

Numerous researchers have examined the adsorption potential of different magnetic nanoparticles, which can be employed as incredibly efficient adsorbents to remove tetracyclines from wastewater [75–82]. Researchers have employed magnetic graphene oxide [62, 69, 124], goethite [125], Fe_3O_4 [126–129], $NiFe_2O_4$ [71], $MnFe_2O_4$ [130], and other nanoparticles in their investigations. According to J. Miao et al. [124], the optimal adsorption of tetracycline and oxytetracycline has been noted at pH = 3.3, whereas magnetic graphene oxide NPs exhibit superior adsorption of chlortetracycline in an alkaline medium. Tetracycline adsorption was carried out using the $Fe_3O_4@SiO2@FeO$ nanocomposite by L. Lian et al. [126], and the impact of pH on adsorption capacity was examined. It was shown that a favourable adsorption efficiency was obtained in the pH range of 3 to 9, which may be related to the fact that $Fe_3O_4@SiO_2@FeO$ has a very low ζ-potential at this pH range. The electrostatic attraction between the negatively charged magnetic sorbent surface and tetracycline molecules occurs, so electron donor groups of tetracycline can easily form stable metal–tetracycline complexes with Fe (II) on the sorbent surface [126].

4.6.3   Adsorption of Minocycline

Pneumonia is among the many bacterial infections that are treated with minocycline, a tetracycline antibiotic. L. Lu et al. [131] created porous magnetic microspheres of $MgFe_2O_4/$-Fe_2O_3, which were then investigated as novel, effective adsorbents for the removal of minocycline from wastewater. The maximum adsorption capacity was 201 mg/g. The study's conclusions showed that adsorption capacity rose with temperature, indicating the endothermic nature of adsorption [131]. The formation of complexes between the functional groups of the antibiotic and the active centres of the adsorbent in the antibiotic–adsorbent system indicated chemisorption. $MgFe_2O_4/$-Fe_2O_3 adsorbents were easily isolated from aqueous solution and could be regenerated using methanol or NaOH solution due to their magnetic properties [131].

4.6.4   Adsorption of Levofloxacin

One of the fluoroquinolone broad-spectrum antibiotics is levofloxacin. Gram-positive and Gram-negative infections, including strains resistant to aminoglycosides, cephalosporins, and/or penicillins, are all susceptible to the action of levofloxacin. Levofloxacin can exist in three different forms: cationic (pH < 6.02), zwitterionic (pH < 6.02 < 8.15), or anionic (pH > 8.5) [132]. Its two dissociation constant values are pKa1 = 6.02 and pKa2 = 8.15. The following magnetic adsorbents are used to remove levofloxacin from water: $Fe_3O_4@SiO_2$ [133], Fe_3O_4–gINPs [134], $NiFe_2O_4$/biochar composites [135], biochar/$MgFe_2O_4$ [136], Ag_3PO_4/rGO/$CoFe_2O_4$ [137], magnetic carbon nanocomposite [138], etc. Levofloxacin adsorption at the surfaces of Fe_3O_4 and $Fe_3O_4@SiO_2$ nanoparticles was investigated by M.H. Al-Jabari et al. [99]. Levofloxacin's highest adsorption capacity ranged from 6.09 to 6.85 mg/g for the magnetic adsorbent. The elimination efficiency that performed the best was 80.2% at pH 6.5. Levofloxacin exists in neutral and zwitterion forms, which is related to this. The electrostatic repulsion of the same kind of charges is related to the removal efficiency decreasing with a change in pH [133].

4.6.5 Adsorption of Enrofloxacin

Enrofloxacin is an antibiotic belonging to the fluoroquinolone class. Due to the dissociating groups -COOH and -NH$_2$ in its structure, enrofloxacin can exist in a variety of forms based on the pH of the solution [139]. The enrofloxacin adsorption mechanisms were studied by G. Peng et al. [140] using a Fe$_3$O$_4$/magnetic montmorillonite nanocomposite. The highest adsorption capacity of 163 mg/g was achieved. Effective adsorption properties were associated with electrostatic interaction. For the purpose of eliminating enrofloxacin, R. Li et al. [141] created a composite of manganese oxide based on magnetic biochar and untreated biochar. The magnetic nanocomposite had a maximum adsorption capacity of 7.19 mg/g. As the pH of the solution increased (between 3.0 and 10.0) and the ionic strength increased (0.001–0.1 M), the adsorption capacity decreased. For the purpose of eliminating enrofloxacin, Z. Gordi et al. [142] created an adsorbent based on a combination of magnetically functionalised graphene oxide and MOF. The enrofloxacin removal efficiency by the magnetically functionalised graphene oxide was 345 mg/g. The efficacy of enrofloxacin elimination rose as the pH of the solution increased to 7.5 and declined above pH = 7.5 (pKa values of carboxyl and amine groups were 5.88 and 7.74, respectively). The electrostatic interaction was the primary adsorption mode. Magnetic carbon nanocomposites were employed by M. Zahoor et al. [143] to extract enrofloxacin from industrial effluent. At pH 6–8, the highest adsorption efficiency was noted. The pH level affects both the surface charge of the magnetic adsorbent and the dissociated form of enrofloxacin in the solution; in an acidic or alkaline environment, the adsorbent and adsorbate surfaces have the same charge signs, and the adsorption capacity is decreased by electrostatic repulsion forces.

4.6.6 Adsorption of Norfloxacin

Norfloxacin belongs to a class of synthetic antibiotics called fluoroquinolones. The pKa dissociation constants of norfloxacin are 6.22 and 8.51. G. Peng et al. [144] utilised magnetic Fe$_3$O$_4$ nanoparticles in conjunction with nitrogen-modified reduced graphene oxide, or N-RGO/Fe$_3$O$_4$, as an adsorbent to remove norfloxacin. A maximum adsorption capacity of 158 mg/g was determined. The two proton-binding centres in norfloxacin—the carboxyl and piperazinyl groups—can exist in different forms based on the pH of the solution, which aids in explaining the adsorption mechanism. Because of electrostatic repulsion, norfloxacin's adsorption is not favourable in acidic solutions, where the cationic form predominates at pH < 6. When pH increases, the surface of the nanoadsorbent and the norfloxacin molecules gain opposite charges, so the adsorption increases [144].

4.6.7 Adsorption of Amoxicillin

Amoxicillin is a broad-spectrum β-lactam antibiotic. Amoxicillin is amphoteric due to its three main functional groups, which are -NH$_2$, -COOH, and -OH [145]. Its three different acid dissociation values are pKa1 = 2.68 for the carboxyl group, pKa2 = 7.49 for the amino group, and pKa3 = 9.63 for the phenolic group [146]. As a result, amoxicillin can exist in an aqueous solution in four different forms (Amox$^+$, Amox, Amox$^-$, and Amox^{2-}) depending on the pH of the solution [146]. M. Pooresmaeil et al. [147] created a magnetic nanocomposite composed of Fe$_3$O$_4$@Cd-MOF@CS to extract amoxicillin from wastewater. Fe$_3$O$_4$@Cd-MOF@CS contains a variety of functional groups that can take part in amoxicillin adsorption. Many hydroxyl and amine functional groups found in chitosan have the ability to bind to antibiotics through hydrogen bonding and electrostatic attraction. The abundance of aromatic rings in the Fe$_3$O$_4$@CdMOF composite offers the chance for pollutant trapping inside the cavities, which further aids in adsorption [147].

4.7 APPLICATIONS OF MAGNETIC ADSORBENTS FOR THE ADSORPTION OF NON-STEROIDAL DRUGS

A class of medications known as non-steroidal anti-inflammatory medicines (NSAIDs) has analgesic, antipyretic (antipyretic) and, at higher dosages, anti-inflammatory properties. The most well-known examples are acetaminophen, keto-profen, diclofenac, ibuprofen, naproxen, and acetylsalicylic acid. Figure 4.4 shows the removal of pharmaceutical contaminants from water using magnetic oxide adsorbents.

4.7.1 ADSORPTION OF IBUPROFEN

Phenylpropionic acid, the source of ibuprofen, has anti-inflammatory, analgesic, and antipyretic properties [148]. As per reference [111], the mean concentration of ibuprofen in Western European surface waters is 0.097 µg/L, whereas the highest quantity ever recorded is 303 µg/L. T.M. Salem Attia et al. synthesised magnetic nanoparticles covered with zeolite (γ-Fe$_2$O$_3$–zeolite) [149]. Zeolites are minerals that have pores the size of molecules and are capable of absorbing and holding onto a wide range of pollutants [150]. At an adsorbent dosage of 1 g/L and an ibuprofen

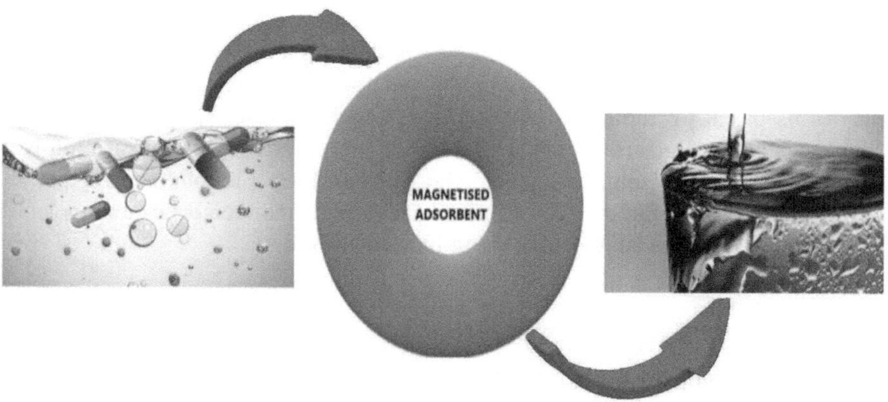

FIGURE 4.4   Removal of pharmaceutical contaminants from water using magnetic oxide adsorbents.

concentration of 0.1 mg/L, the ibuprofen removal efficiency was >95% in less than ten minutes [149]. An $NiFe_2O_4$/activated carbon magnetic composite utilised for ibuprofen adsorption was obtained by A.C. Frohlich et al. [151]. At 328 K, it was discovered that the adsorption was advantageous in an acidic environment. There was a maximum of 261 mg/g for adsorption capacity. Magnetic genipin-crosslinked chitosan/graphene oxide–SO_3H (GC/MGO-SO_3H) was created by Y. Liu et al. [152] as an adsorbent for the elimination of ibuprofen. When temperature increased from 298 to 313 K, the highest adsorption capacity rose from 113 to 138 mg/g. In order to remove ibuprofen from wastewater, G. Wu et al. [153] synthesised magnetic rape straw biomass fibre/β-cyclodextrin/Fe_3O_4. Electrostatic interactions and hydrogen bond formation were involved in the elimination of ibuprofen. The produced adsorbent had a lot of oxygen-containing groups (OH, C=O) on its surface that could form hydrogen bonds with ibuprofen [153].

Ibuprofen, naproxen, diclofenac-Na, and gemfibrozil are a few examples of pharmaceutical and personal care products (PPCPs) that can be removed from aqueous solutions using magnetic nanoparticle-coated zeolite (MNCZ), as described by Attia et al. [154]. To maximise the reaction conditions, several sorption factors including contact time, solution pH, and beginning PPCP concentration were carefully examined. The pH of the adsorptive medium's aqueous phase significantly influenced the removal effectiveness of the PPCPs (ibuprofen, naproxen, diclofenac-Na, and gemfibrozil) under study in a comparable way. Thus, these chemicals were eliminated less effectively as the pH shifts from acidic to basic conditions. Contact time had no discernible impact on MNCZ's ability to adsorb diclofenac-Na. On the other hand, the adsorption of ibuprofen, naproxen, and gemfibrozil was significantly impacted by the duration of contact. For the PPCPs under study, MNCZ had a strong adsorption potential, and the removal efficiency was over 95% in less than ten minutes. When PPCP concentrations were low, the initial concentration did not seem to have a discernible impact on the removal efficiency of the PPCPs under study; nevertheless, as PPCP concentrations increased, the removal of the PPCPs under study, particularly ibuprofen, significantly declined. Because the r^2 fitted with this model is closer to 1 and has a lower SE than that produced by the Langmuir isotherm, the Freundlich isotherm may be the best fit for the adsorption data.

Five kinetic models—simple Elovich, Bangham's model, intra-particle diffusion, pseudo-first order, and pseudo-second order—have been fitted to the experimental data. The model that best fits the experimental data is pseudo-second order. A study comparing the removal efficacy of MNCZ with alternative methods utilised at drinking water treatment plants (DWTPs) was carried out using column adsorption. When comparing MNCZ's removal efficiency to alternative methods for removing PPCPs compounds, greater results were seen.

4.7.2 Adsorption of Acetylsalicylic Acid

Acetylsalicylic acid is a non-steroidal anti-inflammatory medication that has anti-inflammatory, analgesic, and antipyretic effects. Acetylsalicylic acid in Western European surface waters has been found to have an average content of 0.002 µg/L and a high of 0.36 µg/L [111]. Magnetic Fe_3O_4/Douglas fir biochar adsorbents for the removal of acetylsalicylic acid from wastewater were synthesised by A.S. Liyanage et al. [155]. Because of the electrostatic interaction between the negatively charged adsorbate with carboxylate anions and the positively charged adsorbent surface, the maximum removal of acetylsalicylic acid happened at pH = 4–8. Acetylsalicylic acid was extracted from aqueous solutions using a magnetite-core active charcoal–shell matrix, Fe_3O_4@C, by E.A. Moaca et al. [156]. A maximum adsorption capacity of 234 mg/g was achieved. The magnetic matrix demonstrates two benefits: the easy magnetic separation from the liquid medium and the highly porous structure necessary for effective adsorption.

4.7.3 ADSORPTION OF ACETAMINOPHEN

Analgesics like paracetamol are frequently used. Acetaminophen has been found in surface waters of Western Europe at an average value of 0.046 µg/L and at a maximum of 230 µg/L [111]. Acetaminophen adsorption was carried out using an amine-functionalised superparamagnetic silica nanocomposite by S. C. Kollarahithlu et al. [157]. The $NiFe_2O_4$@SiO_2@aminopropyltrimethoxysilane adsorbent contains amino groups ($-NH_2$) that are readily protonated ($-NH_3^+$) and facilitate adsorption in an acidic media (pH range of 4 to 6) via electrostatic interaction. The -OH group of paracetamol and the $-NH_3^+$ of the adsorbent generate hydrogen bonds, which explains this [157]. Acetaminophen was removed from aqueous solutions using rhamnolipid-based chitosan magnetic nanosorbents, which were created by R. Natarajan et al. [158]. The study showed a maximum removal of acetaminophen of 96.7%, adsorption capacity of 96.3 mg/g at 60 min contact time, pH = 5, and temperature of 303 K. Different magnetic adsorbents for antibiotics and non-steroidal anti-inflammatory drugs are shown in Table 4.2.

TABLE 4.2

Magnetic Adsorbent Materials for Antibiotics and Non-Steroidal Anti-Inflammatory Drugs

Magnetic Materials Used for Processes Involved in Decontamination	Pharmaceuticals Present in Water	Processes Involved for Decontamination	Reference
Nano zero-valent iron (nZVI)	Carbamazepine	Adsorption	[159]
Nano zero-valent iron (nZVI) and PEG- and zeolite-supported nZVI	Amoxicillin, ampicillin	Adsorption	[160]
Zero-valent iron (ZVI) carbamazepine chemical degradation	Carbamazepine	Chemical degradation	[161]
Maghemite (Fe_2O_3) core confined in a silica porous layer	Atenolol, gemfibrozil, and sulphamethoxazole	Adsorption	[162]

TABLE 4.2 (*Continued*)

Magnetic Adsorbent Materials for Antibiotics and Non-Steroidal Anti-Inflammatory Drugs

Magnetic Materials Used for Processes Involved in Decontamination	Pharmaceuticals Present in Water	Processes Involved for Decontamination	Reference
MFe_2O_4 (M = Fe, Mn, Co, Zn)	Tetracycline, oxytetracycline, and chlortetracycline	Adsorption	[163]
$MnFe_2O_4$/(activated carbon) AC	Sulphamethoxazole	Adsorption	[164]
$MgFe_2O_4$/γ-Fe_2O_3	Minocycline	Adsorption	[165]
Magnetic graphene oxide	Tetracycline	Adsorption	[124]
Magnetic chitosan-grafted graphene oxide nanocomposite	Ciprofloxacin	Adsorption	[117]
$NiFe_2O_4$/biochar composites	Levofloxacin	Adsorption	[135]
Fe_3O_4/magnetic montmorillonite	Enrofloxacin	Adsorption	[140]
Magnetic molecularly imprinted polymers	Norfloxacin	Adsorption	[166]
Magnetic cadmium-based MOFs modified with chitosan (Fe_3O_4 @Cd-MOF@CS)	Amoxicillin	Adsorption	[146]
Willow sawdust biochar/Fe_3O_4 and pine sawdust biochar/Fe_3O_4	Daptomycin	Adsorption	[167]
Fe_3O_4/activated carbon/chitosan	Erythromycin	Adsorption	[168]
Amine-coated magnetic nanocomposite $NiFe_2O_4$@SiO_2	Streptomycin	Adsorption	[169]
Magnetic Fe_3O_4 NP-coated zeolite	Cephalexin	Adsorption	[170]
Fe_3O_4-chitosan (CTS-MNPs)	Metronidazole	Adsorption	[171]
Fe_3O_4/graphene oxide/citrus peel-derived magnetic bio-nanocomposite	Sparfloxacin	Adsorption	[172]
Activated carbon modified with magnetic Fe_3O_4 NPs	Ceftriaxone	Adsorption	[173]
Bovine serum albumin/Fe_3O_4	Chloramphenicol	Adsorption	[174]
Magnetic genipin-crosslinked chitosan/graphene oxide–SO_3H composite	Ibuprofen	Adsorption	[151]
Fe_3O_4/Douglas fir biochar	Acetylsalicylic acid	Adsorption	[175]
Magnetite nanoparticles modified with β-cyclodextrin, polymer coupled	Acetaminophen	Adsorption	[176]
Fe_3O_4@graphene nanoplatelet nanocomposite	Naproxen	Adsorption	[177]

4.8   FUTURE PERSPECTIVES

The creation of composite adsorbents with magnetic properties for the purification of water is the subject of an increasing amount of research. Samples with specific properties can be created by functionalising and modifying the surfaces of magnetic adsorbents. One of the main advantages of magnetically sensitive adsorbents over ordinary (nonmagnetic) ones is their capacity to be regulated by an external magnetic field. Among the various materials that a magnetic field can penetrate are glass and polymers. Magnetic separation is a non-contact, non-destructive method of removing a wasted adsorbent from water to prevent recontamination with pharmaceuticals or other adsorbates.

The effectiveness of magnetic separation is not significantly affected by temperature, pH, or the amount of adsorbate present in the solution. Therefore, magnetically controlled adsorbents can be used for the adsorption of heavy metals and dyes as well as the purification of pharmaceutical effluents from antibiotics, medications, etc. The use of these materials allows for the replacement of the mechanical separation stage, which is one of the stages in this process that requires the most labour and energy. But developing the ideal treatment technology ought to be a lucrative and economical process that makes it easy to integrate with the facilities that are already in place. After being used in wastewater treatment, traditional adsorbents produce sludge, which causes secondary contamination. Rather, magnetic adsorbents are not a source of this issue.

The process of using magnetic adsorbents to purify water is quite intricate and involves a number of requirements, such as treating low-concentration contaminants, achieving very low concentration limitations at the end of the process, and making sure that after the purification, nanoparticles are removed from the water. Furthermore, the magnetism of nanoparticles results in a significant agglomeration of particles and a decrease in their adsorption ability, making their reuse more challenging. Since most of the studies that were analysed were conducted in laboratories, more work needs to be done to find solutions to a number of problems. These problems include analysing the productivity of magnetic nanoparticles when used in industrial settings, researching the best way to remove pharmaceuticals continuously, examining the toxicity of magnetic nanoparticles after repeated cycles, disposing of magnetic nanoparticles, evaluating their life cycle, and researching the environmental effects of magnetic nanoparticles.

Magnetic adsorbents are attractive because they can be easily removed from the treatment media and because they can be made to change their physicochemical properties when exposed to an external magnetic field created by an electromagnetic field or a permanent magnet. In the second case, heating the magnetic adsorbent offers an extra benefit that might enhance the pollutants' ability to be adsorbed. The applied magnetic field has the ability to modify both the adsorbed molecules and the "magnetic memory effect". When an adsorbent is present, adsorbates can alter their dipole moment, which leads to more agglomeration and increased retention on the adsorbent surface. More investigation is needed into this issue.

It is also important to keep in mind that the regeneration of the magnetic adsorbents is a crucial aspect of their use. Regeneration is impacted by the kind of magnetic adsorbent used as well as the nature of the pollutant. Regeneration is the process of removing contaminants from the adsorbent's surface to enable its reuse. There are several methods for regenerating magnetic adsorbents, including thermal, chemical, biological, and magnetic methods. Thermal regeneration involves heating the adsorbent to a high temperature in order to remove impurities that have been absorbed. During chemical regeneration, the adsorbent is treated chemically, which desorbs the impurities. Contaminants on the adsorbent surface are eliminated or destroyed by the microorganisms involved in the biological regeneration process. A magnetic field is used in another regeneration technique to increase the mobility of absorbed contaminants and expedite their removal from the adsorbent s"rfac'. The adsorbent needs to be thoroughly rinsed in order to remove any contaminants or regeneration ingredients that might have remained after the regeneration. Sufficient regeneration and cleaning procedures are essential to improve the magnetic adsorbent's efficacy after repeated use.

Other significant concerns and challenges for the future development of the described technology include simultaneous adsorption of co-existing drugs and selectivity of the adsorbents to certain pharmaceuticals. To ensure the involvement of all types of adsorption modes, co-existing contaminants should be adsorbed using correctly functionalised magnetic adsorbents. The functional groups can concurrently or selectively adsorb several types of medicines from the solution and produce new adsorption centres. Electrostatic attraction and complex formation are predicted to be the primary processes of concurrent adsorption. As opposed to removing the same medications one at a time, the simultaneous elimination of several medications from the solution may occur under different conditions (such as a higher or lower pH or temperature). For a more thorough understanding of adsorption modes, in situ experimental techniques, artificial intelligence, and quantum chemical modelling should be applied. Therefore, for the continued development of potential magnetic adsorbents, in-depth research on the viability of actual applications in the treatment of water and wastewater is essential.

4.9 CONCLUSIONS AND OUTLOOK FOR THE FUTURE

The current state of knowledge on the removal of some of the most commonly used medications from aquatic environments is presented in this chapter, along with some suggestions for enhancing the adsorption effectiveness of magnetic adsorbents. Since pharmaceuticals have a significant detrimental influence on the environment and negatively affect flora and fauna, pharmaceutical wastewater purification is rising to the top of research priorities for scientists worldwide. It is believed that the most common method of extracting drugs from water is adsorption.

This chapter demonstrates the adsorption of medicines (namely antibiotics and non-steroidal anti-inflammatory drugs) from aqueous environments using various adsorbers with magnetic activity. The review of the literature indicates that magnetic nanoparticles have a high adsorption capacity, which makes them appropriate for wide-scale application. Magnetic adsorbents have many advantages, including the ability to be synthesised using cheap precursors, excellent water stability, biocompatibility, reusability, simple synthesis methods, and surface functionalisation potential. Numerous studies on the elimination of drugs like tetracycline, ciprofloxacin, levofloxacin, diclofenac, and ibuprofen have been published by scientists from throughout the world. The most often studied magnetic adsorbents are Fe_3O_4-based nanoparticles.

Different cations can be added to the structures of magnetite or ferrites to increase their adsorption capacity and surface-active site count, all the while preventing the adsorbent's magnetisation—which is essential for removing used adsorbent from solution—from being greatly decreased. The adsorption capacity of magnetic adsorbents ranges from tens to hundreds of mg/g. It can be amplified several times by creating composite magnetic adsorbents based on graphene oxide, layered hydroxides, charcoal, MOFs/COFs, etc. One disadvantage of magnetic nano-adsorbents is that they tend to agglomerate. However, this can be readily overcome by coating or functionalising the materials with different materials (such as silica, amino acids, surfactants, polymers (polyaniline, polystyrene, polypyrrole, etc.), etc.). The adsorption mode is shown to be significantly influenced by pH, temperature, adsorbent dosage, and adsorbate concentration. We have discussed the relationships between drug–chemical interactions and their ability to adopt cationic, anionic, or zwitterionic forms based on the pH of the aqueous solution. A survey of the literature indicates that the main mechanisms underlying pharmaceutical adsorption on magnetic adsorbents are ion exchange, surface complexations, hydrophobic and electrostatic interactions, π–π interactions, and hydrogen bonds. This suggests that controlling the processes' selectivity—that is, removing only pharmaceuticals from water—is not something that can be done very often.

REFERENCES

[1] Vinayagam V, Murugan S, Kumaresan R, et al. Sustainable adsorbents for the removal of pharmaceuticals from wastewater: A review. Chemosphere. 2022;300, https://doi.org/10.1016/j.chemosphere.2022.134597.

[2] Nannou C, Ofrydopoulou A, Evgenidou E, et al. Lambropoulou, antiviral drugs in aquatic environment and wastewater treatment plants: A review on occurrence, fate, removal and ecotoxicity. Sci Total Environ. 2020;699.

[3] Ahmed MJ, Hameed BH. Removal of emerging pharmaceutical contaminants by adsorption in a fixed-bed column: A review. Ecotoxicol Environ Saf. 2018;149:257–266.

[4] Adewuyi A. Chemically modified biosorbents and their role in the removal of emerging pharmaceutical waste in the water system. Water. 2020;12(6):1551.

[5] Majumder A, Gupta B, Gupta AK. Pharmaceutically active compounds in aqueous environment: A status, toxicity and insights of remediation. Environ Res. 2029;176:108542.

[6] Zhao Y, Zhang C, Yang Z, et al. Global trends and prospects in the removal of pharmaceuticals and personal care products: A bibliometric analysis. J Water Process Eng. 2021;41:102004, https://doi.org/10.1016/j.jwpe.2021.102004.

[7] Khan HK, Rehman MYA, Malik RA. Fate and toxicity of pharmaceuticals in water environment: An insight on their occurrence in South Asia. J Enviorn Manage. 2020a;271:111030.

[8] Priya AK, Gnanasekaran L, Rajendran S, et al. Occurrences and removal of pharmaceutical and personal care products from aquatic systems using advanced treatment-A review. Environ Res. 2022;204:112298, https://doi.org/10.1016/j.envres.2021.112298.

[9] Khan AH, Khan NA, Zubair M, et al. Sustainable green nanoadsorbents for remediation of pharmaceuticals from water and wastewater: A critical review. Environ Res. 2022;204:112243, https://doi.org/10.1016/j.envres.2021.112243.

[10] Bhagyashree T, Sellamuthu B, Ouarda Y, et al. Review on fate and mechanism of removal of pharmaceutical pollutants from wastewater using biological approach. Bioresour Technol. 2017;224:1–12.

[11] Hofman-Caris CHM, Siegers WG, Merlen KVD, et al. Removal of pharmaceuticals from WWTP effluent: Removal of EFOM followed by advanced oxidation. Chem Eng J. 2017;327:514–521.

[12] Bastami TR, Ahmadpour A, Hekmatikar FA. Synthesis of Fe_3O_4/Bi_2WO_6 nanohybrid for the photocatalytic degradation of pharmaceutical ibuprofen under solar light. J Ind Eng Chem. 2017;51:244–254.

[13] Gomes J, Rosa RC, Rui MQF, et al. Application of ozonation for pharmaceuticals and personal care products removal from water. Sci Total Environ. 2017;586:265–283.

[14] Mlunguza NY, Ncube S, Mahlambi PN, et al. Adsorbents and removal strategies of non-steroidal anti-inflammatory drugs from contaminated water bodies. J Environ Chem Eng. 2019;7(3):103142.

[15] Taoufik N, Boumya W, Janani FZ, et al. Removal of emerging pharmaceutical pollutants: A systematic mapping study review. J Environ Chem Eng. 2020;8(5):104251.

[16] Ayati A, Ahmadpour A, Bamoharram FF, et al. A review on catalytic applications of Au nanoparticles in the removal of water pollutant. Chemosphere. 2014;107:163–174.

[17] Ranjbari S, Tanhaei B, Ayati A, et al. Novel Aliquat-336 impregnated chitosan beads for the adsorptive removal of anionic azo dyes. Int J Biol Macromol. 2019;125:989–998.

[18] Zhu X, He M, Sun Y, et al. Insights into the adsorption of Pharmaceuticals and Personal Care Products (PPCPs) on biochar and activated carbon with the aid of machine learning. J Hazard Mater. 2022;423:127060, https://doi.org/10.1016/j.jhazmat.2021.127060.

[19] Magesh N, Renita A A, Siva R, et al. Adsorption behavior of fluoroquinolone(ciprofloxacin) using zinc oxide impregnated activated carbon prepared from jack fruit peel: Kinetics and isotherm studies. Chemosphere. 2022;290:133227, https://doi.org/10.1016/j.chemosphere.2021.133227.

[20] Magesh N, Renita A, Senthil Kumar P, et al. Adsorption of ciprofloxacin from aqueous solution using surface improved tamarind shell as an economical and effective adsorbent. Int J Phytoremediation. 2022;24(3):224–234.

[21] Arabkhani P, Asfaram A. The potential application of bio-based ceramic/organic xerogel derived from the plant sources: A new green adsorbent for removal of antibiotics from pharmaceutical wastewater. J Hazard Mater. 2022;429:128289, https://doi.org/10.1016/j.jhazmat.2022.128289.

[22] Lakshmanan A, Surendran P, Sakthy Priya S, et al. Investigations on structural, optical, dielectric, electronic polarizability, Z-scan and antibacterial properties of Ni/Zn/Fe2O4 nanoparticles fabricated by microwave-assisted combustion method. J Photochem Photobiol A Chem. 2020;402:112794, https://doi.org/10.1016/j.jphotochem.2020.112794.

[23] Lakkaboyana SK, Soontarapa K, Vinaykumar, et al. Preparation of novel chitosan polymeric nanocomposite as an efficient material for the removal of Acid Blue 25 from aqueous environment. Int J Biol Macromol. 2021;168:760–768.

[24] Sathiya Priya A, Geetha D, Karthik K, et al. Investigations on the enhanced photocatalytic activity of (Ag, La) substituted nickel cobaltite spinels. Solid State Sci. 2019;98:105992, https://doi.org/10.1016/j.solidstatesciences.2019.105992.

[25] Carvalho IT, Santos L. Antibiotics in the aquatic environments: A review of the European scenario. Environ Int. 2016;94:736–757.

[26] Reguyal F, Sarmah AK. Adsorption of sulfamethoxazole by magnetic biochar: Effects of pH, ionic strength, natural organic matter and 17 alpha-ethinylestradiol. Sci Total Environ. 2018;628–629:722–730.

[27] Sun PZ, Li YX, Meng T, et al. Removal of sulfonamide antibiotics and human metabolite by biochar and biochar/H2O2 in synthetic urine. Water Res. 2018;147:91–100.

[28] Shikuku VO, Zanella R, Kowenje CO, et al. Single and binary adsorption of sulfonamide antibiotics onto iron-modified clay: Linear and nonlinear isotherms, kinetics, thermodynamics, and mechanistic studies. Appl Water Sci. 2018;8.

[29] Rajapaksha AU, Vithanage M, Lee SS, et al. Steam activation of biochars facilitates kinetics and pH-resilience of sulfamethazine sorption. J Soils and Sediments. 2016;16:889–895.

[30] Premarathna KSD, Rajapaksha AU, Adassoriya N, et al. Clay-biochar composites for sorptive removal of tetracycline antibiotic in aqueous media. J Environ Manag. 2019;238:315–322.

[31] Chayid MA, Ahmecd MJ. Amoxicillin adsorption on microwave prepared activated carbon from Arundo donax Linn: Isotherms, kinetics, and thermodynamics studies. J Environ Chem Eng. 2015;3:1592–1601.

[32] Ahmed MJ. Adsorption of quinolone, tetracycline, and penicillin antibiotics from aqueous solution using activated carbons: Review. Environ Toxicol Pharmacol. 2017;50:1–10.

[33] Ashiq A, Adassooriya NM, Sarkar B, et al. Municipal solid waste biochar-bentonite composite for the removal of antibiotic ciprofloxacin from aqueous media. J Environ Manag. 2019;236:428–435.

[34] Luo J, Li X, Ge C, et al. Sorption of norfloxacin, sulfamerazine and oxytetracycline by KOH-modified biochar under single and ternary systems. Bioresour Technol. 2018;263:385–392.

[35] Chowdhury S, Sikder J, Mandal T, et al. Comprehensive analysis on sorptive uptake of enrofloxacin by activated carbon derived from industrial paper sludge. Sci Total Environ. 2019;665:438–452.

[36] Qin TT, Wang ZW, Xie XY, et al. A novel biochar derived from cauliflower (Brassica oleracea L.) roots could remove norfloxacin and chlortetracycline efficiently. Water Sci Technol. 2017;76:3307–3318.

[37] Martínez-Costa JI, Leyva-Ramos R, Padilla-Ortega E, et al. Antagonistic, synergistic and non-interactive competitive sorption of sulfamethoxazole-trimethoprim and sulfamethoxazole-cadmium (ii) on a hybrid clay nanosorbent. SciTotal Environ. 2018;640–641:1241–1250.

[38] Guo XT, Yin YY, Yang C, et al. Maize straw decorated with sulfide for tylosin removal from the water. Ecotoxicol Environ Safety. 2018;152:16–23.

[39] Kimbell LK, Tong YR, Mayer BK, et al. Biosolids-derived biochar for triclosan removal from wastewater. Environ Eng Sci. 2018;35:513–524.

[40] Liu CH, Chuang YH, Li H, et al. Sorption of lincomycin by manure-derived biochars from water. J Environ Qual. 2016;45:519–527.

[41] Zhang J, Lu M, Wan J, et al. Effects of pH, dissolved humic acid and Cu2+ on the adsorption of norfloxacin on montmorillonite-biochar composite derived from wheat straw. Biochem Eng J. 2018;130:104–112.

[42] Weidemann E, Niinipuu M, Fick J, et al. Using carbonized low-cost materials for removal of chemicals of environmental concern from water. Environ Sci Pollut Res. 2018;25:15793–15801.

[43] García-Mateos FJ, Ruiz-Rosas R, Marqués MD, et al. Removal of paracetamol on biomass-derived activated carbon: Modeling the fixed bed breakthrough curves using batch adsorption experiments. Chem Eng J. 2015;279:18–30.

[44] Mondal S, Patel S, Majumder SK. Naproxen removal capacity enhancement by transforming the activated carbon into a blended composite material. Wat Air and Soil Poll. 2010;231.

[45] Frohlich AC, Foletto EL, Dotto GL. Preparation and characterization of NiFe2O4/activated carbon composite as potential magnetic adsorbent for removal of ibuprofen and ketoprofen pharmaceuticals from aqueous solutions. J Clean Prod. 2019;229:828–837.

[46] Bagheri A, Abu-Danso E, Iqbal J, et al. Modified biochar from Moringa seed powder for the removal of diclofenac from aqueous solution. Environ Sci Pollut Res. 2020;27:7318–7327.

[47] Chakraborty P, Banerjee S, Kumar S, et al. Elucidation of ibuprofen uptake capability of raw and steam activated biochar of Aegle marmelos shell: Isotherm, kinetics, thermodynamics and cost estimation. Process Saf Environ Prot. 2018;118:10–23.

[48] Lessa EF, Nunes ML, Fajardo AR. Chitosan/waste coffee-grounds composite: An efficient and eco-friendly adsorbent for removal of pharmaceutical contaminants from water. Carbohydr Polym. 2018;189:257–266.

[49] Fernandes MJ, Moreira MM, Paiga P, et al. Evaluation of the adsorption potential of biochars prepared from forest and agri-food wastes for the removal of fluoxetine. Bioresour Technol. 2019;292.

[50] Calisto V, Jaria G, Silva CP, et al. Single and multi-component adsorption of psychiatric pharmaceuticals onto alternative and commercial carbons. J Environ Manag. 2017;192:15–24.

[51] Chen J, Liu Y-S, Deng W-J, et al. Removal of steroid hormones and biocides from rural wastewater by an integrated constructed wetland. Sci Total Environ. 2019;660:358–365.

[52] Quesada HB, Alves Baptista AT, Cusioli LF, et al. Surface water pollution by pharmaceuticals and an alternative of removal by low-cost adsorbents. Chemosphere. 2019;222:766–780.

[53] Servos MR, Bennie DT, Burnison BK, et al. Distribution of estrogens, 17 beta-estradiol and estrone, in Canadian municipal wastewater treatment plants. Sci Total Environ. 2005;336:155–170.

[54] Ali MEM, Abd El-Aty AM, Badawy MI, et al. Removal of pharmaceutical pollutants from synthetic wastewater using chemically modified biomass of green alga Scenedesmus obliquus. Ecotoxicol Environ Saf. 2018;151:144–152.

[55] Fu CL, Zhang HL, Xia MZ, et al. The single/co-adsorption characteristics and microscopic adsorption mechanism of biochar-montmorillonite composite adsorbent for pharmaceutical emerging organic contaminant atenolol and lead ions. Ecotoxicol Environ Saf. 2020;187.

[56] Muter O, Perkons I, Bartkevics V. Removal of pharmaceutical residues from wastewater by woodchip-derived biochar. Desalin Water Treat. 2019;159:110–120.

[57] Boudrahem N, Delpeux-Ouldriane S, Khenniche L, et al. Single and mixture adsorption of clofibric acid, tetracycline and paracetamol onto activated carbon developed from cotton cloth residue. Process Saf Environ Pro. 2017;111544–111559.

[58] Liu YY, Blowes DW, Ptacek CJ, et al. Removal of pharmaceutical compounds, artificial sweeteners, and perfluoroalkyl substances from water using a passive treatment system containing zero-valent iron and biochar. Sci Total Environ. 2019;691:165–177.

[59] Hurtado C, Canameras N, Dominguez C, et al. Effect of soil biochar concentration on the mitigation of emerging organic contaminant uptake in lettuce. J Hazard Mater. 2017;323:386–393.

[60] Mondal S, Aikat K, Siddharth K, et al. Optimizing ranitidine hydrochloride uptake of Parthenium hysterophorus derived N-biochar through response surface methodology and artificial neural network. Process Saf Environ Prot. 2017;107:388–401.

[61] Lins PVD, Henrique DC, Ide AH, et al. Evaluation of caffeine adsorption by MgAl-LDH/biochar composite. Environ Sci Pollut Res. 2019;26:31804–31811.

[62] Chen DZ, Xie SS, Chen CQ, et al. Activated biochar derived from pomelo peel as a high-capacity sorbent for removal of carbamazepine from aqueous solution. Rsc Adv. 2017;7:54969–54979.

[63] Prasad NK, Panda D, Singh S, et al. Biocompatible suspension of nanosized γ-Fe_2O_3 synthesized by novel methods. J Appl Phys. 2005;97:10Q903, https://doi.org/10.1063/1.1849056.

[64] Shaw SK, Kailashiya J, Gangwar A, et al. γ-Fe_2O_3 nanoflowers as efficient magnetic hyperthermia and photothermal agent. Appl Surf Sci. 2021;560:150025, https://doi.org/10.1016/j.apsusc.2021.150025.

[65] Kefeni KK, Mamba BB, Msagati TAM. Application of spinel ferrite nanoparticles in water and wastewater treatment: A review. Sep Purif Technol. 2017;188:399–422, https://doi.org/10.1016/J.SEPPUR.2017.07.015.

[66] Paswan SK, Kumar P, Singh RK, et al. Spinel ferrite magnetic nanoparticles. In: Pollut Water Manag. Wiley; 2021. p. 273–305.

[67] Kefeni KK, Msagati TAM, Mamba BB. Ferrite nanoparticles: Synthesis, characterisation and applications in electronic device. Mater Sci Eng B. 2017;215:37–55, https://doi.org/10.1016/J.MSEB.2016.11.002.

[68] Tatarchuk T, Bououdina M, Judith Vijaya J, et al. Spinel ferrite nanoparticles: Synthesis, crystal structure, properties, and perspective applications. In: Fesenko O, Yatsenko L, editors. Nanophysics, Nanomater Interface Stud Appl NANO 2016. Springer Proc. Phys., Springer, 2017. p. 305–325, https://doi.org/10.1007/978-3-319-56422-7_22.

[69] Indira TK, Lakshmi PK. Magnetic nanoparticles: A review. Int J Pharm Sci Nanotechnol. 2010;3:1035.

[70] Datta J, Das M, Sil S, et al. Improvement of charge transport for hydrothermally synthesized $Cd_{0.8}Fe_{0.2}S$ over co-precipitation method: A comparative study of structural, optical and magnetic properties. Mater Sci Semicond Process. 2019;91:133–145.

[71] Jha M, Kumar S, Garg N, et al. Microemulsion based approach for nanospheres assembly into anisotropic nanostructures of $NiMnO_3$ and their magnetic properties. J Solid State Chem. 2018;258:722–727.

[72] Yan W, Wang D, Botte GG. Electrochemical decomposition of urea with Ni-based catalysts. Appl Catal B. 2012;127:221–226.

[73] Chalbot MCG, Kavouras IG. Nuclear magnetic resonance spectroscopy for determining the functional content of organic aerosols: A review. Environ Pollut. 2014;191:232–249.

[74] Grass RN, Stark WJ. Gas phase synthesis of fcc-cobalt nanoparticles. J Mater Chem. 2006;16:1825.

[75] Hicks EM, Zou S, Schatz GC, et al. Controlling plasmon line shapes through diffractive coupling in linear arrays of cylindrical nanoparticles fabricated beam lithography. Nano Lett. 2005;5:1070.

[76] David B, Pizurova N, Schneeweiss O, et al. Magnetic properties of nanometric Fe-based particles obtained by laser-driven pyrolysis. J Phys Chem Solids. 2007;68:1152.

[77] Tung DK, Manh DH, Phong LTH, et al. Iron nanoparticles fabricated by high-energy ball milling for magnetic hyperthermia. J Electron Mater. 2016;45:2644.

[78] Attarad A, Hira Z, Muhammad Z, et al. Synthesis, characterization, applications, and challenges of iron oxide nanoparticles. Nanotechnol Sci Appl. 2016;9:49.

[79] Min-Hung L, Dong-Hwang C. Preparation and characterization of a novel magnetic nano-adsorbent. J Mater Chem. 2002;12:3654–3659, https://doi.org/10.1039/b207158d.

[80] Santhosh C, Daneshvar E, Tripathi KM, et al. Synthesis and characterization of magnetic biochar adsorbents for the removal of Cr(VI) and acid orange 7 dye from

aqueous solution. Environ Sci Pollut Res. 2020;27:32874–32887, https://doi.org/10.1007/s11356-020-09275-1.

[81] Liu J, Yu Y, Zhu S, et al. Synthesis and characterization of a magnetic adsorbent from negatively-valued iron mud for methylene blue adsorption. PLoS One. 2018;13(2):e0191229, https://doi.org/10.1371/journal.pone.0191229.

[82] Jia Y, Xu L, Zhen F, et al. Observation of surface precipitation of arsenate on ferrihydrite. Environ Sci Technol. 2006;40(10):3248–3253.

[83] Zuhaimi NAS, Indran VP, Deraman MA, et al. Reusable gypsum based catalyst for synthesis of glycerol carbonate from glycerol and urea. Appl Catal A: Gen. 2015;502:312–319, http://doi.org/10.1016/j.apcata.2015.06.024.

[84] Khan AH, Khan NA, Zubair M, et al. Sustainable green nanoadsorbents for remediation of pharmaceuticals from water and wastewater: A critical review. Environ Res. 2022;204:112243, https://doi.org/10.1016/j.envres.2021.112243.

[85] Natarajan R, Saikia K, Ponnusamy SK, et al. Understanding the factors affecting adsorption of pharmaceuticals on different adsorbents—A critical literature update. Chemosphere. 2022;287:131958, https://doi.org/10.1016/j.chemosphere.2021.131958.

[86] Rasoulzadeh H, Mohseni-Bandpei A, Hosseini M, et al. Mechanistic investigation of ciprofloxacin recovery by magnetite—Imprinted chitosan nanocomposite: Isotherm, kinetic, thermodynamic and reusability studies. Int J Biol Macromol. 2019;133:712–721, https://doi.org/10.1016/j.ijbiomac.2019.04.139.

[87] Nie J, Zhi D, Zhou Y. Magnetic biochar-based composites for removal of recalcitrant pollutants in water. In: Sorbents Mater. Control. Environ. Pollut. Elsevier; 2021, p. 163–187, https://doi.org/10.1016/B978-0-12-820042-1.00015-8.

[88] Olusegun SJ, Souza TGF, Souza GDO, et al. Iron-based materials for the adsorption and photocatalytic degradation of pharmaceutical drugs: A comprehensive review of the mechanism pathway. J Water Process Eng. 2023;51:103457.

[89] Mangla D, Annu A, Sharma SI. Critical review on adsorptive removal of antibiotics: Present situation, challenges and future perspective. J Hazard Mater. 2022;425:127946, https://doi.org/10.1016/j.jhazmat.2021.127946.

[90] Tang Y, Chen Q, Li W, et al. Engineering magnetic N-doped porous carbon with super-high ciprofloxacin adsorption capacity and wide pH adaptability. J Hazard Mater. 2020;388:122059, https://doi.org/10.1016/j.jhazmat.2020.122059.

[91] Bu F, Huang W, Xian M, et al. Magnetic carboxyl-functionalized covalent organic frameworks for adsorption of quinolones with high capacities, fast kinetics and easy regeneration. J Clean Prod. 2022;336:130485, https://doi.org/10.1016/j.jclepro.2022.130485.

[92] Bao X, Qiang Z, Ling W, et al. Sonohydrothermal synthesis of MFe2O4 magnetic nanoparticles for adsorptive removal of tetracyclines from water. Sep Purif Technol. 2013;117:104–110, https://doi.org/10.1016/j.seppur.2013.03.046.

[93] Yang X, Wan Y, Zheng Y, et al. Surface functional groups of carbon-based adsorbents and their roles in the removal of heavy metals from aqueous solutions: A critical review. Chem Eng J. 2019;366:608–621, https://doi.org/10.1016/j.cej.2019.02.119.

[94] Hu X, Zhao Y, Wang H, et al. Efficient mbly. Int J Environ Res Public Health. 2017;14:1495, https://doi.org/10.3390/ijerph14121495.

[95] Lu L, Liu M, Chen Y, et al. Effective removal of tetracycline antibiotics from wastewater using practically applicable iron (III)-loaded cellulose nano fibres. R Soc Open Sci. 2021;8:210336, https://doi.org/10.1098/rsos.210336.

[96] Igwegbe CA, Oba SN, Aniagor CO, et al. Adsorption of ciprofloxacin from water: A comprehensive review. J Ind Eng Chem. 2021;93:57–77, https://doi.org/10.1016/j.jiec.2020.09.023.

[97] Feizi F, Reguyal F, Antoniou N, et al. Environmental remediation in circular economy: End of life tyre magnetic pyrochars for adsorptive removal of pharmaceuticals from aqueous solution. Sci Total Environ. 2020;739:139855, https://doi.org/10.1016/j.scitotenv.2020.139855.

[98] Shi S, Fan Y, Huang Y. Facile low temperature hydrothermal synthesis of magnetic mesoporous carbon nanocomposite for adsorption removal of ciprofloxacin antibiotics. Ind Eng Chem Res. 2013;52:2604–2612, https://doi.org/10.1021/ie303036e.

[99] Phouthavong V, Yan R, Nijpanich S, et al. Magnetic adsorbents for wastewater treatment: Advancements in their synthesis methods. Materials (Basel). 2022;15:1053, https://doi.org/10.3390/ma15031053.

[100] Wang T, Pan X, Ben W, et al. Adsorptive removal of antibiotics from water using magnetic ion exchange resin. J Environ Sci. 2017;52:111–117, https://doi.org/10.1016/j.jes.2016.03.017.

[101] Gao Y, Li Y, Zhang L, et al. Adsorption and removal of tetracycline antibiotics from aqueous solution by graphene oxide. J Colloid Interface Sci. 2012;368:540–546, https://doi.org/10.1016/j.jcis.2011.11.015.

[102] Okoli CP, Naidoo EB, Ofomaja AE. Role of synthesis process variables on magnetic functionality, thermal stability, and tetracycline adsorption by magnetic starch nanocomposite. Environ Nanotechnol Monit Manag. 2018;9:141–153, https://doi.org/10.1016/j.enmm.2018.02.001.

[103] Fallah Z, Zare EN, Ghomi M, et al. Toxicity and remediation of pharmaceuticals and pesticides using metal oxides and carbon nanomaterials. Chemosphere. 2021;275, https://doi.org/10.1016/j.chemosphere.2021.130055.

[104] Yu B, Bai Y, Ming Z, et al. Adsorption behaviors of tetracycline on magnetic graphene oxide sponge. Mater Chem Phys. 2017;198:283–290, https://doi.org/10.1016/j.matchemphys.2017.05.042.

[105] Yadav P, Yadav A, Labhasetwar PK. Sustainable adsorptive removal of antibiotics from aqueous streams using Fe_3O_4-functionalized MIL101(Fe) chitosan composite beads. Environ Sci Pollut Res. 2022;29:37204–37217, https://doi.org/10.1007/s11356-021-18385-3.

[106] Parashar A, Sikarwar S, Jain R. Removal of pharmaceuticals from wastewater using magnetic iron oxide nanoparticles (IOPs). Int J Environ Anal Chem. 2022;102:117–133, https://doi.org/10.1080/03067319.2020.1716977.

[107] Shearer L, Pap S, Gibb SW. Removal of pharmaceuticals from wastewater: A review of adsorptive approaches, modelling and mechanisms for metformin and macrolides. J Environ Chem Eng. 2022;10:108106, https://doi.org/10.1016/j.jece.2022.108106.

[108] Sparkes D, Enoch DA, Microbiology C, et al. Quinolones. 2022;7:240–254, https://doi.org/10.1016/B978-0-12-820472-6.00171-7.

[109] Palacio DA, Rivas BL, Urbano BF. Ultrafiltration membranes with three watersoluble polyelectrolyte copolymers to remove ciprofloxacin from aqueous systems. Chem Eng J. 2018;351:85–93, https://doi.org/10.1016/j.cej.2018.06.099.

[110] Roca Jalil ME, Baschini M, Sapag K. Influence of pH and antibiotic solubility on the removal of ciprofloxacin from aqueous media using montmorillonite. Appl Clay Sci. 2015;114:69–76, https://doi.org/10.1016/j.clay.2015.05.010.

[111] aus der Beek T, Weber FA, Bergmann A, et al. Pharmaceuticals in the environment-Global occurrences and perspectives. Environ Toxicol Chem. 2016;35:823–835, https://doi.org/10.1002/etc.3339.

[112] Al-Buriahi AK, Al-shaibani MM, Mohamed RMSR, et al. Ciprofloxacin removal from non-clinical environment: A critical review of current methods and future trend prospects. J Water Process Eng. 2022;47:102725, https://doi.org/10.1016/j.jwpe.2022.102725.

[113] Aydin S, Aydin ME, Beduk F, et al. Removal of antibiotics from aqueous solution by using magnetic Fe3O4/red mud-nanoparticles. Sci Total Environ. 2019;670:539–546, https://doi.org/10.1016/j.scitotenv.2019.03.205.

[114] Mahmoud ME, Saad SR, El-Ghanam AM, et al. Developed magnetic Fe_3O_4—MoO_3-AC nanocomposite for effective removal of ciprofloxacin from water. Mater Chem Phys. 2021;257:123454, https://doi.org/10.1016/j.matchemphys.2020.123454.

[115] Zhao J, Liang G, Zhang X, et al. Coating magnetic biochar with humic acid for high efficient removal of fluoroquinolone antibiotics in water. Sci Total Environ. 2019;688:1205–1215, https://doi.org/10.1016/j.scitotenv.2019.06.287.

[116] Asadi-ghalhari M, Kishipour A, Mostafaloo R. Ciprofloxacin removal from aqueous solutions by granular ferric hydroxide: Modeling and optimization. J Trace Elem Miner. 2022;2:100007, https://doi.org/10.1016/j.jtemin.2022.100007.

[117] Wang F, Yang B, Wang H, et al. Removal of ciprofloxacin from aqueous solution by a magnetic chitosan grafted graphene oxide composite. J Mol Liq. 2016;222:188–194, https://doi.org/10.1016/j.molliq.2016.07.037.

[118] Rakshit S, Sarkar D, Elzinga EJ, et al. Mechanisms of ciprofloxacin removal by nano-sized magnetite. J Hazard Mater. 2013;246–247:221–226, https://doi.org/10.1016/j.jhazmat.2012.12.032.

[119] Mao H, Wang S, Lin JY, et al. Modification of a magnetic carbon composite for ciprofloxacin adsorption. J Environ Sci (China). 2016;49:179–188, https://doi.org/10.1016/j.jes.2016.05.048.

[120] Gaur S, Bal AM, Morgannwg T, et al. Tetracyclines. 2022;7:136–153, https://doi.org/10.1016/B978-0-12-820472-6.00185-7.

[121] Rusu A, Buta EL. The development of third-generation tetracycline antibiotics and new perspectives. Pharmaceutics. 2021;13:2085, https://doi.org/10.3390/pharmaceutics13122085.

[122] Gopal G, Alex SA, Chandrasekaran N, et al. A review on tetracycline removal from aqueous systems by advanced treatment techniques. RSC Adv. 2020;10:27081–27095, https://doi.org/10.1039/D0RA04264A.

[123] Ge L, Dong Q, Halsall C, et al. Aqueous multivariate phototransformation kinetics of dissociated tetracycline: Implications for the photochemical fate in surface waters. Environ Sci Pollut Res. 2018;25:15726–15732, https://doi.org/10.1007/s11356-018-1765-0.

[124] Miao J, Wang F, Chen Y, et al. The adsorption performance of tetracyclines on magnetic graphene oxide: A novel antibiotics absorbent. Appl Surf Sci. 2019;475:549–558, https://doi.org/10.1016/J.APSUSC.2019.01.036.

[125] Zhao Y, Geng J, Wang X, et al. Adsorption of tetracycline onto goethite in the presence of metal cations and humic substances. J Colloid Interface Sci. 2011;361:247–251, https://doi.org/10.1016/j.jcis.2011.05.051.

[126] Lian L, Lv J, Wang X, et al. Magnetic solid—Phase extraction of tetracyclines using ferrous oxide coated magnetic silica microspheres from water samples. J Chromatogr A. 2018;1534:1–9, https://doi.org/10.1016/j.chroma.2017.12.041.

[127] Wang Q, Zhang L. Fabricated ultrathin magnetic nitrogen doped graphene tube as efficient and recyclable adsorbent for highly sensitive simultaneous determination of three tetracyclines residues in milk samples. J Chromatogr A. 2018;1568:1–7, https://doi.org/10.1016/j.chroma.2018.07.012.

[128] Ahamad T, Ruksana, Chaudhary AA, et al. Fabrication of $MnFe_2O_4$ nanoparticles embedded chitosan diphenylureaformaldehyde resin for the removal of tetracycline from aqueous solution. Int J Biol Macromol. 2019;134:180–188.

[129] Rouhani M, Ashrafi SD, Taghavi K, et al. Evaluation of tetracycline removal by adsorption method using magnetic iron oxide nanoparticles (Fe_3O_4) and clinoptilolite from aqueous solutions. J Mol Liq. 2022;356:119040, https://doi.org/10.1016/j.molliq.2022.119040.

[130] Hu Z-T, Wang X-F, Xiang S, et al. Self-cleaning Mn-Zn ferrite/biochar adsorbents for effective removal of tetracycline. Sci Total Environ. 2022:157202, https://doi.org/10.1016/j.scitotenv.2022.157202.

[131] Lu L, Li J, Yu J, et al. A hierarchically porous $MgFe_2O_4/\gamma\text{-}Fe_2O_3$ magnetic microspheres for efficient removals of dye and pharmaceutical from water. Chem Eng J. 2016;283:524–534, https://doi.org/10.1016/j.cej.2015.07.081.

[132] Saya L, Malik V, Gautam D, et al. A comprehensive review on recent advances toward sequestration of levofloxacin antibiotic from wastewater. Sci Total Environ. 2022;813:152529, https://doi.org/10.1016/j.scitotenv.2021.152529.

[133] Al-Jabari MH, Sulaiman S, Ali S, et al. Adsorption study of levofloxacin on reusable magnetic nanoparticles: Kinetics and antibacterial activity. J Mol Liq. 2019;291:111249, https://doi.org/10.1016/j.molliq.2019.111249.

[134] Altaf S, Zafar R, Zaman WQ, et al. Removal of levofloxacin from aqueous solution by green synthesized magnetite (Fe_3O_4) nanoparticles using Moringa olifera: Kinetics and reaction mechanism analysis. Ecotoxicol Environ Saf. 2021;226:112826, https://doi.org/10.1016/j.ecoenv.2021.112826.

[135] Xu Z, Xiang Y, Zhou H, et al. Manganese ferrite modified biochar from vinasse for enhanced adsorption of levofloxacin: Effects and mechanisms. Environ Pollut. 2021;272:115968, https://doi.org/10.1016/j.envpol.2020.115968.

[136] Yao B, Luo Z, Du S, et al. Sustainable biochar/$MgFe_2O_4$ adsorbent for levofloxacin removal: Adsorption performances and mechanisms. Bioresour Technol. 2021;340:125698, https://doi.org/10.1016/j.biortech.2021.125698.

[137] Hu Z, Ge M, Guo C. Efficient removal of levofloxacin from different water matrices via simultaneous adsorption and photocatalysis using a magnetic $Ag_3PO_4/$rGO/$CoFe_2O_4$ catalyst. Chemosphere. 2021;268:128834, https://doi.org/10.1016/j.chemosphere.2020.128834.

[138] Ullah A, Zahoor M, Alam S, et al. Separation of levofloxacin from industry effluents using novel magnetic nanocomposite and membranes hybrid processes. Biomed Res Int. 2019, https://doi.org/10.1155/2019/5276841.

[139] Wang S, Wang Z, Hao C, et al. DFT/TDDFT insights into effects of dissociation and metal complexation on photochemical behavior of enrofloxacin in water. Environ Sci Pollut Res. 2018;25:30609–30616, https://doi.org/10.1007/s11356-018-3032-9.

[140] Peng G, Li T, Ai B, et al. Highly efficient removal of enrofloxacin by magnetic montmorillonite via adsorption and persulfate oxidation. Chem Eng J. 2019;360:1119–1127, https://doi.org/10.1016/j.cej.2018.10.190.

[141] Li R, Wang Z, Zhao X, et al. Magnetic biochar-based manganese oxide composite for enhanced fluoroquinolone antibiotic removal from water. Environ Sci Pollut Res. 2018;25:31136–31148, https://doi.org/10.1007/s11356-018-3064-1.

[142] Gordi Z, Ghorbani M, Ahmadian Khakhiyani M. Adsorptive removal of enrofloxacin with magnetic functionalized graphene oxide@ metal—Organic frameworks employing D-optimal mixture design. Water Environ Res. 2020;92:1935–1947, https://doi.org/10.1002/wer.1346.

[143] Zahoor M, Ullah A, Alam S. Removal of enrofloxacin from water through magnetic nanocomposites prepared from pineapple waste biomass. Surf Eng Appl Electrochem. 2019;55:536–547, https://doi.org/10.3103/S1068375519050156.

[144] Peng G, Zhang M, Deng S, et al. Adsorption and catalytic oxidation of pharmaceuticals by nitrogen-doped reduced graphene oxide/Fe_3O_4 nanocomposite. Chem Eng J. 2018;341:361–370, https://doi.org/10.1016/j.cej.2018.02.064.

[145] Anastopoulos I, Pashalidis I, Orfanos AG, et al. Removal of caffeine, nicotine and amoxicillin from (waste)waters by various adsorbents: A review. J Environ Manage. 2020;261, https://doi.org/10.1016/j.jenvman.2020.110236.

[146] Homsirikamol C, Sunsandee N, Pancharoen U, et al. Synergistic extraction of amoxicillin from aqueous solution by using binary mixtures of Aliquat 336, D2EHPA and TBP. Sep Purif Technol. 2016;162:30–36, https://doi.org/10.1016/j.seppur.2016.02.003.

[147] Pooresmaeil M, Namazi H. Chitosan coated Fe_3O_4@Cd-MOF microspheres as an effective adsorbent for the removal of the amoxicillin from aqueous solution. Int J Biol Macromol. 2021;191:108–117, https://doi.org/10.1016/j.ijbiomac.2021.09.071.

[148] Oba SN, Ighalo JO, Aniagor CO, et al. Removal of ibuprofen from aqueous media by adsorption: A comprehensive review. Sci Total Environ. 2021;780:146608, https://doi.org/10.1016/j.scitotenv.2021.146608.

[149] Salem Attia TM, Hu XL, Yin DQ. Synthesized magnetic nanoparticles coated zeolite for the adsorption of pharmaceutical compounds from aqueous solution using batch and column studies. Chemosphere. 2013;93:2076–2085, https://doi.org/10.1016/j.chemosphere.2013.07.046.

[150] Soltys LM, Mironyuk IF, Tatarchuk TR, et al. Zeolite-based composites as slowrelease fertilizers (review). Phys Chem Solid State. 2020;21:89–104, https://doi.org/10.15330/pcss.21.1.89-104.

[151] Frohlich AC, Foletto EL, Dotto GL. Preparation and characterization of $NiFe_2O_4$/activated carbon composite as potential magnetic adsorbent for removal of ibuprofen and ketoprofen pharmaceuticals from aqueous solutions. J Clean Prod. 2019;229:828–837, https://doi.org/10.1016/j.jclepro.2019.05.037.

[152] Liu Y, Liu R, Li M, et al. Removal of pharmaceuticals by novel magnetic genipin-cross-linked chitosan/graphene oxide-SO3H composite. Carbohydr Polym. 2019;220:141–148, https://doi.org/10.1016/j.carbpol.2019.05.060.

[153] Wu G, Liu Q, Wang J, et al. Facile fabrication of rape straw biomass fiber/β-CD/Fe3O4 as adsorbent for effective removal of ibuprofen. Ind Crops Prod. 2021;173:114150, https://doi.org/10.1016/j.indcrop.2021.114150.

[154] Attia TMS, Hu XL, Yin DQ. Synthesized magnetic nanoparticles coated zeolite for the adsorption of pharmaceutical compounds from aqueous solution using batch and column studies. Chemosphere. 2013;93(9):2076–2085.

[155] Liyanage AS, Canaday S, Pittman CU, et al. Rapid remediation of pharmaceuticals from wastewater using magnetic Fe_3O_4/Douglas fir biochar adsorbents. Chemosphere. 2020;258:127336, https://doi.org/10.1016/j.chemosphere.2020.127336.

[156] Moacă EA, Mihali CV, Macaşoi IG, et al. Fe3O4@C matrix with tailorable adsorption capacities for paracetamol and acetylsalicylic acid: Synthesis, characterization, and kinetic modeling. Molecules. 2019;24, https://doi.org/10.3390/molecules24091727.

[157] Chandrashekar Kollarahithlu S, Balakrishnan RM. Adsorption of pharmaceuticals pollutants, ibuprofen, acetaminophen, and streptomycin from the aqueous phase using amine functionalized superparamagnetic silica nanocomposite. J Clean Prod. 2021;294:126155, https://doi.org/10.1016/j.jclepro.2021.126155.

[158] Natarajan R, Anil Kumar M, Vaidyanathan VK. Synthesis and characterization of rhamnolipid based chitosan magnetic nanosorbents for the removal of acetaminophen from aqueous solution. Chemosphere. 2022;288:132532, https://doi.org/10.1016/j.chemosphere.2021.132532.

[159] Shirazi E, Torabian A, Nabi-Bidhendi G. Carbamazepine removal from groundwater: Effectiveness of the TiO2/UV, nanoparticulate zero-valent iron, and Fenton (nZVI/H_2O_2) processes. CLEAN—Soil, Air, Water. 2013;41:1062–1072.

[160] Ghauch A, Tuqan A, Assi HA. Antibiotic removal from water: Elimination of amoxicillin and ampicillin by microscale and nanoscale iron particles. Environ Pollut. 2009;157:1626–1635.

[161] Segura Y, Martinez F, Melero JA. Effective pharmaceutical wastewater degradation by Fenton oxidation with zero-valent iron. Appl Catal B: Environ. 2013;136:64–69.

[162] Bautitz IR, Velosa AC, Nogueira RF. Zero valent iron mediated degradation of the pharmaceutical diazepam. Chemosphere. 2012;88:688–692.

[163] Huang YX, Keller AA. Magnetic nanoparticle adsorbents for emerging organic contaminants. ACS Sustain Chem Eng. 2013;1:731–736.

[164] Bao X, Qiang Z, Ling W, et al. Sonohydrothermal synthesis of MFe_2O_4 magnetic nanoparticles for adsorptive removal of tetracyclines from water. Sep Purif Technol. 2013;117:104–110.

[165] Wan J, Deng HP, Shi J, et al. Synthesized magnetic manganese ferrite nanoparticles on activated carbon for sulfamethoxazole removal. CLEAN—Soil, Air, Water. 2014;42:1199–1207.

[166] Gaho MM, Memon GZ, Memon JUR, et al. Synthesis of novel magnetic molecularly imprinted polymers by solidphase extraction method for removal of norfloxacin. Chinese J Anal Chem. 2022;50, https://doi.org/10.1016/j.cjac.2022.100079.

[167] Ai T, Jiang X, Liu Q, et al. Daptomycin adsorption on magnetic ultrafine wood-based biochars from water: Kinetics, isotherms, and mechanism studies. Bioresour Technol. 2019;273:8–15, https://doi.org/10.1016/j.biortech.2018.10.039.

[168] Danalıoglu ST, Bayazit ŞS, Kerkez Kuyumcu Ö, et al. Efficient removal of antibiotics by a novel magnetic adsorbent: Magnetic Activated Carbon/Chitosan (MACC) nanocomposite. J Mol Liq. 2017;240:589–596, https://doi.org/10.1016/j.molliq.2017.05.131.

[169] Chandrashekar Kollarahithlu S, Balakrishnan RM. Adsorption of pharmaceuticals pollutants, ibuprofen, acetaminophen, and streptomycin from the aqueous phase using amine functionalized superparamagnetic silica nanocomposite. J Clean Prod. 2021;294:126155, https://doi.org/10.1016/j.jclepro.2021.126155.

[170] Mohseni-Bandpi A, Al-Musawi TJ, Ghahramani E, et al. Improvement of zeolite adsorption capacity for cephalexin by coating with magnetic Fe_3O_4 nanoparticles. J Mol Liq. 2016;218:615–624, https://doi.org/10.1016/j.molliq.2016.02.092.

[171] Asgari E, Sheikhmohammadi A, Yeganeh J. Application of the Fe_3O_4-chitosan nanoadsorbent for the adsorption of metronidazole from wastewater: Optimization, kinetic, thermodynamic and equilibrium studies. Int J Biol Macromol. 2020;164:694–706, https://doi.org/10.1016/j.ijbiomac.2020.07.188.

[172] Zhou Y, Cao S, Xi C, et al. A novel Fe_3O_4/graphene oxide/citrus peel-derived bio-char based nanocomposite with enhanced adsorption affinity and sensitivity of ciprofloxacin and sparfloxacin. Bioresour Technol. 2019;292:121951, https://doi.org/10.1016/j.biortech.2019.121951.

[173] Yegane Badi M, Azari A, Pasalari H, et al. Modification of activated carbon with magnetic Fe_3O_4 nanoparticle composite for removal of ceftriaxone from aquatic solutions. J Mol Liq. 2018;261:146–154, https://doi.org/10.1016/j.molliq.2018.04.019.

[174] Zhang B, Zhang H, Li X, et al. Synthesis of BSA/Fe_3O_4 magnetic composite microspheres for adsorption of antibiotics. Mater Sci Eng C. 2013;33:4401–4408, https://doi.org/10.1016/j.msec.2013.06.038.

[175] Liyanage AS, Canaday S, Pittman CU, et al. Rapid remediation of pharmaceuticals from wastewater using magnetic Fe_3O_4/douglas fir biochar adsorbents. Chemosphere. 2020;258:127336, https://doi.org/10.1016/j.chemosphere.2020.127336.

[176] Shi Y, Zhang Y, Cui Y, et al. Magnetite nanoparticles modified β-cyclodextrin polymer coupled with $KMnO_4$ oxidation for adsorption and degradation of acetaminophen. Carbohydr Polym. 2019;222:114972, https://doi.org/10.1016/j.carbpol.2019.114972.

[177] Al-Khateeb LA, Hakami W, Abdel Salam M, et al. Solid phase-fabrication of magnetically separable Fe_3O_4@graphene nanoplatelets nanocomposite for efficient removal of NSAIDs from wastewater: Perception of adsorption kinetics, thermodynamics, and extra-thermodynamics. Anal Chim Acta. 2022:340158, https://doi.org/10.1016/j.aca.2022.340158.

5 Layered Double Hydroxide Adsorbents for Pharmaceutical Removal

Karina Nava-Andrade and
Santiago José Guevara-Martínez

5.1 INTRODUCTION

Layered double hydroxides (LDHs) are a family of anionic clays with a hydrotalcite-like structure. Their chemical composition is represented by the formula $[M^{2+}_{1-x}M^{3+}_x(OH)_2]^{x+}[A^{n-}]_{x/n} \cdot mH_2O$, where M and A are metal cations and anions, respectively. The LDH structure is derived from brucite $[Mg(OH)_2]$, in which the Mg^{2+} ions are bound to six hydroxide ions. These octahedral units form infinite layers, which are stacked on top of each other to generate the three-dimensional structure. As shown in Figure 5.1a, the typical structure of LDHs is obtained by replacing a portion of the divalent cations in the brucite lattice with trivalent cations [1], forming layered structures with fully hydroxylated surfaces (see Figure 5.1b). The trivalent cations give the layers an excess of positive charge, which is balanced by the anions. In addition to the anions, water molecules are found in the interlayer space after crystallization.

A wide range of M^{2+}/M^{3+} combinations can be used to generate the layered structure, and it is even possible to incorporate monovalent and tetravalent cations with values of x between 0.2 and 0.33 to obtain pure phases. To balance positive charges, LDHs can incorporate anions of different chemical natures into the interlayer space by ionic exchange. A peculiar property of the hydrotalcite-like compounds is the so-called memory effect, which allows them to restore their layered structure after heat treatment under mild conditions. During the reconstruction process, chemical species found in the medium can be taken up in the interlayer space.

Due to the capacity to incorporate metal cations and anionic molecules into their structure, LDHs have been widely investigated as adsorbents for environmental applications. Numerous studies support the high adsorption capacity of LDH-based adsorbents for the removal of heavy metals, textile dyes, and pesticides from aqueous environments [2]. In recent years, the promising use of LDH-based adsorbents for the elimination of pharmaceutical pollutants from

DOI: 10.1201/9781003340164-5

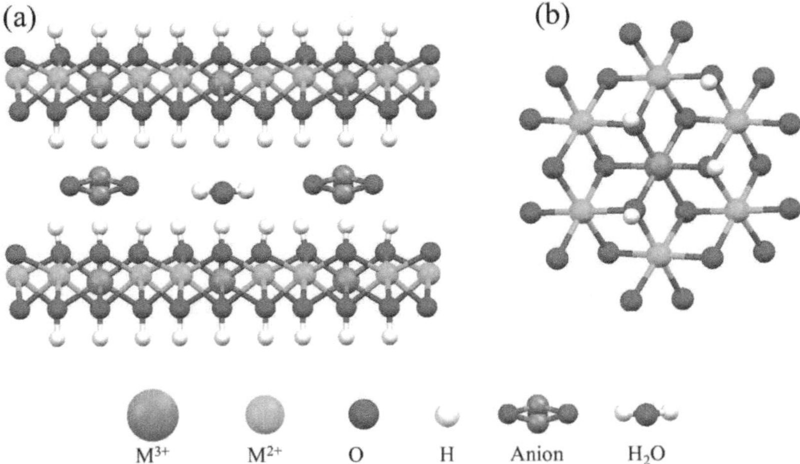

FIGURE 5.1 Representation of the general structure of the LDHs. (a) The side view shows the metal cations attached to OH^- groups forming layers with hydroxylated surfaces, and the presence of water molecules and anions located between the layers. (b) In the top view of the layer, the hexacoordinated metal cations are observed.

wastewater has also been described [3]. Adsorption is a surface phenomenon effective for wastewater treatment; the adsorption capacity of LDHs depends on their chemical composition and microstructure, which are controlled by the synthesis conditions. Hydrotalcite-like compounds are also used to prepare adsorbent materials with supramolecular structures [4].

5.2 PREPARATION OF LDH-BASED ADSORBENTS

Several approaches have been described for the preparation of LDH-based adsorbents, which can be obtained by direct or indirect preparation. Direct preparation involves obtaining the materials, either pure LDHs or assembled with other structures, in a single step. In indirect preparation, the adsorbents are obtained through at least one LDH post-synthesis treatment. The reported strategies for direct and indirect preparation are represented in Figure 5.2 and are described in the following sections.

5.2.1 DIRECT PREPARATION

Precipitation is a commonly used method for the direct synthesis of LDH particles. This method consists of the mixture of an alkaline solution, usually NaOH or NH₄OH, with an acidic solution containing a metallic salt combination. Metal nitrates and chlorides are the most widely used precursor salts since these monovalent anions favor the incorporation of other anionic molecules of interest in the LDH structure [5]. The precipitation can be carried out under constant or variable pH. In the precipitation carried

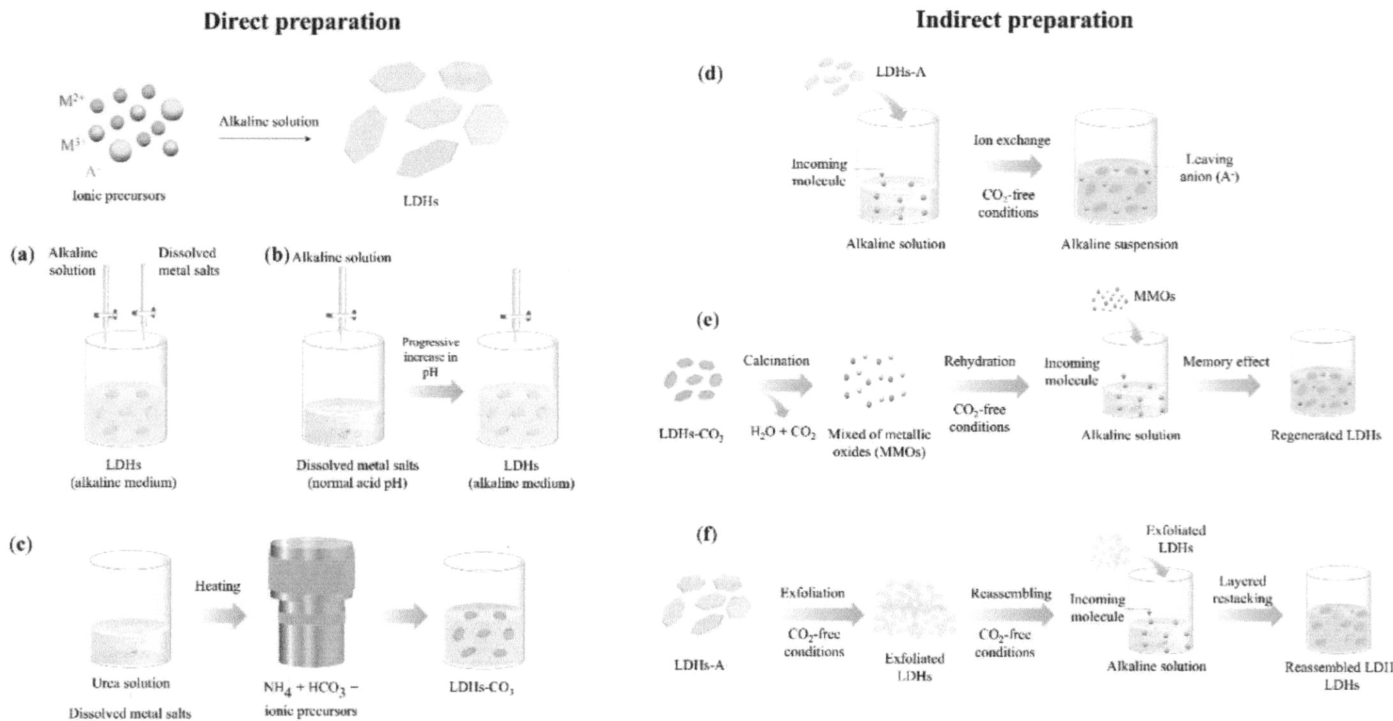

FIGURE 5.2 Direct preparation of LDHs-based adsorbents by (a) coprecipitation, (b) precipitation at variable pH, and (c) urea hydrolysis. Indirect preparation through post-synthesis treatments such as (d) ion exchange, (e) calcination-rehydration, and (f) exfoliation-reassembly processes.

out at constant pH, also called coprecipitation, both solutions are slowly added at the same time to maintain the desired pH constant [6] (see Figure 5.2a), whereas in precipitation at variable pH, the alkaline solution is slowly added to the solution containing the metal salts until the desired pH is reached (see Figure 5.2b).

Urea hydrolysis is also a precipitation method used for the synthesis of LDH particles. Urea is a weak Bronsted base (pKb = 13.8) highly soluble in water. When an aqueous solution of urea and metallic salts is heated above 100 °C for at least 10 h [7], the urea is progressively degraded to CO_3^{2-} and NH_4^+ ions, which provide the required pH for the synthesis of LDHs. Under this methodology, it is only possible to synthesize LDHs with interlayered carbonates, a highly crystalline structure, and an expected chemical composition [8] (see Figure 5.2c). When the molecules to assemble hybrid LDHs do not react with metallic salts, such as biochar or metallic nanoparticles, they can be incorporated into the reaction mixture to directly obtain the hybrid LDH material [9]. Otherwise, preparation approaches that involve more than one step are applied.

5.2.2 Indirect Preparation

Depending on the chemical nature and size of the material to be assembled, LDH-based absorbents can be prepared by the following methodologies:

i. Ion exchange. This method is used for the incorporation of anionic species in the interlayer region. First, LDHs with interlayer monovalent anions are suspended in an alkaline solution containing the incoming anion. Afterwards, the reaction mixture is kept under stirring for hours or days to favor the ionic exchange. Each step must be carried out under a CO_2-free atmosphere to reduce contamination by carbonates that limit the reaction (see Figure 5.2d).

ii. Calcination–rehydration. This process consists of a previous step of LDH calcination at temperatures between 300–500 °C to subsequently suspend the mixture of metallic oxides (MMOs) in an alkaline solution that contains the chemical species to be assembled. This rehydration step promotes the reconstruction of the layered structure by the so-called memory effect (see Figure 5.2e). According to Mascolo and Mascolo [10] preparation by the calcination–rehydration method is simply a direct synthesis of these attractive solids but is not due to the memory effect. This strategy is useful when the incoming molecules are large and cannot be intercalated by ion exchange.

iii. Exfoliation–reassembling. This is a two-step methodology used to capture large anionic species and assemble LDH sheets to polymer matrices or to carbon nanostructures. In the first step, a pristine LDH is treated with formamide or alcohol reflux to break the interactions between the layers and the anions [11, 12]. In the second step, the LDH nanosheets obtained are added to a solution that contains the chemical species to be assembled in order to produce the hybrid LDHs. This methodology also requires a CO_2-free atmosphere to avoid carbonate contamination (see Figure 5.2f).

5.3 LDH-BASED ADSORBENTS FOR THE REMOVAL OF PHARMACEUTICAL POLLUTANTS

5.3.1 LDH Particles

As seen in Table 5.1, the adsorption capacity of LDHs is influenced by their chemical composition. Abdel et al. [13] investigated the influence of the trivalent cation on the absorption capacity of Mg-containing LDHs. MgAl–CO_3 LDHs exhibited a higher percentage of doxycycline removal than MgFe–CO_3 LDHs. Unlike the Fe^{3+} cations, the inclusion of Al^{3+} yields LDH structures with a greater pore volume and better stability in an aqueous solution, which are favorable properties for antibiotic removal. Regarding the anions, the use of LDHs intercalated with monovalent anions exhibited a greater removal efficiency. For example, the MgAl–Cl LDHs [14] and MgAl–CO_3 LDHs [15] studied in diclofenac removal exhibited a maximum adsorption capacity of 0.95 and 0.26 mmol/g, respectively. Monovalent anions such as nitrates and chlorides are easily displaced by anions with a higher charge density. Due to the carbonate's bivalence, these anions are strongly retained in between layers and therefore difficult to replace [2]. Thus, LDHs intercalated with Cl^- or NO_3^- are better adsorbents than those containing interlayered CO_3^{2-}. In addition to the ion exchange process, anions can influence the interactions with the pollutant molecules. Gupta et al. [16] introduced a $(MoS_4)^{2-}$ moiety to the layered structure of CaAl LDHs for the removal of ciprofloxacin (CIP) and ofloxacin (OFL) from contaminated aquatic sources. CaAl–MoS_4 LDHs demonstrated a more effective removal of CIP than OFL due to steric hindrance. The adsorbate–adsorbent interaction was favored by weak hydrogen bonds between S- in LDHs and the NH_2^+ groups in drug molecules, the electrostatic interaction, and the anion exchange mechanisms [16].

Taking advantage of LDHs' ion exchange capacity, Shafiei et al. [22] successfully intercalated calix[4]arene (ACA) into MgAl–NO_3 LDHs and investigated the extractability of the MgAl–ACA LDHs towards aspirin, ibuprofen, and naproxen in human urine and water samples. MgAl–ACA LDHs promoted a more effective interaction with the selected drugs that led to a higher extraction efficiency than that of pure ACA or MgAl–NO_3 LDHs alone. This finding showed that the MgAl–ACA LDH adsorbent could be useful for the sensitive, uncomplicated, and cost-effective determination of nonsteroidal anti-inflammatory drugs (NSAIDs) in real samples.

5.3.2 LDH-Derived MMOs

Several studies corroborate that MMOs' high absorption capacity is obtained by the calcination of LDHs, usually at 500 °C, as observed in Table 5.2. Ghemit et al. [23] reported that the ZnAlFe MMOs obtained by the calcination of ZnFe-CO_3 LDHs at 500 °C showed a higher absorption capacity for diclofenac than their respective LDH precursors. During the rehydration process, diclofenac anions were intercalated into ZnAlFe LDHs. The maximum adsorption capacity for diclofenac on ZnAlFe MMOs was 641.4 mg/g, higher than that reported for commercial activated carbon (76 mg/g), revealing a potential material for the removal of anionic pollutants in the treatment of industrial wastewater [23]. Likewise, it has been described that ZnAl MMOs

TABLE 5.1

LDHs Adsorbents Used in the Adsorptive Removal of Pharmaceutical Pollutants

Adsorbent	Preparation	Surface Area (m²/g)	Drug	Adsorption Conditions			Adsorption Capacity (mg/g)	Isotherm/Kinetics	Regeneration/ Reuse	Reference
				pH	Time (min)	Concentration (mg/L)				
MgAl LDHs	Urea hydrolysis	109	Metronidazole	9	120	40	62.8	Langmuir/Avrami	–	[17]
MgAl LDHs	Coprecipitation	34.8	Tetracycline	7	120	1066	822.2	Freundlich/pseudo-second order	–	[14]
			Diclofenac			710	281.3	Langmuir/pseudo-second order		
MgAl LDHs	Coprecipitation	4.0	Doxycycline	4	60	10–60	100.0	Langmuir/pseudo-second order	–	[13]
MgAl LDHs	Coprecipitation	–	Tetracycline	6.6–7.5	4	30	–	–	–	[18]
CaAl(MoS$_4$) LDHs	Coprecipitation, anionic exchange	60.9	Ciprofloxacin Ofloxacin	6	120	50	707.2 476.7	Langmuir/pseudo-second order	250 °C/3 cycles	[16]
NiAlTi LDHs	Urea hydrolysis	121.8	Tetracycline	9	60	20	238.1	Langmuir/pseudo-second order	Ethanol/5 cycles	[19]
MgAl LDHs	Coprecipitation	–	Amoxicillin	6	120	240	30.4	Sips/pseudo-second order	500 °C/3 cycles	[20]
ZnCoFe LDHs	Coprecipitation	155.6	Oxytetracycline	5	60	25	240.3	Langmuir/pseudo-second order	H$_2$O, 50 °C/7 cycles	[21]
MgAl–ACA LDHs	Coprecipitation, anionic exchange	–	Aspirin Ibuprofen Naproxen	7	10	–	52.8 46.2 52.8	.	Methanol	[22]
MgAl LDHs	Coprecipitation	4.5	Diclofenac	8.5	60	30	60.2	Langmuir/pseudo-second order	–	[15]

TABLE 5.2

LDH-Derived MMO Adsorbents Used in the Adsorptive Removal of Pharmaceutical Pollutants

Adsorbent	Preparation	Surface Area (m²/g)	Drug	Adsorption Conditions			Adsorption Capacity (mg/g)	Isotherm/Kinetics	Regeneration/ Reuse	Reference
				pH	Time (min)	Concentration (mg/L)				
ZnAl MMOs	Coprecipitation, calcination 600 °C	94	Tetracycline	–	240	66	97	–	–	[24]
ZnAlFe MMOs	Coprecipitation, calcination 500 °C	79	Diclofenac	4	180	5–1000	641.5	Langmuir/ pseudo-second order	Ethanol/5 cycles	[23]
ZnAl MMOs	Coprecipitation, calcination 300 °C	–	Salicylic acid	–	600	30	28.4	–	300 °C/2 cycles	[26]
CoFe MMOs	Coprecipitation, calcination 500 °C	348.5	Tetracycline	6	60	40	138.6	–	NaOH/5 cycles	[27]
ZnAl MMOs	Coprecipitation, calcination 500 °C	–	Sulfamethox azole	7	60	50	4314	Freundlich/pseudo- second order	Na_2CO_3/6 cycles	[25]
ZnTiAl MMOs	Coprecipitation, calcination 400 °C	78	Diclofenac Salicylic acid	4 3	120	7.5 3.5	121.1 10.5	Toth/pseudo-second order	–	[28]
ZnAl MMOs	Coprecipitation, calcination 450 °C	214	Diclofenac	7	1440	10	152	Pseudo-First order	–	[29]

demonstrate a high adsorption capacity for tetracycline [24] and sulfamethoxazole antibiotics [25]. Tetracycline was adsorbed through a surface complexation of the A-ring with the Al–OH centers on calcined ZnAl LDHs. The influence of the chemical composition of the LDH precursors on their adsorption properties has also been investigated. Santamaría et al. [3] reported the removal of the drugs diclofenac and salicylic acid (SA) using ZnAl–CO_3 MMOs and ZnAlTi–CO_3 MMOs as adsorbents. According to their results, the titanium incorporation into the structure reduced the ZnAlTi–CO_3 MMOs' adsorption capacity, since it reduced the amount of hydroxyl groups on the layers' surfaces to interact with pollutant molecules.

The molar ratio (r) and the calcination temperature have an enormous influence on the adsorption performance of LDHs. Elhalil et al. [26] synthesized several ZnAl LDHs with molar ratios of 1, 3, and 5 by the coprecipitation method. Subsequently, the obtained LDHs were calcined at 300, 400, 500, and 600 °C and evaluated as adsorbents for the elimination of salicylic acid (SA). The best SA removal efficiency was generated by ZnAl LDHs with a Zn/Al molar ratio of 3, calcined at 300 °C [26]. The absorption of the LDHs improved when increasing the calcination temperature up to 400 °C. This result was attributed to the decomposition of interlayer carbonates and the formation of mixed oxides, capable of reconstructing the LDHs' structure by the incorporation of SA. At 600 °C, spinel phases are formed, which do not favor the reconstruction process.

5.3.3 LDH Hybrids

According to the studies summarized in Table 5.3, the adsorption properties of LDHs and LDH-derived MMOs can be improved by the assembly of other adsorbing agents to increase adsorption sites. In recent years, the functionalization of LDHs with biochar has been a used as a strategy to obtain adsorbents for the removal of pharmaceutical pollutants [30]. Biochar is a carbon-rich organic material generated by the pyrolysis of organic matter [31]. The main methods for the synthesis of biochar–LDH composites are precipitation, hydrothermal treatment, and co-pyrolysis. The synthesis method influences the surface functionalities, elemental composition, crystallographic structure, morphology, and textural properties of the biochar–LDH composite [32]. LDHs and LDH-derived MMOs have been successfully assembled with biochar from kapok fibers [33], bagasse biomass [34], *Syagrus coronata* [35], *Mikania micrantha* Kunth plant [36], and sludge-based biochar [37], resulting in excellent adsorbents for the elimination of drugs from aqueous matrices.

Multi-walled carbon nanotubes (MWCNs) are another type of carbon structure with excellent adsorption properties, and their assembly with LDHs has yielded hybrid materials with a high capacity for drug removal. For example, MgAl LDHs enhanced the removal percentage of doxycycline from 52 to 92% when assembled with MWCNs [13], possibly due to the increase in surface area from 4 to 92 m²/g after the modification with the MWCNTs, resulting in an increment in the number of active sites. In addition, it has been described that LDHs supported on graphene oxide and chitosan exhibit an antibiotic removal efficiency of 98%. It has also been described that LDHs supported on graphene oxide and polymers, such as chitosan [43] and polyacrylic acid [44], showed high efficiency in the removal of antibiotics.

TABLE 5.3

LDH Hybrids Used in the Adsorptive Removal of Pharmaceutical Pollutants

Adsorbent	Preparation	Surface Area (m^2/g)	Drug	Adsorption Conditions			Adsorption Capacity (mg/g)	Isotherm/Kinetics	Regeneration/ Reuse	Reference
				pH	Time (min)	Concentration (mg/L)				
SiO_2@LDHs	Coprecipitation, ultrasound aging	61.1	Diclofenac Ibuprofen Folic acid	–	90	10	430 400 332	–	Co^{2+}–ozone/ 4 cycles	[38]
MgAl MMOs/ γ-AlO(OH)/C	Urea hydrolysis, calcination 500 °C	287	Minocycline	7	420	200	447	Langmuir/pseudo-second order	500 °C/4 cycles	[33]
MgAl MMOs/ Biochar	Coprecipitation, calcination 450 °C	–	Tetracycline	6	1440	100	1002	Langmuir/pseudo-second order	NaOH/5 cycles	[34]
MgFe-MMOs/C	Coprecipitation, calcination 400 °C	117	Tetracycline	9	180	375	195.3	Langmuir/pseudo-second order	Na_2CO_3/5 cycles	[39]
BiOBr/ZnAlFe LDHs/C	Urea hydrolysis, chemical deposition	157.4	Doxycycline	–	420	200	227.4	Pseudo-second order	–	[40]
MgAl LDHs/C	Coprecipitation	151.2	Diclofenac	5.5	1440	30	2114	Langmuir/pseudo-second order	NaOH or NaCl/5 cycles	[30]
NiFe LDHs/ $Cu(OH)_2$	Electrosynthesis	2	Ketoconazole Clotrimazole Miconazole	5	20	0.25	142.5 198.8 67.5	–	Acetonitrile	[41]

NiMgAl LDHs/ CuFe$_2$O$_4$	Urea hydrolysis	–	Oxytetracycline	4	300	250	192	Langmuir/pseudo-second order	NaOH or acetone or ethanol/ 3 cycles	[42]
MgAl LDHs/ Biochar	Coprecipitation	168	Diclofenac	5.6	420	500–1000	135.8	Sips/pseudo-second order	–	[35]
NiZrAl LDHs/ graphene oxide/ chitosan	Hydrothermal method, ex situ polymerization	128	Nalidixic acid	8	7	15	277.8	Freundlich/pseudo-second order	Propanol, hydrochloric acid/5 cycles	[43]
ZnFe LDHs/ Biochar	Coprecipitation	54.1	Oxytetracycline	2–10	1440	50	426.6	Pseudo-second order	NaOH/5 cycles	[36]
CoFe LDHs/ Biochar	Coprecipitation	108	Ciprofloxacin	5	2880	10	–	–	Ethanol/5 cycles	[37]
MnO$_2$@NiFe@ Diatomite	Hydrothermal method	126.9	Tetracycline	4	270	20	268.6	Langmuir/pseudo-second order	–	[4]
Graphene oxide/ NiAl LDHs/poly acrylic acid	Hydrothermal method	–	Tetracycline	6	18.5	110	887.5	Freundlich/pseudo-second order	5 cycles	[44]
MgFe LDHs	Solvothermal method	–	Oxytetracycline	7	120	50	–	Pseudo-second order	–	[45]
CMC/CoNiFe-LDHs/ZIF-8	Hydrothermal reaction, immersion	388.5	Ciprofloxacin	6	300	70	1397.5	Langmuir/pseudo-second order	Methanol- ethyl acetate/5 cycles	[46]

The capacity of LDHs for the removal of pharmaceutical pollutants has also been improved by the assembly with metallic oxides [4, 43], in some cases generating adsorbents with hierarchical structures. Chen et al. [38] described the synthesis of monodispersed MgAl LDH spheres with a hierarchical structure using a facile ultrasound-assisted method. On the negatively charged surface of the previously synthesized SiO_2 spheres, MgAl LDHs nanosheets were deposited by electrostatic attractions. The subsequent aging process was assisted by ultrasound, and a hierarchical structure was obtained when the oriented LDHs nanosheets formed and grew vertically on the SiO_2 surface. The thickness of the coating was around 70–100 nm with a particle size of 670–770 nm. These 3D LDH spheres demonstrated good effectiveness in the elimination of diclofenac (758 mg/g), ibuprofen (400 mg/g), and folic acid (332 mg/g).

LDHs' adsorption properties have been attractive for drug extraction design and quantification systems from aqueous matrices [9, 47]. Di et al. proposed a sensitive and green method for the determination of ketoprofen, naproxen, and tolmetin in water samples, using Fe_3O_4@ZnAl LDHs as adsorbent. The methodology combined dispersive micro-solid phase extraction and homogeneous liquid–liquid microextraction (HLLME). In the first stage, Fe_3O_4@ZnAl LDHs was added to the drug-spiked water samples. Then, the adsorbent was magnetically separated and dissolved by adding HCl for the next HLLME procedure. Due to their dissolving ability, these materials end the need for harmful chemical solvents during the elution process [9].

Significant benefits come from combining LDH adsorbents with cellulose derivatives to create composite powders that are flexible and moldable. Raicopol et al. [48] prepared composite membranes from cellulose acetate (CA) and MgAl LDHs with interlayered sodium dodecyl sulfate for the removal of diclofenac and tetracycline. The membrane prepared with 4 wt.% MgAl LDHs increased the adsorption capacity of diclofenac to ten times that of the cellulose acetate. This improvement was ascribed to the electrostatic interactions between the positively charged MgAl LDH layers and the negatively charged drug molecule. A small increase in the removal rate was observed for tetracycline, and this was attributed to hydrogen bond interactions between the drug and MgAl LDHs [48]. In another investigation, Li et al. [46] successfully prepared a malleable cellulose aerogel using carboxymethylcellulose (CMC), CoNiFe LDHs, and zeolitic imidazolate frameworks (ZIF-8). The product displayed a high ciprofloxacin adsorption capacity up to 1397.5 mg/g (Li et al., 2023). The CMC/CoNiFe-LDHs/ZIF-8 functioned as a source framework for the in situ synthesis of ZIF-8 with a high mass loading. The assembly of ZIF-8 provided secondary active sites.

5.4 MECHANISMS OF REMOVAL

The removal mechanism using LDH-based adsorbents is determined by the chemical nature of the drug and the specific composition of the LDHs. The main adsorption mechanisms described for LDH-based adsorbents are shown in Figure 5.3. The absorption of anionic drugs on LDHs is promoted by electrostatic interactions between the anion and the positively charged layers (see Figure 5.3a). Small drug molecules can be captured through intercalation by displacing host anions in the LDH adsorber by ion exchange [14] (see Figure 5.3b). Large anions are retained only on the external surfaces of the LDHs. As expected, the absorption of anionic drugs is favored at an alkaline pH, at which deprotonated species predominate [19]. Also, neutral drugs have been removed by LDH adsorbents, since their structures generally contain heteroatoms such as oxygen and nitrogen, which interact with the layer's hydroxyls through non-electrostatic interactions such as hydrogen bonding, surface complexation, van der Waals forces, and hydrophobic interactions [13, 18] (see Figures 5.3c–5.3f).

For LDH-derived MMOs, the adsorption mechanism involves ion exchange and reconstruction of the interlayer structure due to the so-called memory effect. When the LDHs are mildly calcined with a complete decomposition, the resulting MMOs are more or less crystalline with increased surface area due to crystal shuttering and to the creation of a micro-crack maze of the LDHs' precursor [10]. For example,

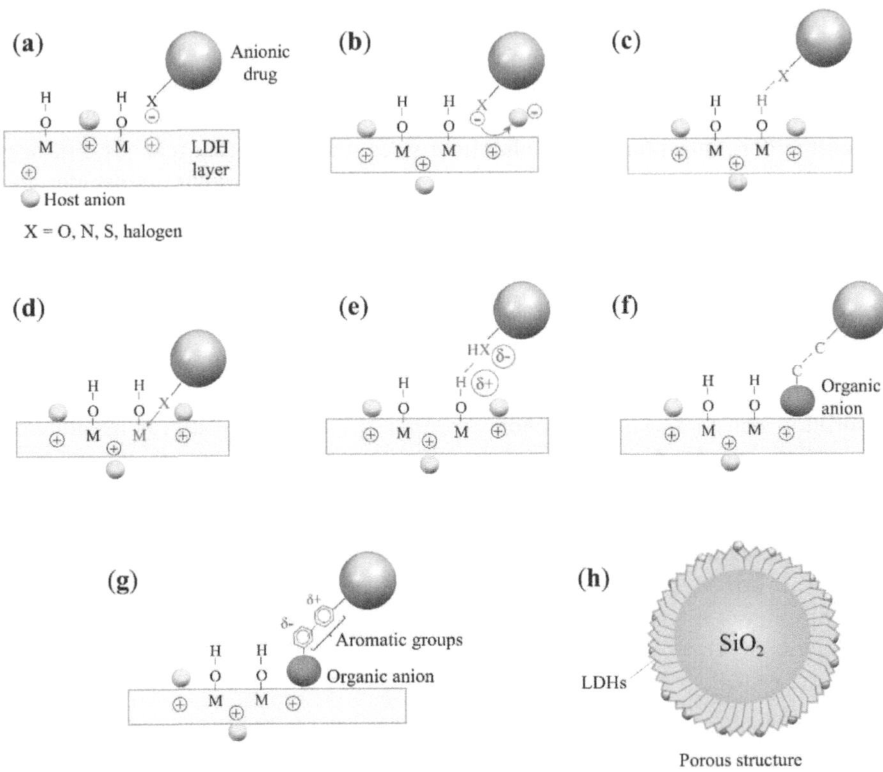

FIGURE 5.3 Representation of the adsorption mechanisms described for LDHs-based adsorbents: (a) electrostatic interactions, (b) ion exchange, (c) hydrogen bonds, (d) complex formation, (e) Van der Waals forces, (f) hydrophobic interactions, (g) π–π interactions, and (h) intraparticle diffusion–adsorption.

the Brunauer–Emmett–Teller (BET) surface of MgAl–CO$_3$ LDHs increases from 31 m^2/g to 214 m^2/g after being calcined at 400 °C [29]. In adsorption tests, the pollutant is found in an aqueous matrix. These conditions favor the rehydration of the MMOs and the uptake of the contaminant for the recovery of the interlayer structure. Depending on its size and charge density, the contaminant may be intercalated or retained uniquely on the hydroxylated surfaces of the reconstructed LDHs.

Regarding the LDHs modified with carbon structures, the adsorption mechanism is based on π–π interactions (see Figure 5.3g). Hence, this type of material favors the elimination of organic pollutants that contain aromatic rings in their molecular structure, for example, antibiotics. In LDH hybrids with hierarchically porous structures, the removal of contaminants occurs mainly by physisorption and diffusion, where the three-dimensional hierarchical structure functions for molecule transport (see Figure 5.3h). In general, the adsorption mechanism depends on the structure of the drug and the surface of the adsorbent. For example, it was described that the adsorption of ciprofloxacin by CMC/CoNiFe-LDHs/ZIF-8 is favored by electrostatic interactions, hydrogen bonding, metal complexation, and π–π bonds [46].

The adsorption of solute molecules on LDH surfaces has been described mainly using the Langmuir and Freundlich isotherms, as observed in Tables 5.1–5.3. Most pharmaceutical pollutant adsorption processes are fitted to the Langmuir model, which describes the formation of a monomolecular layer on the surface of the adsorbent and assumes that all adsorption sites are homogeneous and independent. When the concentration of the solute increases, the adsorption reaches a peak as the surface becomes saturated and cannot hold any more solute molecules in a monolayer [49]. The Langmuir model is described by Equation (5.1):

$$q_e = \frac{K_L \cdot q_{max} \cdot C_e}{1 + K_L \cdot C_e} \tag{5.1}$$

where C_e is the solute concentration, q_e is the amount of solute adsorbed under same equilibrium conditions, q_{max} is the maximum adsorbed, and K_L is the Langmuir adsorption constant. This mathematical model is especially useful in situations where there is limited competition between molecules for adsorption sites and provides information about the interaction between molecules in the medium and on the adsorbent surface [3].

Alternatively, the Freundlich model describes the adsorption of solutes on solid surfaces. It provides a relationship between the amount of solute adsorbed and the concentration of the solute in the medium [49]. The Freundlich model is expressed by Equation (5.2):

$$q_e = K_F \cdot C_e^{\left(\frac{1}{n}\right)} \tag{5.2}$$

where q_e is the amount adsorbed at equilibrium conditions, K_F is the Freundlich adsorption capacity of the adsorbent, C_e is the equilibrium concentration of the adsorbate, and n is the Freundlich exponent, which indicates the intensity of adsorption. This model is useful for describing situations where adsorption can occur in multiple layers and on heterogeneous surfaces. High n values indicate strong adsorption and the ability of the surface to retain the solute. It is a valuable tool to understand the interaction between solute molecules in the medium and on the surface of the adsorbent.

On the other hand, almost all reported kinetic studies for the adsorption of drugs by LDH-based adsorbents describe that the process follows a pseudo-second-order model. This suggests that the removal of the drug occurs by a chemisorption process, wherein the pollutants probably interact with the metal cations in the layers, forming a metal–drug complex. A pseudo-first-order model describes a physisorption process [3].

5.5 RECOVERY OF LDH-BASED ADSORBENTS

An advantage offered by LDH-based adsorbents is their suitable regeneration and reuse procedures. These materials are commonly regenerated by thermal decomposition, advanced oxidation processes, and the desorption method. By high-temperature thermal decomposition, drugs adsorbed on LDH-derived MMO adsorbents are eliminated mainly as H_2O and CO_2 molecules. In advanced oxidation processes, the retained pollutants on the adsorbent surfaces are degraded by strong oxidizing agents such as ozone [38]. And in the regeneration of LDH-based absorbents by desorption, an alkaline solution is required such as NaOH [30], organic solvents [46], or a solution containing an excess of anions [19, 40]. The efficiency of the process depends on the desorption agent and the adsorbent material. After three regeneration cycles of the $CuFe_2O_4$/NiMgAl LDHs used for oxytetracycline removal, NaOH aqueous solution had a better desorption efficiency than ethanol and acetone [42]. Li et al. evaluated CH_3COOH, CH_3OH/CH_3COOH, CH_3OH/HCl, HCl, NaCl, NaOH, and CH_3OH as desorption agents for the removal of retained ciprofloxacin on CMC/CoNiFe-LDHs/ZIF-8 [46]. The CH_3OH/CH_3COOH system (ratio 9:1) was the best eluent with, a desorption percentage of 96%.

It has also been reported that the development of LDH-based adsorbents with magnetic properties facilitates the recovery of the adsorbent material from the treated water by applying an external magnetic field. This simplifies the separation process and allows the reuse of the adsorbents, thus reducing costs and minimizing the generation of waste [50].

5.6 FACTORS INFLUENCING PHARMACEUTICAL POLLUTANT ADSORPTION ON LDHS

5.6.1 pH

The medium's pH greatly influences the removal of pharmaceutical contaminants since it determines the mechanism and efficiency of adsorption of LDH-based adsorbents. Depending on the pH, the drug molecule can be found as a cationic, neutral, or anionic species, each interacting differently with the adsorbent. The surface charge of the adsorbent depends on the point of zero charge (pH_{pzc}), a feature that indicates the pH at which the material's surface is electrically neutral. LDHs have a pH_{pzc} of 7–8 [3, 14], and under ambient pH conditions, LDHs possess positively charged surfaces [19, 51], facilitating the adsorption of anionic drugs. In acidic solutions, LDH-based adsorbents exhibit a low percentage of drug removal due to partial dissolution of LDHs and electrostatic repulsion between the positive charges of the drug and the layers [26, 35]. A low percentage of drug removal has also been observed in solutions with pH values greater than 9, since the negative charge of the LDH's surface does not allow the retention of anionic drugs due to electrostatic repulsion [23]. Regarding the MMOs, their absorption capacity decreases as the pH of the medium increases, since there is competition between the anionic drug, the carbonates, and the hydroxyl groups for the positive sites generated during the reconstruction [25].

5.6.2 Temperature

Drug adsorption using LDH-based materials can be endothermic or exothermic. In an endothermic process, the adsorption of pollutants increases when increasing the temperature by favoring the mobility of drug molecules and their affinity for the LDH adsorbent surfaces [21, 43]. In an exothermic adsorption process, contaminant removal decreases by increasing the temperature, since thermal agitation can remove contaminant molecules weakly retained on the LDHs surface [25].

5.6.3 Drug Concentration

Several investigations corroborate that the highest adsorption capacity of LDHs in solutions with concentrations of pharmaceutical pollutants is around 100 mg/mL [17, 42]. When the concentration of the drugs is higher than 200 mg/mL, the percentage of removal decreases due to the saturation of adsorption sites found on the adsorbent's surface [3, 19]. This does not represent a disadvantage for the employment of LDH-based adsorbents since the concentration of pharmaceutical pollutants in environmental waters and wastewaters rarely exceeds 1 mg/mL [7].

5.6.4 Competitive Compounds

Real wastewater contains various salts that can affect the adsorption capacity of LDH-based adsorbents to remove pharmaceutical pollutants. Multivalent anions and anionic drugs can compete for adsorption sites, reducing drug removal efficiency [24, 40]. Particularly, carbonates generated by the dissolution of atmospheric CO_2 show a high affinity for LDH surfaces. Ifebajo et al. reported that the tetracycline removal capacity using CoFe LDH-derived MMOs decreases from 96.6% to 84–87.3% in the presence of Cl^- and NO_3^- ions [27]. Anionic drugs can also act as ligands and react with metal cations found in the medium to form coordination compounds [34], therefore affecting the removal efficiency. Nevertheless, experiments carried out with environmental water samples revealed that the ions contained in the medium did not significantly affect the removal process of the pharmaceutical pollutant [19, 40].

Competitive adsorption caused by natural organic matter (NOM), such as humic acid and fulvic acid, is a problem when applying adsorbents in natural water and wastewater. Smata and Yoshimura investigated the effect of NOM on the adsorption of oxytetracycline (OTC) by MgFe LDHs to evaluate their potential application in real environmental conditions [45]. The MgFe LDHs were prepared by the one-step solvothermal technique with four different Fe/Mg ratios (1:1, 1:2, 1:3, and 1:4). Except for the LDHs with the highest Fe/Mg ratio, the MgFe LDHs obtained exhibited a selectivity in OTC removal due to the small interlaminar space, which limited the intercalation of compounds present in the NOM. Moreover, the results revealed that NOM did not considerably influence the removal efficiency of OTC.

5.7 CHALLENGES AND FUTURE PERSPECTIVES

LDH-based materials have been extensively studied for the removal of contaminants from aqueous matrices and wastewater by adsorption processes. Interest in LDH particles as adsorbents is due to their versatile chemical composition, tunable microstructure, excellent ionic exchange capacity, high surface area, high porosity, and low toxicity. Although the adsorption capacity of LDHs is widely demonstrated, their application as adsorbents for the elimination of drugs and other organic pollutants in real wastewater treatment systems remains a challenge. Most investigations report the capacity of LDHs for the removal of a single pollutant, and in some cases, the simultaneous absorption of two types of pollutants has also been demonstrated. However, actual industrial effluents contain a variety of pollutants in wastewater, evidencing the need to investigate the potential of LDHs for the adsorption of multiple pollutants found in wastewater, including emerging contaminants that have been poorly explored such as hormones.

In addition to their adsorption capacity, there is also the need to investigate the antimicrobial properties of LDH adsorbents to obtain efficient systems for the simultaneous elimination of chemical and biological pollutants. In addition, LDH adsorption studies could also provide a cost analysis of LDH-based adsorbent synthesis, recycling, and regeneration processes, showing the potential use of each material as an adsorbent in real wastewater treatment. Similarly, it is essential to investigate the ecotoxicity of this type of material, evaluating and minimizing any potential environmental impacts associated with the synthesis, use, and disposal of LDHs in wastewater treatment. It is also necessary to verify the stability and effectiveness of LDH-based adsorbents in various conditions and aqueous matrices for their applicability in real situations. In general, wastewater can have a wide range of pH values, depending on the wastewater and the processes to which it has been subjected. Under highly acidic or alkaline conditions, such as those found in industrial wastewater, the partial or complete collapse of the LDH structure occurs, releasing metal anions and cations into the aqueous medium.

Finally, the adsorption process can be combined with other approaches, such as photocatalysis or electrochemistry, to develop emerging technologies for the effective implementation of LDHs in advanced water treatment systems on a larger scale. Addressing these challenges can lead to the effective and sustainable development of LDH-based technologies for the removal of drugs from wastewater, thus contributing to the preservation of water quality and environmental protection.

REFERENCES

[1] Evans DG, Slade RCT. Structural aspects of layered double hydroxides. In: Duan X, Evans DG, editors. Layered Double Hydroxides. Springer; 2006. p. 1–87.

[2] Johnston AL, Lester E, Williams O, et al. Understanding layered double hydroxide properties as sorbent materials for removing organic pollutants from environmental waters. J Environ Chem Eng. 2021;9:105197.

[3] Santamaría L, Vicente MA, Korili SA, et al. Progress in the removal of pharmaceutical compounds from aqueous solution using layered double hydroxides as adsorbents: A review. J Environ Chem Eng. 2020;8:104577.

[4] Dai X, Rao J, Bao Z, et al. Magnetic double-core@shell MnO_2@NiFe@DE as a multi-functional scavenger for efficient removal of tetracycline, anionic and cationic dyes. J Colloid Interface Sci. 2022;628:769–783.

[5] Theiss FL, Ayoko GA, Frost RL. Synthesis of layered double hydroxides containing Mg^{2+}, Zn^{2+}, Ca^{2+} and Al^{3+} layer cations by co-precipitation methods—A review. Appl Surf Sci. 2016;383:200–213.

[6] Bukhtiyarova MV. A review on effect of synthesis conditions on the formation of layered double hydroxides. J Solid State Chem. 2019;269:494–506.

[7] Nava AK, Carbajal-Arízaga GG, Obregón S, et al. Layered double hydroxides and related hybrid materials for removal of pharmaceutical pollutants from water. J Environ Manage. 2021;288:112399.

[8] Liu J, Song J, Xiao H, et al. Synthesis and thermal properties of ZnAl layered double hydroxide by urea hydrolysis. Powder Technol. 2014;253:41–45.

[9] Di X, Zhao X, Guo X. Dispersive micro-solid phase extraction combined with switchable hydrophilicity solvent-based homogeneous liquid-liquid microextraction for enrichment of non-steroidal anti-inflammatory drugs in environmental water samples. J Chromatogr A. 2020;1634:461677.

[10] Mascolo G, Mascolo MC. On the synthesis of layered double hydroxides (LDHs) by reconstruction method based on the "memory effect." Microporous Mesoporous Mater. 2015;214:246–248.

[11] Mao N, Zhou CH, Tong DS, et al. Exfoliation of layered double hydroxide solids into functional nanosheets. Appl Clay Sci. 2017;144:60–78.

[12] Wang H, Xu W, Su L, et al. Ultra-adsorption enhancing peroxymonosulfate activation by ultrathin NiAl-layered double hydroxides for efficient degradation of sulfonamide antibiotics. J Clean Prod. 2022;369:133277.

[13] Abdel MSA, Mahmoud RK, Mohamed NA, et al. Synthesis and characterisation of LDH-type anionic nanomaterials for the effective removal of doxycycline from aqueous media. Water Environ J. 2019; 1–19.

[14] Li E, Liao L, Lv G, et al. The interactions between three typical PPCPs and LDH. Front Chem. 2018;6:1–9.

[15] Mkaddem H, Rosales E, Pazos M, et al. Anti-inflammatory drug diclofenac removal by a synthesized MgAl layered double hydroxide. J Mol Liq. 2022;359:119207.

[16] Gupta K, Huo JB, Yang JCE, et al. $(MoS_4)^{2-}$ intercalated CAMoS4·LDH material for the efficient and facile sequestration of antibiotics from aqueous solution. Chem Eng J. 2019;355:637–649.

[17] Sepehr MN, Al-Musawi TJ, Ghahramani E, et al. Adsorption performance of magnesium/aluminum layered double hydroxide nanoparticles for metronidazole from aqueous solution. Arab J Chem. 2017;10:611–623.

[18] Panplado K, Subsadsana M, Srijaranai S, et al. Rapid removal and efficient recovery of tetracycline antibiotics in aqueous solution using layered double hydroxide components in an in situ-adsorption process. Crystals. 2019;9:1–10.

[19] Rathee G, Singh N, Chandra R. Simultaneous elimination of dyes and antibiotic with a hydrothermally generated NiAlTi layered double hydroxide adsorbent. ACS Omega. 2020;5:2368–2377.

[20] Elhaci A, Labed F, Khenifi A, et al. MgAl-layered double hydroxide for amoxicillin removal from aqueous media. Int J Environ Anal Chem. 2021;101:2876–2898.

[21] Kamal W, Mahmoud R, Allah AE, et al. Insights into synergistic utilization of residual of ternary layered double hydroxide after oxytetracycline as a potential catalyst for methanol electrooxidation. Chem Eng Res Des. 2022;188:249–264.

[22] Shafiei NS, Hosseinzadeh R, Ghani M. Solid-phase extraction of nonsteroidal anti-inflammatory drugs in urine and water samples using acidic calix[4]arene intercalated in LDH followed by quantification via HPLC-UV. Microchem J. 2022;183:107985.

[23] Ghemit R, Boutahala M, Kahoul A. Removal of diclofenac from water with calcined ZnAlFe-CO$_3$ layered double hydroxides: Effect of contact time, concentration, pH and temperature. Desalin Water Treat. 2017;83:75–85.

[24] Starukh GM, Oranska OI, Levytska SI. Reconstruction of calcined Zn-Al layered double hydroxides during tetracycline adsorpsion. Odesa Natl Univ Herald Chem. 2015;20:82–93.

[25] Mourid EH, Lakraimi M, Benaziz L, et al. Wastewater treatment test by removal of the sulfamethoxazole antibiotic by a calcined layered double hydroxide. Appl Clay Sci. 2019;168:87–95.

[26] Elhalil A, Farnane M, Machrouhi A, et al. Effects of molar ratio and calcination temperature on the adsorption performance of Zn/Al layered double hydroxide nanoparticles in the removal of pharmaceutical pollutants. J Sci Adv Mater Devices. 2018;3:188–195.

[27] Ifebajo AO, Oladipo AA, Gazi M. Efficient removal of tetracycline by CoO/CuFe$_2$O$_4$ derived from layered double hydroxides. Environ Chem Lett. 2019;17:487–494.

[28] Santamaría L, López-Aizpún M, García-Padial M, et al. Zn-Ti-Al layered double hydroxides synthesized from aluminum saline slag wastes as efficient drug adsorbents. Appl Clay Sci. 2020;187:105486.

[29] Amamra S, Elkolli H, Benguerba Y, et al. Synthesis and characterization of layered double hydroxides aimed at encapsulation of sodium diclofenac: Theoretical and experimental study. J Mol Liq. 2021;338:116677.

[30] Lins PVS, Henrique DC, Ide AH, et al. Adsorption of a non-steroidal anti-inflammatory drug onto MgAl/LDH-activated carbon composite—Experimental investigation and statistical physics modeling. Colloids Surfaces A Physicochem Eng Asp. 2020;586:124217.

[31] Vithanage M, Ashiq A, Ramanayaka S, et al. Implications of layered double hydroxides assembled biochar composite in adsorptive removal of contaminants: Current status and future perspectives. Sci Total Environ. 2020;737:139718.

[32] Zubair M, Ihsanullah I, Abdul Aziz H, et al. Sustainable wastewater treatment by biochar/layered double hydroxide composites: Progress, challenges, and outlook. Bioresour Technol. 2021;319:124128.

[33] Li J, Zhang N, Ng DHL. Synthesis of a 3D hierarchical structure of γ-AlO(OH)/Mg-Al-LDH/C and its performance in organic dyes and antibiotics adsorption. J Mater Chem A. 2015;3:21106–21115.

[34] Tan X, Liu S, Liu Y, et al. One-pot synthesis of carbon supported calcined-Mg/Al layered double hydroxides for antibiotic removal by slow pyrolysis of biomass wasteOne-pot synthesis of carbon supported calcined-Mg/Al layered double hydroxides for antibiotic removal by slow pyrolysis. Sci Rep. 2016;6:1–12.

[35] de Souza dos Santos GE, Ide AH, Duarte JLS, et al. Adsorption of anti-inflammatory drug diclofenac by MgAl/layered double hydroxide supported on Syagrus coronata biochar. Powder Technol. 2020;364:229–240.

[36] Li X, Gan T, Zhang J, et al. High-capacity removal of oxytetracycline hydrochloride from wastewater via Mikania micrantha Kunth-derived biochar modified by Zn/Fe-layered double hydroxide. Bioresour Technol. 2022;361:127646.

[37] Zheng D, Wu M, Zheng E, et al. Adsorption and oxidation of ciprofloxacin by a novel layered double hydroxides modified sludge biochar. J Colloid Interface Sci. 2022;625:596–605.

[38] Chen C, Wang P, Lim TT, et al. A facile synthesis of monodispersed hierarchical layered double hydroxide on silica spheres for efficient removal of pharmaceuticals from water. J Mater Chem A. 2013;1:3877–3880.

[39] Wu H, Gao H, Yang Q, et al. Removal of typical organic contaminants with a recyclable calcined chitosan-supported layered double hydroxide adsorbent: Kinetics and equilibrium isotherms. J Chem Eng Data. 2018;63:159–168.

[40] Bing X, Jian X, Chu J, et al. Hierarchically porous BiOBr/ZnAl$_{1.8}$Fe$_{0.2}$O$_4$ and its excellent adsorption and photocatalysis activity. Mater Res Bull. 2019;110:1–12.

[41] Barabi A, Seidi S, Rouhollahi A, et al. Electrochemically synthesized NiFe layered double hydroxide modified Cu(OH)$_2$ needle-shaped nanoarrays: A novel sorbent for thin-film solid phase microextraction of antifungal drugs. Anal Chim Acta. 2020;1131:90–101.

[42] Eniola JO, Kumar R, Mohamed OA, et al. Synthesis and characterization of CuFe$_2$O$_4$/NiMgAl-LDH composite for the efficient removal of oxytetracycline antibiotic. J Saudi Chem Soc. 2020;24:139–150.

[43] Radmehr S, Hosseini Sabzevari M, Ghaedi M, et al. Adsorption of nalidixic acid antibiotic using a renewable adsorbent based on Graphene oxide from simulated wastewater. J Environ Chem Eng. 2021;9:105975.

[44] Omidi MH, Azqhandi MHA, Ghalami-Choobar B. Synthesis, characterization, and application of graphene oxide/layered double hydroxide/poly acrylic acid nanocomposite (LDH-rGO-PAA NC) for tetracycline removal: A comprehensive chemometric study. Chemosphere. 2022;308:136007.

[45] Smata A, Yoshimura C. Size-based selectivity of magnetic-layered double hydroxide (MLDH) for efficient removal of oxytetracycline in the presence of natural organic matter. Appl Clay Sci. 2023;232:106775.

[46] Li C, Wang F, Xu X, et al. A high-capacity malleable cellulose aerogel with layered double hydroxide decorating ZIF-8 for efficient adsorption of ciprofloxacin. Chem Eng J. 2023;455:140841.

[47] Mohiuddin I, Grover A, Aulakh JS, et al. Starch-Mg/Al layered double hydroxide composites as an efficient solid phase extraction sorbent for non-steroidal anti-inflammatory drugs as environmental pollutants. J Hazard Mater. 2021;401:123782.

[48] Raicopol MD, Andronescu C, Voicu SI, et al. Cellulose acetate/layered double hydroxide adsorptive membranes for efficient removal of pharmaceutical environmental contaminants. Carbohydr Polym. 2019;214:204–212.

[49] Obregón S, Mendoza-Reséndez R, Luna C. Facile synthesis of ultrafine akaganeite nanoparticles for the removal of hexavalent chromium: Adsorption properties, isotherm and kinetics. J Nanosci Nanotechnol. 2017;17:4471–4479.

[50] Sajid M, Ihsanullah I. Magnetic layered double hydroxide-based composites as sustainable adsorbent materials for water treatment applications: Progress, challenges, and outlook. Sci Total Environ. 2023;880:163299.

[51] Oliveira C, Lima DLD, Silva CP, et al. Photodegradation of sulfamethoxazole in environmental samples: The role of pH, organic matter and salinity. Sci Total Environ. 2019;648:1403–1410.

6 Carbon Materials as Adsorbents and Catalysts

Akeem Adeyemi Oladipo, Ezinne Favour Ogulewe, Hoda Ansari, Abimbola Aleshinloye, and Mustafa Gazi

6.1 INTRODUCTION: PRESENCE AND PERSISTENCE OF PHARMACEUTICALS IN THE ENVIRONMENT

Over the past few decades, the world population has increased, leading to a rise in the investment in healthcare. This has resulted in advancements in research and development, as well as an increase in animal husbandry. The global market has also become more pervasive, while industrialized countries have seen their societies age. As a result, there has been a significant increase in the consumption of pharmaceuticals for diagnostic, preventive, and therapeutic purposes. However, there is growing concern about the occurrence of pharmaceuticals in environmental media, which are classified as "contaminants of emerging concern" or "emerging contaminants." This has led many researchers to study how to remove these contaminants from aqueous solutions [1–5].

Every day, people around the world consume thousands of pharmaceuticals. While biological pharmaceuticals like vitamins, antibodies, insulin, vaccines, and other protein-based drugs are usually metabolized by humans or biodegrade quickly and are not considered an environmental concern, non-biological substances like synthetic hormones, antibiotics, and non-steroid anti-inflammatory drugs are naturally excreted from the body because they are not completely metabolized by organisms and can contribute to the presence of pharmaceutically active compounds and their byproducts in the environment [6].

Recent research has revealed that over 700 different pharmaceutical substances have been found in various environmental sources, including groundwater, wastewater, drinking water, sediments, surface water, and living organisms [2, 7–11]. Moreover, pharmaceuticals can enter the environment through improper disposal of medications, hospital effluents, and wastewater discharges from manufacturing facilities. Unfortunately, wastewater treatment plants do not completely eliminate most of these compounds, making them persistent pollutants in our environment.

Pharmaceuticals have several other potential pathways to enter receiving waters, including stormwater runoff, septic systems, aquaculture farms, untreated landscape leachate, livestock farms, and sewer overflows during rainfall. Once they enter the aquatic environment, they can undergo biotic and abiotic transformation

DOI: 10.1201/9781003340164-6

processes but may accumulate in the ecosystem and harm non-target organisms through intermediate metabolites. Additionally, pharmaceuticals can enter the food chain and ultimately affect human health. The properties of pharmaceuticals, such as their polarity, stability, volatility, degradation ability, absorbability, and solubility, can vary significantly, and they can be biologically effective even at extremely trace concentrations.

Wastewater contains a multitude of pharmaceuticals such as non-steroidal anti-inflammatory drugs, antibiotics, analgesics, β-blockers, psychoactive compounds, antiretroviral drugs, endocrine disruptors, and cancer treatments. Antibiotics, in particular, pose a major concern due to their persistence, partial metabolism, and ability to easily spread throughout ecosystems, as noted by Mukhtar et al. [12]. Furthermore, it is important to note that antibiotics can be harmful to organisms due to their potency at low concentrations. Additionally, when antibiotics are combined with other drugs or xenobiotic compounds, there can be a synergistic effect that further exacerbates their toxic effects.

The overuse of antibiotics in the environment can lead to the enrichment of antibiotic-resistant bacteria or resistance genes, which can be transmitted from the environment to humans. The presence of antibiotic resistance genes has been associated with treatment failure in human medicine, leading to longer periods of illness, increased morbidity, and higher mortality rates [13, 14]. It is concerning to learn that in the E.U., 33,000 people pass away each year due to antibiotic-resistant bacteria infections, and the situation is similarly alarming in the U.S., where over 2.8 million people contract antibiotic-resistant infections annually, resulting in more than 35,000 fatalities [14–16]. Recent studies by Kariuki et al. [17] and Murray et al. [18] reveal that antimicrobial resistance is linked to over 700,000 deaths each year in Africa, with an estimated 4.1 million fatalities projected by 2050. The economic and healthcare systems suffer greatly from the impact of antibiotic-resistant infections.

Figure 6.1 presents a summary of the classification of frequently found pharmaceuticals in the environment, their effects, and pathways. It is important to note that the prevalence and amount of these pharmaceutically active substances in the environment, particularly in wastewater treatment plant influents, can differ based on factors like the demographics of the population served by the plant, consumption habits, climate, and water usage patterns. Several wastewater treatment plants are not equipped to effectively remove persistent micropollutants like pharmaceuticals. As a result, several technologies, such as adsorption and catalytic degradation, are being employed as primary, secondary, or optional tertiary or advanced treatment processes to eliminate these harmful substances from water and wastewater.

This is an important step towards ensuring the safety and purity of the water supply. This chapter will predominantly centre around the employment of carbon-based materials, both as adsorbents and catalysts, for water treatment in the presence of pharmaceuticals. The various features of carbon-based materials will be explored in detail, encompassing the aspects that affect their efficacy, mechanisms of removal, obstacles encountered, and the latest advancements in research to enhance their performance. The final aim is to pave the way for the commercialization of carbon-based water treatment technologies.

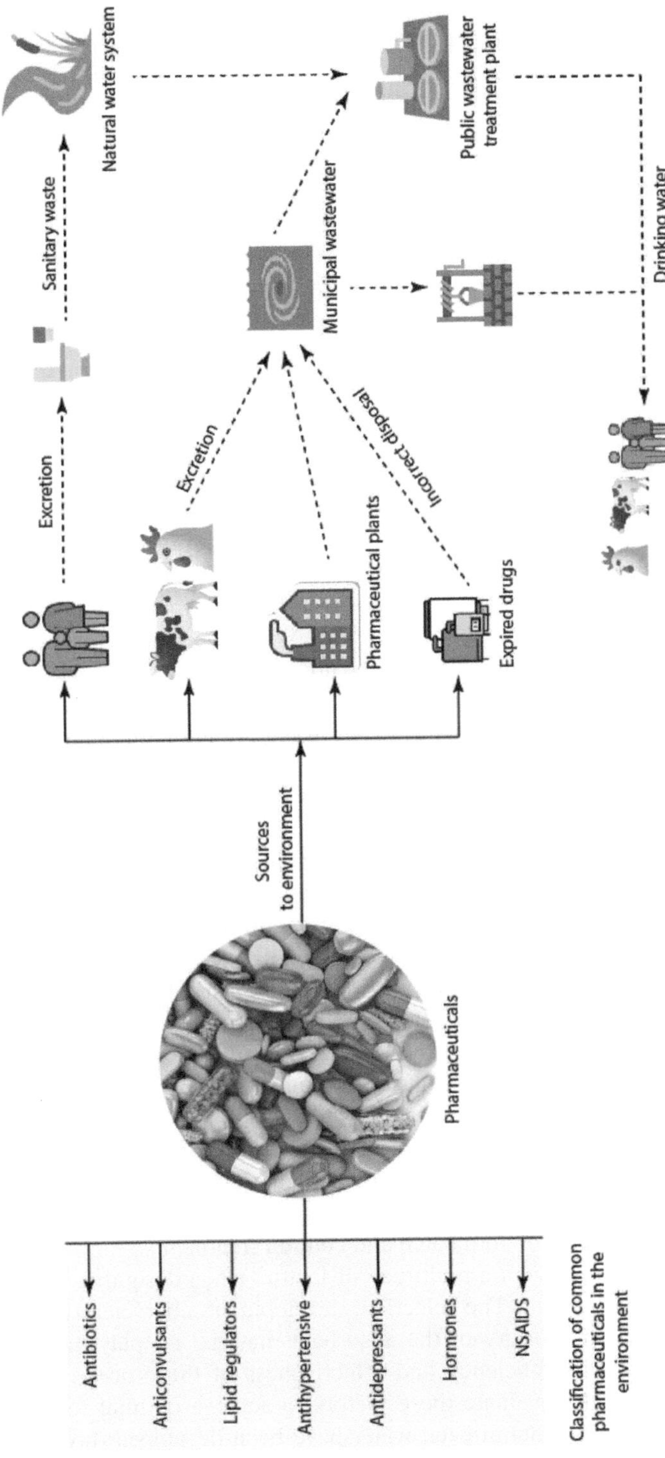

FIGURE 6.1 Classification, pathways, and distribution of pharmaceuticals in the environment.

6.2 REMOVAL OF PHARMACEUTICAL COMPOUNDS FROM WATER: ADSORPTION AND CATALYSIS

The elimination of pharmaceutical substances from water is a crucial matter that demands the exploration and adoption of effective treatment techniques. This is because elevated concentrations of active pharmaceutical compounds or their byproducts in environmental media can endanger human health and the ecosystem. Although not exhaustive in its coverage of pharmaceutical removal treatments in wastewater, this chapter emphasizes the importance of selecting a practical, cost-effective, and efficient approach. Additionally, it is vital to take into account the potential long-term effects of water treatment methods on the environment and living organisms' health.

The adsorption process is a surface-based technique that can effectively remove pharmaceuticals from contaminated water resources. This method involves the use of an adsorbent material to which the pharmaceuticals adhere through various physicochemical processes, such as hydrogen bond formation, surface complexation, ion exchange, changes in functional groups, van der Waals interactions, electron donor–acceptor interactions, diffusion, coordination, and electrostatic interactions. Adsorption is a more effective and cost-efficient approach to removing pharmaceutical pollutants from water due to their low concentrations. Additionally, sustainable adsorbents have excellent regeneration properties, making adsorption a preferred method industrially. Several studies, including those by Ahmed and Hameed [19], Quesada et al. [20], Natarajan et al. [21], and Kang et al. [22], have demonstrated that adsorption is a highly suitable technology for removing pharmaceuticals from wastewater. Moreover, Oladipo and Gazi [23], Oladipo et al. [24], and Rathi et al. [25] have also concluded that adsorption is a reliable and efficient method for this purpose.

There are two operation modes used in the process of adsorption: column treatment and batch treatment. Column treatment involves employing continuous fixed or packed beds, pulsed and fluidized beds, or continuous moving bed systems. On the other hand, batch treatment involves using static or stirred tanks. Batch adsorption treatment is an effective way to remove various pollutants from both real and synthetic wastewaters when the effluent's pollution load is low and the quantity is small. However, when dealing with a higher pollution load, fixed or packed bed columns are commonly used to remove a larger quantity of pollutants from wastewater.

To compare batch and column adsorption procedures, numerous scientists have used adsorption capacity, percentage removal, and other characteristics [19, 26–27]. Figure 6.2 shows some of the salient characteristics, parameters, benefits, and performance assessments of both batch and column treatments.

It is imperative to consider a multitude of factors when designing an adsorption process for water treatment. The selection, development, characterization, adsorption, and regeneration capacity of the adsorbent material all play pivotal roles in determining the overall efficiency and effectiveness of the process. Therefore, it is crucial to thoroughly evaluate these factors to achieve optimal results. Several adsorbents for treating contaminated water have been developed through research

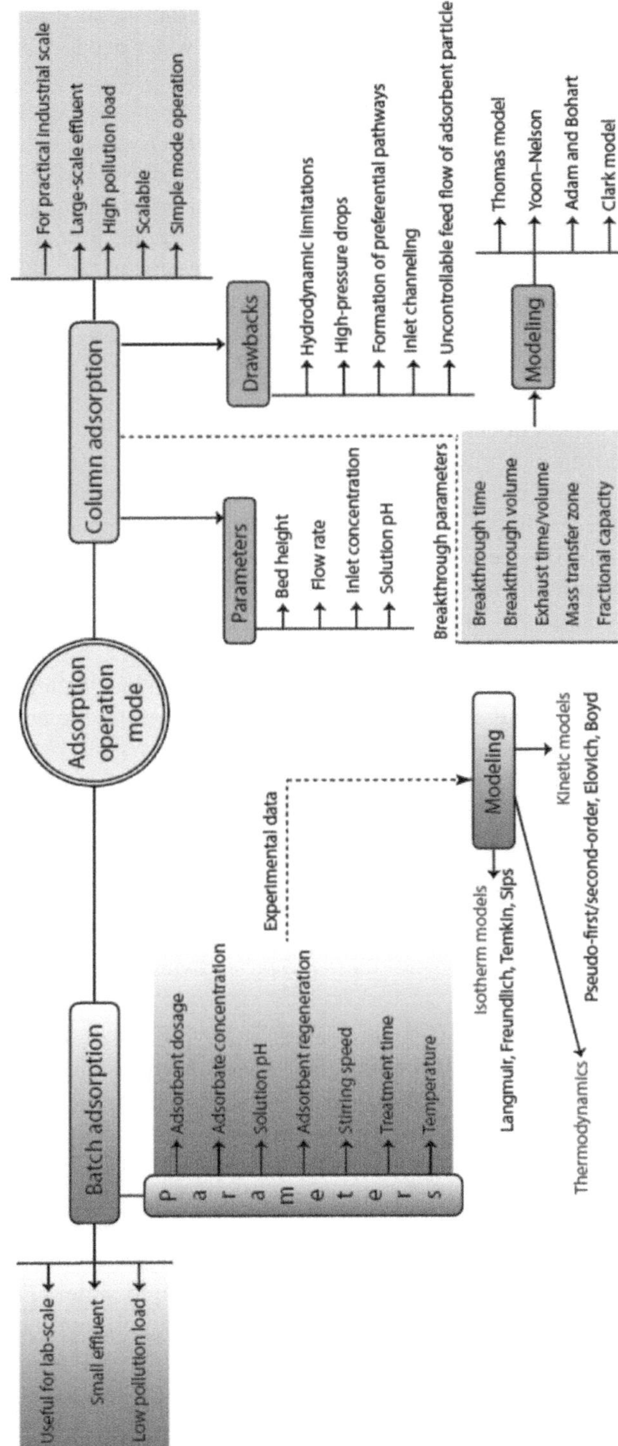

FIGURE 6.2 Adsorption operation mode and key parameters for batch and column treatments.

investigations; nevertheless, it should be noted that an efficient adsorbent for effective adsorption of pharmaceuticals from contaminated water needs to have the following characteristics:

- **High removal efficiency and adsorption capacity**: To treat contaminated water properly, an adsorbent needs to have a high removal rate of pharmaceutical pollutants from the water and a high adsorption capacity per gram of adsorbent. This will not only shorten the adsorption process but also reduce the amount of adsorbent required for the treatment.
- **Economical, sustainable, and available**: As the adsorbent material accounts for nearly 85% of the operating expenses, it must be cost-effective and sustainable to obtain or develop, as large amounts of it must be transported to treatment plants.
- **Good physicochemical and morphological properties**: To effectively remove pharmaceutically active pollutants, adsorbents must have a large surface area, appropriate functional groups, suitable textural features, and sufficient pore volume. Adsorption is a surface-based phenomenon that requires interaction between the adsorbent and the pollutants.
- **High mechanical and chemical stability**: The adsorbent used in columns for continuous water treatment needs to be mechanically stable to avoid high-pressure drops and preferential pathways. Additionally, it should possess a high degree of chemical stability, since different water matrices have unique chemical properties that can affect the adsorbent's performance.
- **Fast adsorption and desorption kinetics**: The properties of the adsorbent material should enable rapid achievement of high removal efficiency and adsorption capacity. In addition, the desorption kinetics must be sufficiently rapid to permit the regeneration and subsequent reuse of the adsorbent. As a result, the size and operational costs of the water treatment plant will be impacted if the duration of the complete adsorption process is shorter.

Adsorbents play a crucial role in water treatment processes. However, many expensive and inexpensive adsorbents fail to meet the minimum requirements for an effective adsorbent. As a result, researchers are constantly developing new adsorbent materials to achieve the properties necessary for effective adsorption processes. Recent research by Xiang et al. [28], Ashiq et al. [29], and Nabgan et al. [30], as well as industrial performance, has shown that carbon-derived materials such as activated carbons, chars, nanomaterials, and biochars, as well as graphene oxide and quantum dots, are gaining interest as efficient adsorbent materials for treating pharmaceutically contaminated water. The majority of pharmaceuticals are classified as persistent organic pollutants due to their stable chemical structure. Therefore, researchers are developing catalysts that can demineralize these pollutants.

Carbon-based nanomaterials such as graphene, quantum dots, nanofibers, nanohorns, and carbon nanotubes have been used as catalysts for the degradation of pharmaceuticals in water. These materials have shown promising results as electrocatalysts and photocatalysts for the degradation of pharmaceuticals. The catalytic capacities of carbonaceous materials, which are metal-free and environmentally benign catalysts, are determined by various factors such as their high surface area, black body properties, defects in graphitic structures, degree of graphitization, and high charge carrier mobility [31].

Although carbon-based materials have been extensively studied as semiconductor supports, dopants, and hybridizing materials, they can be employed directly as photocatalysts or electrocatalysts. Studies have shown that carbon-based materials, like graphite felt and carbon sponge, are highly effective as electrocatalysts for the degradation of pharmaceuticals. These materials offer several benefits, including a wide range of potential, strong electronic conduction, low resistance, and residual currents. Additionally, their high surface area and porosity provide numerous sites for redox reactions and exceptional electrolytic efficiency. Researchers Liu et al. [32], Ganiyu et al. [33], and Huong et al. [34] have all found these materials to be effective in electrocatalysis.

The photocatalytic removal of pharmaceuticals from contaminated water using carbonaceous materials, like graphene quantum dots with a small band gap, involves a process called photocatalytic oxidation (advanced oxidation processes, AOPs). This process occurs when the graphene quantum dots absorb visible light directly, which produces photo-generated holes and electrons [31, 35]. These charge carriers can then initiate various reactions that create three active species: superoxide ($O_2^{\cdot-}$) radicals produced by photo-induced electrons via the reduction of surface-adsorbed oxygen, photo-excited holes (h^+), and hydroxyl ($OH^\cdot$) radicals from the oxidation of H_2O by holes. These radicals can react with the pharmaceuticals, creating intermediate species that eventually lead to complete mineralization.

In carbonaceous electrocatalytic materials such as graphite, the electrocatalytic oxidation of pharmaceuticals primarily operates through the adsorption of H_2O onto the active edges of the anodic material. This leads to the oxidation of H_2O to produce surface-adsorbed hydroxyl radicals ($OH^\cdot$) [36], which are then converted into a higher oxidation state [37]. This process results in the selective oxidation of pharmaceuticals. Electrocatalysts like graphene, graphite, and carbon nanotubes possess carbon surfaces that are appropriate for adsorbing various chemical species. Due to their high conductivity, these surfaces can facilitate specific electrocatalytic reactions [38]. Carbon-based electrocatalytic electrodes are believed to be capable of serving as efficient anodes and cathodes for the degradation of pharmaceutical pollutants via oxidation and reduction. Additionally, Figure 6.3 showcases the distinctive traits of materials that can be utilized as efficient photocatalysts, electrocatalysts, and photoelectrocatalysts. In the following sections, the characteristic properties of selected carbonaceous materials will be discussed in relation to their application as adsorbents and catalysts for removing pharmaceuticals from wastewater and various industrial effluents.

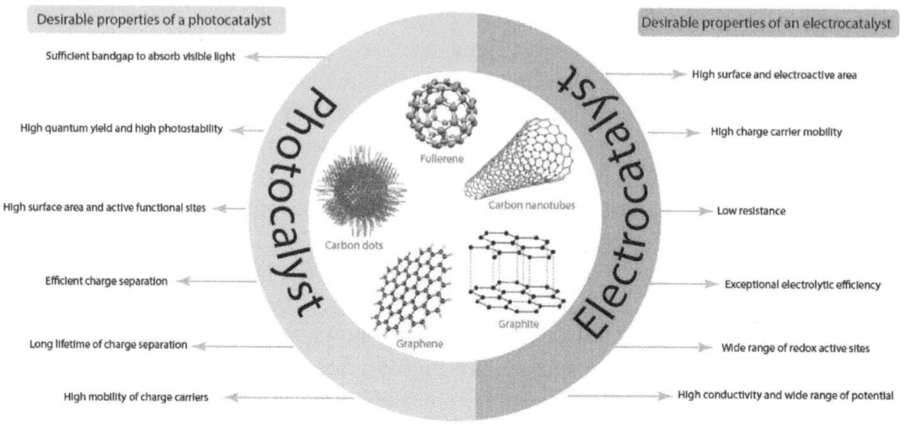

FIGURE 6.3   Desirable properties of effective catalysts for the removal of organic pollutants.

6.3 CARBON-BASED MATERIALS: CHARACTERISTICS AND FORMS

Due to its distinctive electronic structure, carbon, possessing four valence electrons (two electrons in each of the 2s and 2p subshells), can take on diverse forms. This structure permits the establishment of stable chemical bonds in varying arrangements by blending the 2s and 2p orbitals, leading to the creation of single, double, or triple bonds. Carbon exhibits a remarkable ability to self-bond, resulting in the formation of straight or branched, cyclic or acyclic chains. Furthermore, it can form bonds with other non-metallic elements. As a result of the linear combination of valence carbon orbitals, carbon can assume a wide range of structures in one, two, and three dimensions [39].

Carbon is truly an intriguing element with three distinct naturally occurring allotropes: amorphous carbon (such as carbon black, charcoal, coke, and more), crystalline graphite, and diamond. The crystalline forms of carbon are particularly fascinating, as they are formed by carbon atoms with sp^2 hybridization and display a wide range of unique structures and properties. Researchers tend to favour crystalline allotropes of carbon over amorphous forms like carbon black since the latter are less modifiable and harder to alter. Additionally, carbonaceous materials exist in diverse, stable forms, ranging from naturally occurring bulk structures like activated carbon and biochar to discrete structures such as single-walled carbon nanotubes, multi-walled carbon nanotubes, graphene, and hierarchical porous carbon.

Carbon-based materials offer numerous advantages such as simple synthesis procedures, affordability and accessibility of raw materials, and exceptional adsorption abilities thanks to their vast specific surface area, extensive porosity, and well-defined and adaptable pore structure that includes micropores, mesopores, and macropores, as well as various surface functional groups (-NH$_2$, -OH, C=O, Si–H, C–O, -COOH, -CO$_3$, etc.). Consequently, these materials hold immense potential for applications in pharmaceutical removal. A diverse range of carbon-based materials exists, each boasting distinctive physical and chemical properties. These materials are highly sought after as adsorbents and catalysts for the elimination of pharmaceuticals. Among the most frequently utilized carbon-based materials are activated carbon, biochar, hydrochar, pyrochar, graphene, graphite, carbon nanotubes, carbon dots, and hierarchical porous carbon. The mechanisms by which these materials interact with pharmaceuticals involve hydrogen bonding, electrostatic interaction, π–π interaction, hydrophobic interaction, and pore filling and trapping. The carbon-based materials are classified into two types: pristine and engineered. The engineered ones can be further categorized into composites such as carbon–clay, carbon–polymer, carbon–metal oxide, carbon–metal nanoparticles, and more, as well as modified carbon-based materials.

6.3.1 Carbon-Based Adsorbents for the Removal of Pharmaceuticals from Water

Activated carbon: As illustrated in Figure 6.4, activated carbon is a black, solid substance with an exceptionally porous structure and an expansive surface area. While it can be derived from numerous carbon-based materials, the characteristics of activated carbon may vary depending on the source and the activation technique employed. Various factors affect the activation process, such as the activation temperature, raw material type, activation time, activator type, activation agents, and activator flow.

These factors can be altered to adjust the physicochemical properties of the resulting activated carbon. The most commonly utilized raw materials for producing activated carbon are animal bones; biomass; fruit peels, pits, and shells; wood; charcoal; peat; and paper mill waste. Activated carbon is a highly beneficial adsorbent for purifying water and air on an industrial scale due to its advantageous physicochemical properties. It's worth noting that the band gap of activated carbon may vary

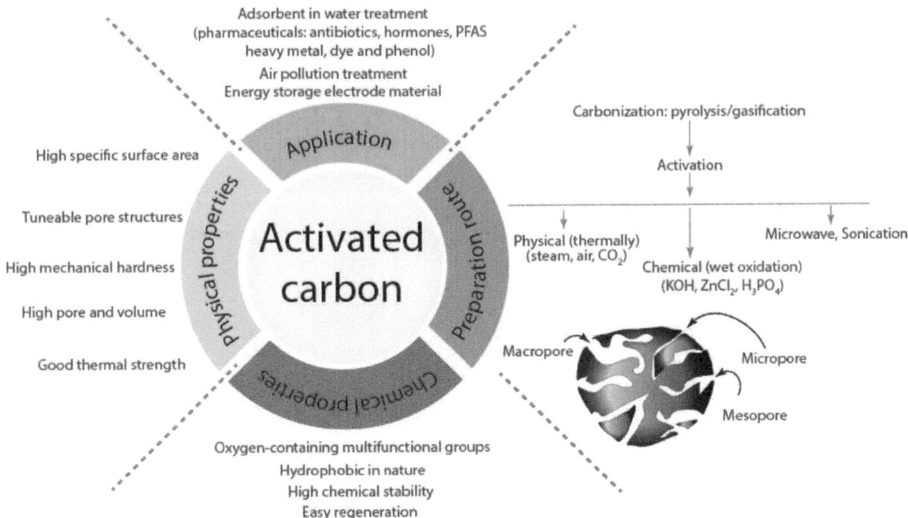

FIGURE 6.4 Characteristics and suitability of activated carbon for the adsorption of pharmaceuticals.

based on the method of synthesis and degree of activation. Studies have shown that activated carbon can have an energy gap between 2.80 and 3.15 eV when created through physical and chemical activation methods [40].

Another study has found that activated carbon fibres have band gaps ranging from 0.5 to 1.5 eV [41]. It should be noted, however, that activated carbon is typically not used as a photocatalyst due to its inability to absorb light and create electron–hole pairs. To improve its efficiency and performance, researchers have explored various pathways, including identifying cost-effective raw materials and optimal activation methods, as well as enhancing its adsorption capabilities.

Yazidi et al. [42] conducted a study where they produced activated carbon from durian shell powder using the physical activation method. After the pyrolysis process, CO_2 was used as the activating agent at a temperature of 550°C. The team then investigated the adsorption capacity and mechanism of amoxicillin and tetracycline antibiotics on the durian shell-activated carbon in both single and binary systems. They found that the activated carbon had an adsorption capacity of 106–142.7 mg/g in single antibiotic systems and 64.9–147.8 mg/g in binary systems.

In their recent study, Sousa et al. [43] utilized microwave pyrolysis at 105°C for 24 hours in an oven to produce pyrolyzed carbon from paper mill sludge. The resulting product was chemically activated with alkaline KOH, a method chosen for its ability to produce activated carbon with a high surface area and narrow pore size distribution while minimizing environmental contamination and corrosion. The researchers then tested the sludge-based activated carbon as an adsorbent for amoxicillin and sulphamethoxazole antibiotics, achieving a removal efficiency of 72–85% under optimized conditions. According to a recent study by Wang et al. [44], activated carbon with a hierarchal structure was successfully developed using peeled and dried loofah.

The goal of the research was to create a material capable of efficiently adsorbing multi-level antibiotic pollutants. The authors discovered that the activated carbon made from loofah possessed a well-organized and hierarchal microstructure, with nanoscale protrusions covering the surface, as depicted in Figure 6.5. This distinctive structure was found to greatly enhance the material's adsorption capacities, specifically for norfloxacin, ofloxacin, and tetracycline. Table 6.1 provides a summary of

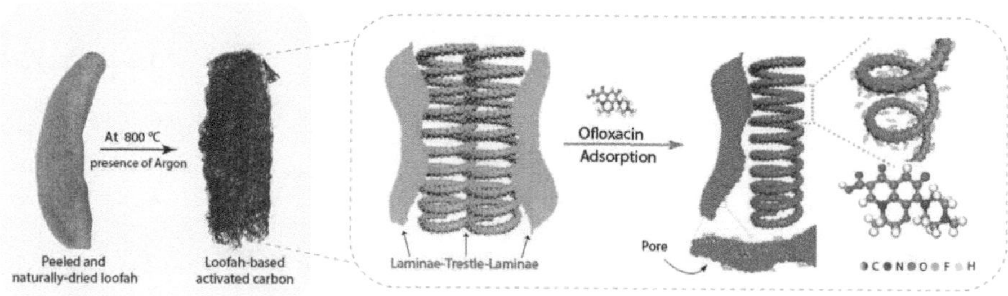

FIGURE 6.5 Characteristics and suitability of activated carbon for the adsorption of pharmaceuticals.

Source: (Modified from "Loofah activated carbon with hierarchical structures for high-efficiency adsorption of multi-level antibiotic pollutants", Wang et al. [44], with permission).

TABLE 6.1

Summary of Recently Reported Activated Carbon Applications for the Adsorption of Pharmaceuticals

Precursor Materials	Preparation Routes	Pharmaceuticals	Removal Conditions	Comments	References
Vegetal: corncob, rice hulls, corn straw, or date stone	Commercial (preparation routes not given)	Amoxicillin, enrofloxacin, sulphadiazine, and trimethoprim	28–67% removal (61.12–137.22 mg/g) in 1 h by 100 mg/L activated carbon	Surface area: 745.4 m^2/g. π–π interactions influence adsorption Removal: Langmuir, Freundlich isotherms, and pseudo-second-order kinetics	[45]
Prosopis juliflora	Pre-treated with H_2SO_4, pyrolysis, and chemical activation with KOH	Sulphadiazine, metronidazole, and tetracycline	85–97% removal (18.48–28.81 mg/g) using 1 g/L of activated carbon under a pH range of 6–8	The data fit the pseudo-second-order kinetic model and Langmuir isotherm. Multicomponent systems showed both antagonistic and synergistic adsorption, influenced by surface functional groups (O–H, N–O, C–O).	[46]
Olive biomass	Muffle furnace: Pyrolysis at 600°C and chemical activation with $ZnCl_2$. Microwave-induced: five cycles of 80 s at 1200 W.	Amoxicillin	91.96–94.43% removal (166.96–237.02 mg/g) at 25°C using 30 mg/50 mL at pH 7	Activated carbon has a specific surface area of 1742 m^2/g and follows Avrami nonlinear kinetics and the Liu isotherm model. The adsorption process is driven by hydrophobic interactions, polar interactions of oxygen-containing functional groups, polar groups of activated carbon, and hydrogen bonds.	[47]
Azolla filiculoides	Pyrolysis at 600°C for 4 h and activated by $ZnCl_2$	Azithromycin	97.9% (374 mg/g) removal in 75 min using 1 mg/mL activated carbon at pH = 9	The specific surface area: 484.1 m^2/g. Experimental data fit the Freundlich isotherm and pseudo-first-order kinetics.	[48]
Macroalgae: *Enteromorpha clathrate*	Pyrolysis at 500°C and chemically activated with NaOH at 800°C	Tetracycline	95.78% removal (381.6 mg/g) at pH 5 using 0.2 g activated carbon	Activated carbon had a mesoporous structure and a surface area of 71,238.5 m^2/g. Electrostatic interactions, π–π interactions, and hydrogen bonding influenced antibiotic adsorption. Langmuir and pseudo-second-order kinetics models were suitable to describe the removal mechanism.	[49]

(Continued)

TABLE 6.1 (*Continued*)

Summary of Recently Reported Activated Carbon Applications for the Adsorption of Pharmaceuticals

Precursor Materials	Preparation Routes	Pharmaceuticals	Removal Conditions	Comments	References
Semi-coke	After pyrolysis, it was activated by KOH and calcinated for 3 h.	Tetracycline	302.99 mg/g adsorption capacity using 30 mg of activated carbon at pH 6	Activated carbon, with a mesoporous structure and 1365.8 m^2/g surface area, adsorbs antibiotics due to its high specific surface area, surface functional groups, π–π bond, and hydrophobicity. Langmuir and pseudo-second-order kinetics models describe the removal mechanism, and it has great potential for treating wastewater.	[50]
Pawpaw seeds	Microwave synthesis for 10 min at 540 W and activated with HCl	Ciprofloxacin and tetracycline	88–295 mg/g adsorption capacity using 0.05 g activated carbon	Adsorption occurs via physical processes such as hydrogen bonding, electrostatic interactions, π–π interactions, and hydrophobic interactions. Freundlich and pseudo-second-order kinetics models describe the removal mechanism.	[51]
Paper mill sludge	Pyrolysis at 800°C and chemically activated with KOH	Sulphamethoxazole, carbamazepine, venlafaxine, lorazepam, paroxetine, piroxicam	65–96% removal in various pharmaceutical effluents.	The activated carbon was improved by adding amine and sulphonic groups. The removal mechanism was through aromatic–aromatic interactions via π–π electron donor–acceptor complexes between the activated carbon and the antibiotic.	[52]
Waterfowl feather and *Phragmites australis*	Pyrolyzed at 450°C for 1 h and chemically activated using 85 wt% H_3PO_4 solution	Amoxicillin and cephalexin	66.44–92.59% removal at pH = 5.5 and pH = 7	The main adsorption mechanisms were covalent bonds and electrostatic attraction. The pseudo-second-order kinetics, Langmuir, and Freundlich isotherms were used to correlate with the experimental results.	[53]
Garlic skin	Carbonization at 600°C and chemically activated using KOH	Tetracycline, doxycycline, oxytetracycline, and chlortetracycline	1393.8–1221.4 mg/g adsorption capacity using 10 mg/L of activated carbon	Surface area: 3686 m^2/g Adsorption is influenced by pore filling, hydrogen bonding, π–π, and electrostatic interactions. Data fit the Freundlich isotherm and pseudo-first-order kinetics.	[54]

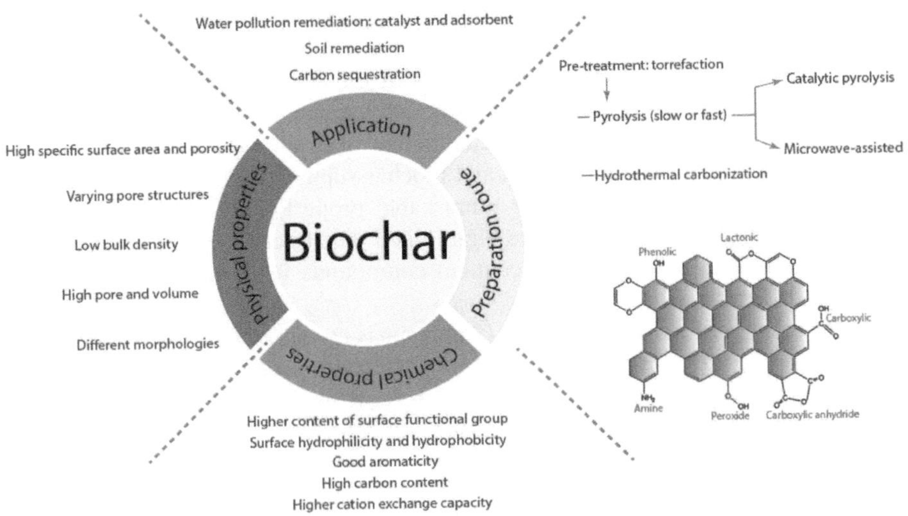

FIGURE 6.6 Characteristics and suitability of biochar for the adsorption of pharmaceuticals.

the pharmaceuticals that have been adsorbed on activated carbon recently reported in the literature.

Biochar: This is a captivating, lightweight, pyrogenic-carbon-rich solid material that has garnered attention for its adsorbent properties. It is produced by heating biomass in an environment with low oxygen. One of the notable benefits of biochar is its cost effectiveness, making it a more accessible option than non-renewable activated carbon. Furthermore, biochar has been demonstrated to be a more effective and renewable choice, making it a potential adsorbent and modifier for various environmental and industrial applications. It is worth noting that the characteristics of biochar shown in Figure 6.6 are heavily influenced by the source of the feedstock and the conditions used to prepare it [55, 56].

For instance, carrying out pyrolysis at temperatures over 600°C results in a material that gradually loses its functional groups, leading to a more condensed and persistent polycyclic aromatic structure. Conversely, pyrolyzing at lower temperatures produces biochar with a higher carbon content, greater surface area, increased porosity, and better adsorption properties [57]. The physicochemical properties of biochar are critical in determining its capacity for adsorption, efficiency, and mechanisms. Recent studies by Layek et al. [57], Mukherjee et al. [58], and Law et al. [56] indicate that modified biochar with improved physicochemical properties is particularly effective in removing a broad spectrum of pollutants from wastewater due to its enhanced adsorbability.

The process of biochar adsorption is multifaceted and influenced by several factors, including biochar characteristics, the type of pollutant present, and solution chemistry [59]. Specifically, organic pollutant adsorption by biochar is primarily driven by hydrogen bonding, hydrophobic effects, electrostatic interactions, and π–π interactions. Meanwhile, metal ion adsorption is mainly governed by surface complexation, adsorption–reduction, ion exchange, and co-precipitation on biochar [60].

The distinct photoelectric characteristics of biochar make it a desirable material for doping with photocatalysts. Additionally, it can serve as a cathode electrocatalyst support for environmental remediation and water purification. Scientific interest in biochar-based adsorbents and biochar-supported photo-/electrocatalysts has greatly increased due to their remarkable properties and customizable surface chemistry. Recent studies have highlighted the effectiveness of biochar-based adsorbents in eliminating pharmaceutical compounds from aqueous matrices, as outlined in Table 6.2.

Graphene and its derivatives: Graphene is a remarkable two-dimensional sheet made up of a single layer of sp^2 hybridized carbon atoms arranged in a hexagonal honeycomb aromatic ring structure. It is considered the parent of all graphitic forms; for instance, a rolled-up graphene layer becomes a one-dimensional carbon nanotube, folded-up graphene forms zero-dimensional fullerene, and three-dimensional graphite sheets generate graphene. The graphene's unique intrinsic properties have captured the attention of the scientific and engineering communities worldwide. These characteristics include a high specific surface area of around 2650 m^2/g, the ability to adjust its structure, high electronic conductivity and electron mobility, an π-conjugated system, exceptional mechanical strength of approximately 39.9 N/m, optical transparency of roughly 98%, and good thermal and chemical stability [71, 72].

The characteristics outlined contribute to graphene's exceptional performance in adsorption and photocatalysis activities. Graphene materials are commonly utilized in aqueous solutions, but their synthesis on a large scale can be difficult, and graphene itself is hydrophobic. The techniques used during synthesis have a significant impact on the quality and performance of graphene. Among the techniques employed are chemical vapour deposition, exfoliation, and organic synthesis. These synthesis techniques can be approached in two ways: top down, which minimizes van der Waals forces of attraction inside the graphite layers, and bottom up, which integrates unique molecular units [73]. To enhance the solubility and functionality of graphene, various methods have been developed to modify it, resulting in graphene derivatives like graphene oxide (GO) and reduced graphene oxide (rGO). GO can be produced through a top-down approach from graphite, such as the Hummers' method Hummers and Offeman [74], or bottom-up synthesis methods like chemical vapour deposition.

Graphene oxide is a form of graphene that has undergone a high level of oxidation, resulting in the presence of numerous oxygen functional groups such as hydroxyl, carboxyl, carbonyl, and epoxy. This alteration of the sp^2 bonding network in GO leads to unique properties and a hydrophilic behaviour that is largely determined by the degree of oxidation [71, 75]. Due to their considerable charge and hydrophilicity, GO sheets are easily dispersed. Reduced graphene oxide (rGO) is obtained from the chemical or thermal treatment of graphene oxide. While GO and rGO differ significantly in their structural and chemical properties, the primary distinction lies in the C/O ratio within their structures. While GO has a very low C/O ratio, rGO boasts a substantially higher ratio, with oxygen content nearly reaching zero.

TABLE 6.2

Summary of Recently Reported Biochar Application for the Adsorption of Pharmaceuticals

Precursor Materials	Preparation Routes	Pharmaceuticals	Removal Conditions	Comments	References
Spent coffee grounds, cattle manure, and biosolids	Pyrolysis at 550°C for 90 min	Tetracycline, sulphamethoxazole, erythromycin, ampicillin, clarithromycin, trimethoprim, and ofloxacin	70–100% removal in 90 min using 10 g/L biochar at pH 8–10.5	Surface area: 1.53–11.48 m²/g. The material is free from any visible pores and is made up of closely packed elongated structures. Each biochar applied at a low dosage exhibited impressive efficacy, removing over 70% of antibiotics and reaching 100% in some cases. However, neither ofloxacin nor sulphamethoxazole was eliminated. The Freundlich model was the most suitable for the adsorption results, except for sulphamethoxazole. Antibiotic adsorption was primarily driven by surface complexation, H-bonding, π–π electron-donor–acceptor interactions, and pore-filling effects.	[61]
Sunflower seed husk	Pyrolyzed at 600°C for 2 h and activated by 85% phosphoric acid	Tetracycline, ciprofloxacin, ibuprofen, and sulphamethoxazole	251.1–429.3 mg/g adsorption capacity using 25 mg/L of biochar under a pH range of 3–11	The specific surface area is 378.8 m²/g and exhibits a mesoporous structure. The data fit the pseudo-second-order kinetic model and Langmuir isotherm. Multicomponent systems show both antagonistic and synergistic adsorption, influenced by surface functional groups (O–H, N–O, C–O).	[62]
Grapevine cane, holm tree biochar, and eucalyptus	Pyrolysis at 500°C for 8 h	Venlafaxine, trazodone, and fluoxetine	0.99–8.68 mg/g adsorption capacity using 0.25 g/25 mL using both batch and column adsorption systems with simulated and real pharmaceutical wastewater	The primary mechanism in the adsorption process is the interaction between the donor and acceptor electrons. The multicomponent system demonstrated a synergistic behaviour for trazodone and an antagonistic behaviour for venlafaxine and fluoxetine pharmaceuticals after the column adsorption experimental data were simulated using the Thomas, Yoon–Nelson, and Yan models.	[63]

(Continued)

TABLE 6.2 (*Continued*)

Summary of Recently Reported Biochar Application for the Adsorption of Pharmaceuticals

Precursor Materials	Preparation Routes	Pharmaceuticals	Removal Conditions	Comments	References
Pharmaceutical sludge	Impregnation method and dry mixing method followed by activation with NaOH and pyrolysis at 600°C for 1 h	Tetracycline	Maximum adsorption capacities of 361.8–379.8 mg/g at 25°C using 20 mg/50 mL antibiotic solution	The antibiotic's adsorption was primarily influenced by the pore-filling effect, π–π conjugation, and hydrogen bonding. The adsorption process accelerated as a result of the temperature rise.	[64]
Pharmaceutical sludge	Pyrolysis at 800°C for 1.5 h and chemically activated with KOH, $ZnCl_2$, and CO_2	Levofloxacin	99.9% removal (159.3 mg/g) using 10 mg of $ZnCl_2$-activated biochar in the presence of 40 mg/L of antibiotic at 35 °C	The $ZnCl_2$-activated biochar had a microporous structure, contained rich acidic functional groups, and had a surface area of 534.91 m^2/g. Surface complexation, electrostatic interactions, hydrogen bonding and π–π interactions dominated the antibiotic adsorption mechanism.	[65]
Penicillin fermentation residue and sludge	The raw material was pyrolyzed at 400–800°C	Penicillin and antibiotic resistance genes	93.3–98.5% (23.26–44.1 mg/g) removal efficiency using 80 mg/50 mL of biochar at an optimal pH 11	The adsorption of penicillin utilizing biochar derived from fermentation sludge was based on chemisorption, which has a low selectivity and a high sensitivity to pH and other environmental factors. In contrast, physical adsorption, which is more selective and less impacted by pH and other environmental factors, was the basis for the pharmaceuticals' removal from the biochar made from fermentation residue.	[66]

Wood chip	Crushed wood chip was oven-dried at 105°C and pyrolyzed at 300–600°C.	Ciprofloxacin and norfloxacin	Batch adsorption capacity is 4.23–7.59 mg/g and photocatalytic degradation efficiency is 88–88.4% using 0.25 g biochar.	The primary adsorption processes for norfloxacin (NOR) and ciprofloxacin (CIP) are the polar interactions and the π–π interactions. The primary active species for the oxidative destruction of CIP and NOR in the photocatalytic process are h^+ and OH·. Both batch and fixed-bed adsorption techniques were used; the fixed-bed method produced a greater removal capacity. CIP's maximum fixed-bed adsorption capacity was 9.82 mg/g, greater than NOR's (7.19 mg/g) maximum.	[67]
Rice straw	Pyrolyzed at 450°C and chemically activated with KOH.	Sulphamethazine, oxytetracycline hydrochloride, and amoxicillin	7.67–51.77 mg/g removal using 3.2 g/L biochar at pH 3	The adsorption process was heterogeneous and mostly controlled by chemisorption, as batch adsorption tests clarified. The biochar surface's -OH groups may act as the main active sites for the removal of antibiotics, according to the density functional theory.	[68]
Bagasse, bamboo, and hickory chips	Oven-dried at 80°C, pyrolyzed at 300°C–600°C for 1.5 h, and ball milled for 12 h	Sulphamethoxazole and sulphapyridine	83.3–89.6% removal using 10 mg of biochar at pH = 5.	Several processes, such as hydrogen bonding, electrostatic interaction, π–π interaction, and hydrophobic interaction, governed sulphonamide batch adsorption on the biochars. The adsorption was pH dependent because of the significance of electrostatic interaction.	[69]
Walnut shells	Pre-treated and dried at 110°C for 12 h; pyrolyzed at 600°C and chemically activated using 69% nitric acid	Sulphadiazine, sulphamethazine, and sulphachloropyridazine	32–46 mg/g adsorption capacity using 20 mg/L of biochar at pH 4–6	Sulphonamides were primarily adsorbed onto the biochar by chemisorption, as suggested by the fact that the kinetics followed the Elovich model and the isotherm conformed to Freundlich. Weak acidic solutions were more favourable for the adsorption of antibiotics, and the primary interactions were electrostatic, Lewis acid–base, hydrophobic, and H-bond.	[70]

According to Table 6.3, it can be observed that graphene and its derivatives possess high efficacy in adsorbing pharmaceutical-based pollutants from various aqueous media. The unique chemical structures and functional groups of graphene and its derivatives lead to different adsorption capacities for pharmaceutical-based pollutants. Generally, pharmaceutical molecules with aromatic rings can interact with graphene in several ways, such as π–π interaction between antibiotics and graphene's aromatic ring's π electron, hydrogen bonding between antibiotics and GO's carboxyl or hydroxyl ion, hydrophobic interaction between antibiotics and graphene-based adsorbents' hydrophobic groups, and electrostatic interactions between antibiotics and active functional groups on the adsorbent. By utilizing van der Waals or π–π electron donor–acceptor interactions, graphene's polarized surface enables it to interact with organic molecules like antibiotics. According to Ghorbani et al. [76], the adsorbent and adsorbate achieved a rapid equilibrium adsorption time due to the enhanced mass transfer rate caused by the graphene-based materials' shallow thickness.

In a recent work published by Wernke et al. [77], the antibiotic cephalexin was adsorbed in an aqueous solution using GO that was synthesized using a modified version of Hummers' technique. As per the scientists' findings, pH 7 was the most favourable for the cephalexin adsorption procedure, yielding an adsorption capacity of 164.35 mg/g. This translates to a 65.73% antibiotic removal rate utilizing 0.01 g of mass of GO. A modified Hummers' method-prepared GO was employed as both an adsorbent and a radical initiator in a subsequent work by Moreira et al. [78] to explore the adsorption and degradation of the antibiotic norfloxacin in water. According to the authors, the π–π interaction and van der Waals force primarily dominated the norfloxacin adsorption mechanism onto the GO. The oxygen functional groups on the surface of the GO contribute to some of the adsorption process, which was also aided by the interaction of the hydrogen bond donor and acceptor.

Additionally, Carrales-Alvarado et al. [79] investigated the adsorption of graphite, graphene oxide, and reduced graphene oxide-based carbon materials on the antibiotics trimethoprim and metronidazole. The maximum adsorption uptakes for both antibiotics were 190 mg/g for metronidazole and 218 mg/g for trimethoprim at pH 7 and 10, respectively, for the graphene oxide (GO) adsorbents, which the authors said showed outstanding adsorption capabilities. When the pH was lowered from 10 to 7, the ability of materials based on graphene oxide to absorb trimethoprim did not change appreciably. The π–π dispersive interactions were primarily attributed to the adsorption of both antibiotics on the graphite and reduced graphene oxide-based materials. While typical graphene-based adsorbents have a large specific surface area and high adsorption capacity, there are several disadvantages to employing pristine graphene and its derivatives for the removal of organic contaminants from aqueous environments. For example, graphene nanosheets can be stacked more easily with one another through π–π stacking and van der Waals interactions. These interactions impact site exposure, hinder rapid mass movement, and restrict the adsorption efficiency of graphene nanosheets. Similarly, low dispersion and agglomeration in aqueous solution and the quantity of surface functional groups on graphene oxide alone limit its adsorption capability [80, 81].

Additionally, GO's higher hydrophilic and highly dispersible qualities may hinder the removal of organic contaminants from aqueous environments. Several studies are being conducted to enhance GO and rGO's adsorption capabilities.

TABLE 6.3

Graphene and Carbon Nanotube-Based Materials for the Removal of Pharmaceuticals

Carbon-Based Materials	Adsorption/ Catalysis	Pharmaceuticals	Removal	Comments	References
Nitrogen-doped graphene	Catalyst	Sulphachlorpyridazine	99% removal efficiency achieved with 60% mineralization	To improve the mineralization of sulphachlorpyridazine, carbocatalysis facilitated the activation of peroxymonosulphate. The addition of Nitrogen-doped reduced graphene oxide to the system boosted the removal rate of sulphachlorpyridazine, which in turn improved the removal of all organic matter.	→
N, O co-doped carbon nanotubes (Co@N-O-CNTs-2)	Catalyst	Sulphamethoxazole	100% removal efficiency and 35.6% mineralization	On the carbon nanotubes (CNTs), the co-doping of nitrogen and oxygen produced more active sites. peroxymonosulfate (PMS) can be efficiently activated by cobalt embedded oxygen and nitrogen co-doped CNTs-2 to facilitate the decomposition of sulphamethoxazole. Over the decomposition of sulphamethoxazole, radicals and surface-bound reactive species predominated. Both sulphate and hydroxyl radicals had a minor effect on sulphamethoxazole degradation, while singlet oxygen had a minor impact.	[91]
Graphene oxide	Adsorbent	Metronidazole and trimethoprim	The maximum adsorption is 190 and 218 mg/g for metronidazole and trimethoprim at pH 7 and pH 10, respectively.	$\pi-\pi$ dispersive interactions were primarily responsible for the adsorption of both antibiotics. The results of the study showed that the graphitic materials exhibited an activated carbon-like adsorption process.	[78]
Graphene oxide	Adsorbent	Norfloxacin	A removal capacity of 374.9 ± 29.8 mg/g and removal efficiency of 94.8%	Van der Waals force and the $\pi-\pi$ interaction dominated the norfloxacin adsorption process onto Graphene oxide (GO). Since GO had functional groups that carried oxygen, the interaction between the donor and acceptor of hydrogen bonds also played a role in the adsorption.	[77]

These include incorporating additional materials to develop hybrid composites with better performance, modifying the synthesis parameters to modulate the morphology and structure of the graphene-based adsorbent to result in more active adsorption sites, and surface functionalization with particular functional groups.

6.3.2 CARBON-BASED CATALYSTS FOR THE REMOVAL OF PHARMACEUTICALS FROM WATER

Given the many appealing qualities of carbon-based materials that have already been mentioned, they have been used as catalysts to remove pharmaceuticals from industrial effluents. In particular, graphene-based nanomaterials, carbon nanotubes, and quantum dots generated from carbon have been employed extensively in the catalytic decomposition of antibiotics. The extremely reactive radicals produced by advanced oxidation processes (AOPs) are capable of degrading antibiotics into less harmful byproducts. Examples of these radicals are hydroxyl radicals ($HO^•$), sulphate radicals ($SO_4^{•-}$), and superoxide radicals ($O_2^{•-}$). Persulphate-based AOPs are promising among the various AOPs; they can be triggered by carbon materials to generate reactive radicals [72, 82, 83].

The degradation of pharmaceutical pollutants by carbon-activated persulphate AOPs can be categorized into three main pathways: adsorption, free radical, and non-radical (direct oxidation, singlet oxygen, and electron transfer) [84, 85]. These activation processes frequently take place in a reaction concurrently. According to Gao et al. [86], the degree of graphitization, surface functional groups, porous structure, and surface electrochemical impedance of carbon-based materials impact the activation of persulphate. These factors are associated with the raw materials, preparation techniques, and modification methodologies employed. According to the following equations (6.1–6.6), carbon materials' active sites can catalyze the activation of persulphate to generate sulphate radicals [86, 87]. Between pH values of 2 and 7, the primary active species ($SO_4^{•-}$) can combine with -OH to generate $HO^•$, and at between pH values of 7 and 10, it can coexist with $HO^•$. When dissolved oxygen is present, $O_2^{•-}$ may be generated as an intermediate product. During this time, $O_2^{•-}$ can recombine in a variety of ways to yield $^1O_2$. The degradation of pollutants is initiated by all of these reactive oxygen species.

$$S_2O_8^{2-} + Carbon\ surface - OH \rightarrow SO_4^{•-} + HSO_4^{-} + Carbon\ surface - O^• \tag{6.1}$$

$$HSO_5^{-} + Carbon\ surface - \pi \rightarrow SO_4^{•-} + OH^{-} + Carbon\ surface - \pi^{+} \tag{6.2}$$

$$SO_4^{•-} + OH^{-} \rightarrow HO^• + SO_4^{2-} \tag{6.3}$$

$$O_2 + e^{-} \rightarrow O_2^{•-} \tag{6.4}$$

$$O_2^{•-} + HO^• \rightarrow {}^1O_2 + OH^{-} \tag{6.5}$$

$$O_2^{•-},\ HO^•,\ SO_4^{•-} + \text{pollutants} \rightarrow Intermediate\ products \rightarrow CO_2 + H_2O \tag{6.6}$$

It is worth noting that the roles of adsorption, non-radical pathways, and free radical pathways can vary in different catalytic systems utilizing persulphate, particularly when utilizing diverse carbon materials. These materials boast active sites on their surface that can alter degradation pathways and removal rates, as noted by Ushani et al. [87]. Furthermore, the carbon matrix with ketonic groups and sp^2 hybridized orbitals is capable of transferring electrons from antibiotic pollutants to persulphate, which yields the production of $SO_4^{\cdot-}$ and $HO^\cdot$ [86]. The presence of intrinsic and external defects in carbon materials can result in differences in electric density and regional electrons, ultimately functioning as active sites that initiate both radical and non-radical pathways.

Recent studies have indicated that carbon materials exhibiting high defect levels, extensive surface areas, minimal oxygen content, and an abundance of surface functional groups exhibit robust reactivity with persulphate in free radical reactions. As a result, $SO_4^{\cdot-}$ or $OH^\cdot$ radicals are generated, which can be effectively utilized to oxidize organic pollutants [88]. It has been discovered through recent reports by Ma et al. [72] and Akbar et al. [89] that reactive oxygen species are not produced in a non-radical system. Instead, a ternary system consisting of a pollutant (acting as an electron donor), a carbon catalyst (acting as an electron bridge), and persulphate (acting as an electron acceptor) undergo electron abstraction or electron shuttle. The non-radical pathway has demonstrated remarkable efficacy across a wide pH spectrum and showcases superior selectivity towards organic contaminants with electron-rich targets, in comparison to the radical oxidation pathway. As such, the use of non-radical-based oxidation reactions can present a highly efficient method for treating pharmaceutical-laden wastewater of diverse compositions.

Carbon nanotubes: CNTs show great potential as a carbon-based catalyst for removing pharmaceutical pollutants in water and wastewater treatment. CNTs comprised a rolled-up graphene sheet, which is a hexagonal lattice of carbon atoms possessing sp2 hybridization. They have a length of a micron and a diameter of a nanometre, and they exist in three different geometries: chiral, zigzag, and armchair. The optical, electrical, and mechanical properties of CNTs are influenced by their chirality, among other factors [88, 90]. CNTs are classified into two main groups based on the number of graphite layers they contain. SWCNTs are made up of a single hexagonal carbon atom sheet rolled into a tube with a diameter ranging from 0.4 to 3 nm. In contrast, MWCNTs contain multiple hexagonal sheets of carbon atoms that are rolled up with a distance of about 0.34 nm between each sheet. CNTs have some remarkable properties that make them highly effective in the realm of environmental science. These properties include excellent electrical and thermal conductivity, a large surface area, great strength, a high pore capacity, and specific structural and functional characteristics such as a high aspect ratio, electron emission, optical absorption, and low energy loss. CNTs are particularly good at adsorption because of the oxygen-containing functional groups on their surface, as well as their strong interaction with organic molecules through non-covalent forces. Additionally, CNTs are promising materials for supporting heterogeneous catalysis due to their exceptional physical properties, such as excellent electron conductivity, good chemical inertness, and relatively high oxidation stability.

According to a recent study conducted by Pham et al. [88], the efficacy of two representative carbon catalysts, namely granular activated carbon (GAC) and carbon nanotubes (CNTs), was analyzed with regard to their ability to degrade acetaminophen through persulphate (PS) activation. The study revealed that the GAC possessed a superior quantity of surface functional groups and a larger surface area in comparison to CNT. However, the sp2 and C=O groups of the CNTs were found to be more efficient than GAC. Furthermore, both materials exhibited similar structural defect levels and surface charging characteristics. The researchers conducted a study on the degradation of acetaminophen using CNTs, GAC, and PS. Results showed that when used alone, PS or CNTs achieved only low levels of removal, but when combined, complete removal efficiency was achieved. The behaviour of acetaminophen degradation and PS decomposition in each system reflected two distinct reaction mechanisms. The GAC/PS system relied on a radical mechanism, while the CNT/PS system utilized a non-radical mechanism. The GAC/PS system boasts impressive oxidizing power, thanks to its oxidative radicals. By activating PS on its surface, sulphate ($SO_4^{\bullet-}$) or hydroxyl ($OH^{\bullet}$) radicals are generated, allowing for superior pollutant degradation. On the other hand, the CNT/PS system exhibits a remarkable PS decomposition rate in the presence of acetaminophen. This suggests that the CNT/PS system's enhanced oxidative power is due to electron transport processes or non-radical pathways, with CNT serving as an electron transfer bridge, PS as an electron acceptor, and acetaminophen as an electron donor.

A study by Wang and Wang [91] found that Co@N-O-CNTs, created by embedding cobalt into oxygen- and nitrogen-co-doped CNTs, were effective in activating peroxymonosulphate for sulphamethoxazole degradation. The results demonstrated that Co@N-O-CNTs had superior catalytic activity, producing bulk and surface-bound radicals that contributed to the degradation of sulphamethoxazole. The degradation rate reached 0.247 min^{-1}, resulting in 35.6% mineralization.

Carbon-derived quantum dots: C-dots, or carbon dots, are tiny spherical particles with a diameter of under 10 nanometres. They were initially discovered in 2004 while purifying single-walled carbon nanotubes. C-dots are composed of sp^2/sp^3 hybridized carbon atoms, giving them exceptional properties such as the capacity to absorb light across a broad range, easy modifiability, light emission, electron mediation, chemical stability, high solubility in water, photostability, and non-toxicity [89, 92]. These distinct characteristics have resulted in their extensive usage in a variety of fields, including photocatalysis and other technologies. Carbon dots come in an array of structural cores, including partially crystalline, amorphous, and crystalline. Typically, crystalline carbon dots are synthesized through top-down synthetic approaches using graphene and are referred to as "graphene quantum dots" (GQDs). Crystalline graphitic core carbon dots obtained through bottom-up synthetic techniques are also classified as GQDs.

While C-dots are occasionally referred to as quantum dots in the literature, they generally differ from traditional semiconductor quantum dots by displaying distinct electronic quantum states. Unlike typical quantum dots, the absorption behaviour of C-dots is not solely determined by their size, as their bandgap and corresponding absorption/emission wavelengths are not directly proportional to their size. The existence of genuine quantum confinement in C-dots remains a point of debate, with conclusive evidence yet to be presented. There are not many published reports that focus on the use of bare carbon dots for the photodegradation of antibiotics. Most of the research studies are based on C-dots hybrids. Additionally, there are only a few reports that discuss the utilization of C-dots as active light-harvesting materials that generate electron–hole pairs transferred to other catalysts.

According to a recent study by Hetefi et al. [93], maltose-based graphene quantum dots were used as a highly effective photocatalyst for the photodegradation of imipramine in wastewater samples. The best removal efficiency of 80% was achieved under optimal conditions, and the photocatalytic material synthesized showed good stability and reusability, with only a 15% decrease in photoactivity after the fourth recycling run. According to a study conducted by Zhao et al. [94], they used tubular g-C_3N_4 that was modified with carbon quantum dots (CQDs) to enhance the photocatalytic activity, specifically for the removal of carbamazepine. The results of the study showed that they achieved 100% removal of carbamazepine. Incorporating CQDs increased the carbamazepine degradation kinetics of tubular g-C_3N_4 by over five times.

In a study by Sharma et al. [95], hybrid trimetallic La/Cu/Zr C-dots were synthesized and tested for their ability to break down the ampicillin antibiotic using visible light. The production of these nanoparticles led to a decrease in the bandgap of the bare C-dots from 2.55 to 1.56 eV, which boosted light absorption in the visible spectrum and enhanced the photocatalytic properties of the composite. After 4 hours of photodegradation, the La/Cu/Zr–C-dot hybrid was able to eliminate 96% of the ampicillin antibiotic. It is noteworthy that even without exposure to light, adsorption alone (in the dark for 1 hour) removed 88% of the ampicillin antibiotic before 4 hours of photodegradation.

These findings suggest that these materials could be instrumental in decreasing the concentration of antibiotics in wastewater.

6.4 CONCLUSIONS AND OUTLOOK

The production of pharmaceuticals around the world is on the rise, which has led to a growing concern about the presence of pharmaceutical residues and metabolites in various environmental media. The overuse of antibiotics has significantly increased the level of these pollutants in the environment beyond what is considered safe, which is an alarming situation. Thus, it has become essential to find effective ways to remove these emerging pollutants from water and wastewater. Carbon-based materials have emerged as a promising solution for the remediation of antibiotic-contaminated waters. They are affordable, environmentally friendly, have excellent adsorption capacity and catalytic activity. Carbon-based materials are non-metallic and chemically inert, which makes them ideal for the photodegradation of pollutants from aqueous environments.

However, the effectiveness of carbon-based materials in adsorbing antibiotics depends on several factors, including their properties, the physical and chemical features of pharmaceutical pollutants, and the chemistry of the background solution. Carbon-based materials are classified into two types: pristine and engineered. The engineered ones can be further categorized into composites such as carbon–clay, carbon–polymer, carbon–metal oxide, carbon–metal nanoparticles, and more, as well as modified carbon-based materials. Activated carbon is one of the commonly used carbon-based materials for removing antibiotics. Activated carbons are available in various forms and have unique surface properties. The physical and chemical surface attributes of activated carbon, such as surface area, pore size, pore volume, and functional groups, are modified to provide different surface qualities and removal effectiveness. Biochar is another carbon-based material that has been utilized as an adsorbent for water contaminants. It has a large specific surface area, porous structure, surface functional groups, and high mineral content. The adsorption capacity of biochar is influenced by its properties, such as the type of feedstock, pyrolysis temperature, and activation process.

Recently, a new type of carbon-based material known as carbon-based quantum dots possesses unique optical and electronic properties, including the ability to convert low-energy photons into high-energy ones, as well as a bivalent redox character. They can function as both photocatalysts and photosensitizers and can absorb light across a broad spectrum of wavelengths. In addition, studies have shown that these quantum dots are effective at removing pharmaceutical residues from wastewater under varying irradiation conditions. Graphene, a fascinating carbon-based substance, boasts a porous structure and ample surface area, making it an ideal contender for adsorption. Its adsorbents facilitate swift and effective absorption, allowing for the diffusion or surface reaction of antibiotics. Additionally, graphene oxide, a derivative of graphene, has been closely investigated as an effective adsorbent for antibiotics.

Although significant progress has been achieved in using carbon-based materials for eliminating pharmaceutical pollutants through photodegradation, there are still various hurdles to overcome. These include the need for cost-effective and easily scalable synthesis methods, refining the structure and surface properties of carbon-based materials, and understanding the mechanisms that govern the adsorption and photodegradation processes. Conquering these obstacles is vital for driving the field forward and creating effective and environmentally conscious solutions for eliminating pharmaceutical pollutants from water and wastewater.

REFERENCES

[1] Scaria J, Gopinath A, Ranjith N, et al. Carbonaceous materials as effective adsorbents and catalysts for the removal of emerging contaminants from water. J Clean Prod. 2022;350.

[2] Vinayagam V, Murugan S, Kumaresan R, et al. Sustainable adsorbents for the removal of pharmaceuticals from wastewater: A review. Chemosphere. 2022;300.

[3] Ricky R, Shanthakumar S. Phycoremediation integrated approach for the removal of pharmaceuticals and personal care products from wastewater—A review. J Environ Manage. 2022;302.

[4] Oladipo AA, Mustafa FS. Bismuth-based nanostructured photocatalysts for the remediation of antibiotics and organic dyes. Beilstein J Nanotechnol. 2023;14: 291–321.

[5] Oladipo AA, Mustafa FS, Ezugwu ON, et al. Efficient removal of antibiotic in single and binary mixture of nickel by electrocoagulation process: Hydrogen generation and cost analysis. Chemosphere. 2022;300.

[6] Ortúzar M, Esterhuizen M, Olicón-Hernández DR, et al. Pharmaceutical pollution in aquatic environments: A concise review of environmental impacts and bioremediation systems. Front Microbiol. 2022;13.

[7] Krahnstöver T, Santos N, Georges K, et al. Low-carbon technologies to remove organic micropollutants from wastewater: A focus on pharmaceuticals. Sustain. 2022;14.

[8] Waleng NJ, Nomngongo PN. Occurrence of pharmaceuticals in the environmental waters: African and Asian perspectives. Environ Chem Ecotoxicol. 2022;4:50–66.

[9] Maculewicz J, Kowalska D, Świacka K, et al. Transformation products of pharmaceuticals in the environment: Their fate, (eco)toxicity and bioaccumulation potential. Sci Total Environ. 2022;802.

[10] Bavumiragira JP, Ge J, Yin H. Fate and transport of pharmaceuticals in water systems: A processes review. Sci Total Environ. 2022;823.

[11] Darvishi P, Mousavi SA, Mahmoudi A, et al. A comprehensive review on the removal of antibiotics from water and wastewater using carbon nanotubes: Synthesis, performance, and future challenges. Environ Sci Water Res Technol. 2022;9:11–37.

[12] Mukhtar A, Manzoor M, Gul I, et al. Phytotoxicity of different antibiotics to rice and stress alleviation upon application of organic amendments. Chemosphere. 2020;258.

[13] Khan HK, Rehman MYA, Malik RN. Fate and toxicity of pharmaceuticals in water environment: An insight on their occurrence in South Asia. J Environ Manage. 2020;271.

[14] Polianciuc SI, Gurzău AE, Kiss B, et al. Antibiotics in the environment: Causes and consequences. Med Pharm Reports. 2020;93:231–240.

[15] CDC. Antibiotic/Antimicrobial Resistance (AR/AMR); 2019. Available from: www. cdc.gov/drugresistance/biggest-threats.html.

[16] ECDPC. 33000 People Die Every Year Due to Infections with Antibiotic-Resistant Bacteria; 2018. Available from: www.ecdc.europa.eu/en/news-events/33000-people-die-every-year-due-infections-antibiotic-resistant-bacteria.

[17] Kariuki S, Kering K, Wairimu C, et al. Antimicrobial resistance rates and surveillance in sub-saharan Africa: Where are we now? Infect Drug Resist. 2022;15:3589–3609.

[18] Murray CJ, Ikuta KS, Sharara F, et al. Global burden of bacterial antimicrobial resistance in 2019: A systematic analysis. Lancet. 2022;399:629–655.

[19] Ahmed MJ, Hameed BH. Removal of emerging pharmaceutical contaminants by adsorption in a fixed-bed column: A review. Ecotoxicol Environ Saf. 2018;149: 257–266.

[20] Quesada HB, Baptista ATA, Cusioli LF, et al. Surface water pollution by pharmaceuticals and an alternative of removal by low-cost adsorbents: A review. Chemosphere. 2019;222:766–780.

[21] Natarajan R, Saikia K, Ponnusamy SK, et al. Understanding the factors affecting adsorption of pharmaceuticals on different adsorbents—A critical literature update. Chemosphere. 2022;287.

[22] Kang J, Zhou L, Duan X, et al. Catalytic degradation of antibiotics by metal-free catalysis over nitrogen-doped graphene. Catal Today. 2020;357:341–349.

[23] Oladipo AA, Gazi M. Application of hydroxyapatite-based nanoceramics in wastewater treatment: Synthesis, characterization, and optimization. Sol-gel Based Nanoceramic Mater Prep Prop Appl. 2017;231–251.

[24] Oladipo AA, Ifebajo AO, Vaziri R. Green adsorbents for removal of antibiotics, pesticides and endocrine disruptors. In: Crini G, Lichtfouse E, editors. Green Adsorbents Pollut Remov Innov Mater: Springer International Publishing Ag; 2018. p. 327–351.

[25] Rathi BS, Kumar PS, Show PL. A review on effective removal of emerging contaminants from aquatic systems: Current trends and scope for further research. J Hazard Mater. 2021;409.

[26] Oladipo AA, Gazi M. Uptake of Ni2+ and rhodamine B by nano-hydroxyapatite/alginate composite beads: Batch and continuous-flow systems. Toxicol Environ Chem. 2016;98:189–203.

[27] Patel H. Comparison of batch and fixed bed column adsorption: A critical review. Int J Environ Sci Technol. 2022;19:10409–10426.

[28] Xiang Y, Xu Z, Wei Y, et al. Carbon-based materials as adsorbent for antibiotics removal: Mechanisms and influencing factors. J Environ Manage. 2019;237:128–138.

[29] Ashiq A, Vithanage M, Sarkar B, et al. Carbon-based adsorbents for fluoroquinolone removal from water and wastewater: A critical review. Environ Res. 2021;197.

[30] Nabgan W, Jalil AA, Nabgan B, et al. A state of the art overview of carbon-based composites applications for detecting and eliminating pharmaceuticals containing wastewater. Chemosphere. 2022;288.

[31] Mestre AS, Carvalho AP. Photocatalytic degradation of pharmaceuticals carbamazepine, diclofenac, and sulfamethoxazole by semiconductor and carbon materials: A review. Molecules. 2019;24.

[32] Liu X, Yang D, Zhou Y, et al. Electrocatalytic properties of N-doped graphite felt in electro-Fenton process and degradation mechanism of levofloxacin. Chemosphere. 2017;182:306–315.

[33] Ganiyu SO, de Araújo MJG, de Araújo Costa ECT, et al. Design of highly efficient porous carbon foam cathode for electro-Fenton degradation of antimicrobial sulfanilamide. Appl Catal B Environ. 2021;283.

[34] Huong Le TX, Bechelany M, Cretin M. Carbon felt based-electrodes for energy and environmental applications: A review. Carbon N Y. 2017;122:564–591.

[35] Yuan A, Lei H, Xi F, et al. Graphene quantum dots decorated graphitic carbon nitride nanorods for photocatalytic removal of antibiotics. J Colloid Interface Sci. 2019;548:56–65.

[36] Meng F, Wang Y, Chen Z, et al. Synthesis of CQDs@FeOOH nanoneedles with abundant active edges for efficient electro-catalytic degradation of levofloxacin: Degradation mechanism and toxicity assessment. Appl Catal B Environ. 2021;282.

[37] Choudhary V, Vellingiri K, Thayyil MI, et al. Removal of antibiotics from aqueous solutions by electrocatalytic degradation. Environ Sci Nano. 2021;8:1133–1176.

[38] McCreery RL. Advanced carbon electrode materials for molecular electrochemistry. Chem Rev. 2008;108:2646–2687.

[39] Sabzehmeidani MM, Mahnaee S, Ghaedi M, et al. Carbon based materials: A review of adsorbents for inorganic and organic compounds. Mater Adv. 2021;2:598–627.

[40] Shahcheragh SK, Bagheri Mohagheghi MM, Shirpay A. Effect of physical and chemical activation methods on the structure, optical absorbance, band gap and urbach energy of porous activated carbon. SN Appl Sci. 2023;5.

[41] Li C, Yang J, Zhang L, et al. Carbon-based membrane materials and applications in water and wastewater treatment: A review. Environ Chem Lett. 2021;19:1457–1475.

[42] Yazidi A, Atrous M, Edi Soetaredjo F, et al. Adsorption of amoxicillin and tetracycline on activated carbon prepared from durian shell in single and binary systems: Experimental study and modeling analysis. Chem Eng J. 2020;379.

[43] Sousa É, Rocha L, Jaria G, et al. Optimizing microwave-assisted production of waste-based activated carbons for the removal of antibiotics from water. Sci Total Environ. 2021;752.

[44] Wang Z, Wang G, Li W, et al. Loofah activated carbon with hierarchical structures for high-efficiency adsorption of multi-level antibiotic pollutants. Appl Surf Sci. 2021;550.

[45] Berges J, Moles S, Ormad MP, et al. Antibiotics removal from aquatic environments: Adsorption of enrofloxacin, trimethoprim, sulfadiazine, and amoxicillin on vegetal powdered activated carbon. Environ Sci Pollut Res. 2021;28:8442–8452.

[46] Manjunath SV, Singh Baghel R, Kumar M. Antagonistic and synergistic analysis of antibiotic adsorption on Prosopis juliflora activated carbon in multicomponent systems. Chem Eng J. 2020;381.

[47] Rodrigues DLC, Machado FM, Osório AG, et al. Adsorption of amoxicillin onto high surface area—Activated carbons based on olive biomass: Kinetic and equilibrium studies. Environ Sci Pollut Res. 2020;27:41394–41404.

[48] Balarak D, Mahvi AH, Shahbaksh S, et al. Adsorptive removal of azithromycin antibiotic from aqueous solution by azolla filiculoides-based activated porous carbon. Nanomaterials. 2021;11.

[49] Wei M, Marrakchi F, Yuan C, et al. Adsorption modeling, thermodynamics, and DFT simulation of tetracycline onto mesoporous and high-surface-area NaOH-activated macroalgae carbon. J Hazard Mater. 2022;425.

[50] Wang J, Lei S, Liang L. Preparation of porous activated carbon from semi-coke by high temperature activation with KOH for the high-efficiency adsorption of aqueous tetracycline. Appl Surf Sci. 2020;530.

[51] Egbedina AO, Adebowale KO, Olu-Owolabi BI, et al. Microwave synthesized carbon materials as low-cost and efficient adsorbents for the removal of antibiotics in single and binary systems. Arab J Sci Eng. 2022;47:5755–5765.

[52] Jaria G, Lourenço MAO, Silva CP, et al. Effect of the surface functionalization of a waste-derived activated carbon on pharmaceuticals' adsorption from water. J Mol Liq. 2020;299.

[53] Yu J, Kang Y, Yin W, et al. Removal of antibiotics from aqueous solutions by a carbon adsorbent derived from protein-waste-doped biomass. ACS Omega. 2020;5: 19187–19193.

[54] Liu S, Pan M, Feng Z, et al. Ultra-high adsorption of tetracycline antibiotics on garlic skin-derived porous biomass carbon with high surface area. New J Chem. 2020;44:1097–1106.

[55] Shirvanimoghaddam K, Czech B, Abdikheibari S, et al. Microwave synthesis of biochar for environmental applications. J Anal Appl Pyrolysis. 2022;161.

[56] Law XN, Cheah WY, Chew KW, et al. Microalgal-based biochar in wastewater remediation: Its synthesis, characterization and applications. Environ Res. 2022;204.

[57] Layek J, Narzari R, Hazarika S, et al. Prospects of biochar for sustainable agriculture and carbon sequestration: An overview for Eastern Himalayas. Sustain. 2022;14.

[58] Mukherjee A, Patra BR, Podder J, et al. Synthesis of biochar from lignocellulosic biomass for diverse industrial applications and energy harvesting: Effects of pyrolysis conditions on the physicochemical properties of biochar. Front Mater. 2022;9.

[59] Ambaye TG, Vaccari M, van Hullebusch ED, et al. Mechanisms and adsorption capacities of biochar for the removal of organic and inorganic pollutants from industrial wastewater. Int J Environ Sci Technol. 2021;18:3273–3294.

[60] Lu Y, Cai Y, Zhang S, et al. Application of biochar-based photocatalysts for adsorption-(photo)degradation/reduction of environmental contaminants: Mechanism, challenges and perspective. Biochar. 2022;4.

[61] Stylianou M, Christou A, Michael C, et al. Adsorption and removal of seven antibiotic compounds present in water with the use of biochar derived from the pyrolysis of organic waste feedstocks. J Environ Chem Eng. 2021;9:105868.

[62] Nguyen TKT, Nguyen TB, Chen WH, et al. Phosphoric acid-activated biochar derived from sunflower seed husk: Selective antibiotic adsorption behavior and mechanism. Bioresour Technol. 2023;371.

[63] Puga A, Moreira MM, Pazos M, et al. Continuous adsorption studies of pharmaceuticals in multicomponent mixtures by agroforestry biochar. J Environ Chem Eng. 2022;10.

[64] Liu H, Xu G, Li G. Preparation of porous biochar based on pharmaceutical sludge activated by NaOH and its application in the adsorption of tetracycline. J Colloid Interface Sci. 2021;587:271–278.

[65] Wu Q, Zhang Y, Cui M hua, et al. Pyrolyzing pharmaceutical sludge to biochar as an efficient adsorbent for deep removal of fluoroquinolone antibiotics from pharmaceutical wastewater: Performance and mechanism. J Hazard Mater. 2022;426.

[66] Wang Q, Zhang Z, Xu G, et al. Pyrolysis of penicillin fermentation residue and sludge to produce biochar: Antibiotic resistance genes destruction and biochar application in the adsorption of penicillin in water. J Hazard Mater. 2021;413.

[67] Cheng N, Wang B, Chen M, et al. Adsorption and photocatalytic degradation of quinolone antibiotics from wastewater using functionalized biochar. Environ Pollut. 2023;336.

[68] Zhao H, Wang Z, Liang Y, et al. Adsorptive decontamination of antibiotics from livestock wastewater by using alkaline-modified biochar. Environ Res. 2023;226.

[69] Huang J, Zimmerman AR, Chen H, et al. Ball milled biochar effectively removes sulfamethoxazole and sulfapyridine antibiotics from water and wastewater. Environ Pollut. 2020;258.

[70] Geng X, Lv S, Yang J, et al. Carboxyl-functionalized biochar derived from walnut shells with enhanced aqueous adsorption of sulfonamide antibiotics. J Environ Manage. 2021;280.

[71] Wang X, Yin R, Zeng L, et al. A review of graphene-based nanomaterials for removal of antibiotics from aqueous environments. Environ Pollut. 2019;253:100–110.

[72] Minale M, Gu Z, Guadie A, et al. Application of graphene-based materials for removal of tetracyclines using adsorption and photocatalytic-degradation: A review. J Environ Manage. 2020;276.

[73] Ma R, Xue Y, Ma Q, et al. Recent advances in carbon-based materials for adsorptive and photocatalytic antibiotic removal. Nanomaterials. 2022;12.

[74] Hummers Jr WS, Offeman RE. Preparation of graphitic oxide. J Am Chem Soc. 1958;80(6):1339.

[75] Razaq A, Bibi F, Zheng X, et al. Review on graphene-, Graphene oxide-, Reduced graphene oxide-based flexible composites: From fabrication to applications. Materials (Basel). 2022;15.

[76] Ghorbani M, Seyedin O, Aghamohammadhassan M. Adsorptive removal of lead (II) ion from water and wastewater media using carbon-based nanomaterials as unique sorbents: A review. J Environ Manage. 2020;254.

[77] Wernke G, Shimabuku-Biadola QL, dos Santos TRT, et al. Adsorption of cephalexin in aqueous media by graphene oxide: Kinetics, isotherm, and thermodynamics. Environ Sci Pollut Res. 2020;27:4725–4736.

[78] Moreira VR, Lebron YAR, da Silva MM, et al. Graphene oxide in the remediation of norfloxacin from aqueous matrix: Simultaneous adsorption and degradation process. Environ Sci Pollut Res. 2020;27:34513–34528.

[79] Carrales-Alvarado DH, Rodríguez-Ramos I, Leyva-Ramos R, et al. Effect of surface area and physical—Chemical properties of graphite and graphene-based materials on their adsorption capacity towards metronidazole and trimethoprim antibiotics in aqueous solution. Chem Eng J. 2020;402.

[80] Yang W, Cao M. Study on the difference in adsorption performance of graphene oxide and carboxylated graphene oxide for Cu(II), Pb(II) respectively and mechanism analysis. Diam Relat Mater. 2022;129.

[81] Kong Q, Shi X, Ma W, et al. Strategies to improve the adsorption properties of graphene-based adsorbent towards heavy metal ions and their compound pollutants: A review. J Hazard Mater. 2021;415:125690.

[82] Sendão RMS, Esteves da Silva JCG, Pinto da Silva L. Photocatalytic removal of pharmaceutical water pollutants by TiO2—Carbon dots nanocomposites: A review. Chemosphere. 2022;301.

[83] He Y, Zhao Z, Wang T, et al. Carbon-based materials as efficient adsorbents for the removal of antibiotics: The real contributions of carbon edge sites. Chemosphere. 2023;344:140341.

[84] Qi Y, Ge B, Zhang Y, et al. Three-dimensional porous graphene-like biochar derived from enteromorpha as a persulfate activator for sulfamethoxazole degradation: Role of graphitic N and radicals transformation. J Hazard Mater. 2020;399.

[85] Peng Y, Tong W, Xie Y, et al. Yeast biomass-induced Co2P/biochar composite for sulfonamide antibiotics degradation through peroxymonosulfate activation. Environ Pollut. 2021;268.

[86] Gao Y, Wang Q, Ji G, et al. Degradation of antibiotic pollutants by persulfate activated with various carbon materials. Chem Eng J. 2022;429.

[87] Ushani U, Lu X, Wang J, et al. Sulfate radicals-based advanced oxidation technology in various environmental remediation: A state-of-the—Art review. Chem Eng J. 2020;402.

[88] Pham VL, Kim DG, Ko SO. Advanced oxidative degradation of acetaminophen by carbon catalysts: Radical vs non-radical pathways. Environ Res. 2020;188.

[89] Akbar K, Moretti E, Vomiero A. Carbon dots for photocatalytic degradation of aqueous pollutants: Recent advancements. Adv Opt Mater. 2021;9.

[90] Cruz-Cruz A, Rivas-Sanchez A, Gallareta-Olivares G, et al. Carbon-based materials: Adsorptive removal of antibiotics from water. Water Emerg Contam Nanoplastics. 2023;2:2.

[91] Wang S, Wang J. Degradation of sulfamethoxazole using peroxymonosulfate activated by cobalt embedded into N, O co-doped carbon nanotubes. Sep Purif Technol. 2021;277.

[92] Tepliakov NV, Kundelev EV, Khavlyuk PD, et al. Sp2-sp3-hybridized atomic domains determine optical features of carbon dots. ACS Nano. 2019;13:10737–10744.

[93] Hatefi R, Mashinchian-Moradi A, Younesi H, et al. Graphene quantum dots based on maltose as a high yield photocatalyst for efficient photodegradation of imipramine in wastewater samples. J Environ Heal Sci Eng. 2020;18:1531–1540.

[94] Zhao C, Liao Z, Liu W, et al. Carbon quantum dots modified tubular g-C3N4 with enhanced photocatalytic activity for carbamazepine elimination: Mechanisms, degradation pathway and DFT calculation. J Hazard Mater. 2020;381.

[95] Sharma G, Bhogal S, Naushad M, et al. Microwave assisted fabrication of La/Cu/Zr/carbon dots trimetallic nanocomposites with their adsorptional vs photocatalytic efficiency for remediation of persistent organic pollutants. J Photochem Photobiol A Chem. 2017;347:235–243.

[96] Kang Z, Jia X, Zhang Y, et al. A review on application of biochar in the removal of pharmaceutical pollutants through adsorption and persulfate-based AOPs. Sustain. 2022;14.

7 Layered Double Hydroxide (LDH)-Based Materials Applied in Advanced Photocatalytic Oxidation of Pharmaceuticals

Aydin Hassani, Paria Eghbali, and Farshid Ghanbari

7.1 INTRODUCTION

Rapid urbanization and industrialization over recent decades have increasingly contributed to water pollution, thereby giving rise to environmental challenges at the global level. At present, one of the biggest concerns is water pollution coming from pharmaceuticals. As global population growth is leading to upward trends in the production and consumption of such substances [1, 2], bioaccumulation is thus occurring in food chains, to the extent that many low-concentration pharmaceuticals, ranging from ng/L to mg/L, are being detected in natural and domestic sources of water, which seems to be potentially hazardous [3, 4].

Since pharmaceuticals are unsafe even at low concentrations, their persistence, bioaccumulation, and toxicity can threaten aquatic life. Moreover, long-term human exposure to these substances can result in hormonal imbalances and types of cancer [5, 6]. Most pharmaceuticals are refractory, which means they cannot be effectively removed in traditional wastewater treatment plants [2, 7].

An assortment of water remediation methods, such as electrochemical processes, biological techniques, precipitation/coagulation/flocculation, ion exchange, membrane filtration process, reverse osmosis, aerobic/anaerobic microbial removal, and adsorption have been correspondingly developed to remove such substances from aqueous solutions [8]. These methods, however, have been facing some limitations, including low removal efficiency, high cost, being time consuming, excessive sludge generation, and being merely suitable for small-scale remediation but not large-scale applications [8, 9].

Against this background, there is a pressing need to develop eco-friendly and efficient methods for removing pharmaceuticals from aquatic environments. During

DOI: 10.1201/9781003340164-7

the last few decades, chemical and environmental engineers have thus attempted to employ advanced oxidation processes (AOPs) to effectively remove such substances from polluted water [10, 11], since these processes have been acknowledged as very effective and efficient methods for producing highly reactive oxygen compounds like hydroxyl radicals ($^{\bullet}$OH) with E^0 = 1.8–2.7 V and sulfate radicals ($SO_4^{\bullet-}$) with E^0 = 2.5–3.1 V to destroy pharmaceuticals or mineralize them favorably [12–14].

Furthermore, AOPs have their outstanding benefits, including a relatively rapid reactivity rate, plus the ability to treat multiple pollutants simultaneously. Furthermore, they can reduce the toxicity of pollutants and even mineralize them completely [15, 16]. Typical AOPs often consist of oxidizing agents such as hydrogen peroxide (H_2O_2), ozone (O_3), or the Fenton processes in combination with ultraviolet (UV) light irradiation. Despite this, the use of chemicals in the aforementioned processes is likely to impede to their practicality because of the ongoing requirements for a consistent supply of chemicals [17, 18]. To address these drawbacks, photocatalytic processes have recently been utilized as one of the most promising methods for water remediation [19–21]. Photocatalysis processes have several benefits, including high conversion efficiency, environmental friendliness, low cost, and energy saving [13, 22]. A fundamental aspect of photocatalytic efficiency is the understanding of the materials used. To date, a wide variety of semiconductor-based photocatalysts such as TiO_2 [23], ZnO [24], Fe_2O_3 [25], WO_3 [26], CdS [27], Bi_2O_3 [28], and g-C_3N_4 [29] have been employed for the heterogeneous photocatalysis of pharmaceuticals in wastewater. Nevertheless, the photocatalytic performance of the aforementioned photocatalysts is likely to be affected by several factors, such as bandgap width, fast charge recombination of photogenerated electron–hole pairs, and inadequate visible-light absorption [30]. To deal with these serious obstacles, many advanced materials, such as photocatalytic hybrid ones, have been developed in recent years [31, 32].

Recently, two-dimensional (2D) semiconductor photocatalysts seem to be promising materials for wastewater treatment because of their distinctive 2D structure, high abundance in nature, and eco-friendliness [31, 33]. Among them, a family of 2D materials, namely, layered double hydroxides (LDHs), known as hydrotalcite, has been of particular research interest. LDHs also have a variety of chemical, physical, and functional properties that make them applicable in various fields, including biomedicine [34, 35], catalysis, energy, electronics, etc. [36]. Such compounds have been gaining much attention for being exploited as photocatalysts thanks to their excellent physiochemical properties, such as a special layered structure with large surface area, adjustable bandgap, high anion exchange capacity, wide light absorption range, convenient synthesis, low cost, and remarkable recyclability [37–39]. In this chapter, a strong attempt is made to address the significance of LDHs and hybrid LDHs during the removal of pharmaceuticals from aquatic environments. The methods utilized for the preparation of LDHs and the structural properties of LDHs are introduced. The latest achievements in the utilization of LDHs in pharmaceutical wastewater treatment, with much emphasis on photocatalytic and hybrid processes, are then delineated. In addition, the potential future advancements and the associated perspectives are explained.

7.2 LDH CHEMISTRY

7.2.1 FUNDAMENTALS

LDHs are a versatile class of 2D inorganic layered nanomaterials that are known as natural or synthetic anionic hydrotalcite-like clays. They contain positive and negative charges of solvent molecules between their interlayers. One type of LDH is hydrotalcite (with the formula of $Mg_6Al_2(OH)_{16}CO_3.4H_2O$), which is a member of the LDH family and contains high water content, resembling talc [40]. The general formula of LDHs is $[M^{II}_{1-x}M^{III}_x (OH)_2]^{x+}(A^{n-})_{x/n} \cdot yH_2O$, where M^{II} and M^{III} represent divalent (such as Mg^{2+}, Zn^{2+}, or Ni^{2+}) and trivalent metal cations (such as Al^{3+}, Cr^{3+}, Ga^{3+}, V^{3+}, Fe^{3+}, Mn^{3+}, etc.), respectively. The molar ratio of $M^{III}/(M^{II} + M^{III})$ varies between 0.17, 0.33, and 1, depending on the compound [12, 41]. A^{n-} denotes the hydrated interlayer anion and charge density of the LDH layers (e.g., Cl^-, CO_3^{2-}, NO_3^-, SO_4^{2-}), which helps balance the internal positive charges. LDHs can also contain M^+ and M^{4+} cations such as lithium ions (Li^+) and titanium ions (Ti^{4+}), and tetravalent cations such as Zn^{4+} and Sn^{4+} can be incorporated into LDH layers [12]. In LDHs, divalent cations replace trivalent ones and produce a net positive charge on LDH sheets that equilibrates through the anions in the space between the interlayers. The lamellar structure, wide chemical composition, variable charge density, ion exchange properties, water swelling, and rheological properties of LDHs make them of interest in research in various applications, such as catalysts and catalyst supports, photocatalyst materials, chemical adsorbents, and anion exchangers.

7.2.2 STRUCTURE

Comparing the structure of layered double hydroxides (LDHs) with brucite $[Mg(OH)_2]$ is the easiest way to investigate LDHs. Brucite has a hexagonal structure with a close packing of hydroxide ions (OH^-) in alternate octahedral places that are occupied by Mg^{2+} ions. The hydroxide layers in brucite are neutral and stacked on top of each other by the van der Waals gravitational forces, held by a basal gap of approximately 0.48 nm. Substituting some divalent ions in the structure of brucite with trivalent ions leads to the arrangement and mixture of positively charged metal hydroxides in layers ($[M^{II}_{1-x}M^{III}_x (OH)_2]^{x+}$) and the intercalation of anions in the interlayer spaces. This action counterbalances the residual positive charge on the metal hydroxide layers and then simulates the LDH structures. Water molecules in the interlayer spaces further stabilize the crystal structure of LDHs by making strong hydrogen bonds between the metal hydroxide layers and anions. These intercalations in an interlamellar space increase the lamellar distance from 0.48 nm in brucite to approximately 0.77 nm in hydrotalcite [12]. Figure 7.1 shows the general lamellar structure of LDHs. The lamellar structure can contain a wide range of divalent and trivalent metal cations. The type of stacking of the layers in LDHs creates two different polymorphs, namely rhombohedral and hexagonal.

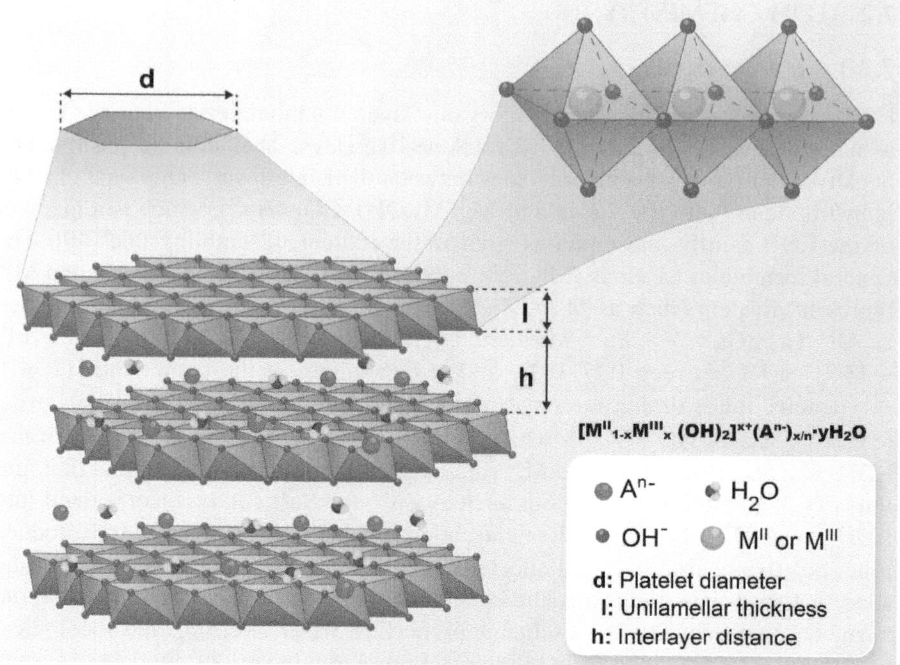

FIGURE 7.1   Schematic representation of the structure of LDHs.

Source: Reproduced with modification from Ref. [42] with permission from ACS.

7.2.3 Synthesis Strategies

LDHs can be synthesized by two methods: top-to-bottom layering and bottom-to-top direct synthesis. The exfoliation of LDHs is an important step in producing positively charged thin platelets and atomically thick multilayers. The process of exfoliation has two main steps. First, macromolecules are used to entangle the layered materials, which increases the distance between the layers. This is followed by an exfoliation step to obtain the exfoliated nanolayers. However, the exfoliation of LDHs is complicated due to the high charge density in their layers, high anion content, and interlaminar hydrogen bonding, which leads to the tight stacking of lamellae [41]. Moreover, the dispersion of nanosheets obtained from their exfoliation is unstable, and they can easily self-assemble to restore the layered structure of LDHs [43]. Nevertheless, efficient approaches have been developed to prepare LDH nanosheets, such as laser ablation, which can complete the exfoliation process in minutes. Various methods can be used to synthesize LDHs, and the selection of the method depends on the cations in the hydroxide layers, the intercalated anions, and the desired physicochemical properties, such as phase purity, porosity, crystallinity, and morphology, as well as the electronic and optical characteristics of the final substances. Direct methods include co-precipitation, the sol–gel process, urea hydrolysis, the molten salt synthesis method, electrochemical synthesis, and in situ film growth. On the other hand, indirect methods include anion exchange, reconstruction by the memory effect, and delamination followed by restacking.

Co-precipitation and hydrothermal methods are commonly used to synthesize LDH-based materials due to their simplicity, low cost, and reliability [12, 44]. Co-precipitation is the most effective method for preparing LDHs, with a pH range of 7–10 depending on the metal ions. In this method, a mixed aqueous solution of divalent and trivalent ion salts is added to water, co-precipitated with dilute solutions of $NaHCO_3$ and/or NaOH at a controlled temperature (60–70 °C), and shaken under inert gas protection [44]. During the hydrothermal technique, urea and metal salts are combined in an autoclave and heated under high pressure and temperature to produce LDHs with large flakes, high crystallinity, and narrow particle size distribution with hexagonal structures. The particle size of LDHs can be controlled by the urea-to-metal salt ratio, reactant concentration, aging time, and reaction temperature [44]. The sol–gel process is a simple, inexpensive preparation technique and an efficient wet-chemical method of obtaining high-purity metal oxides from LDH precursors using condensation and hydrolysis [45]. The sol–gel transition occurs when certain metal precursors, such as metal-based alkoxides or acetyl acetonides, are hydrolyzed into metal hydroxides in an ethanol-water environment with the help of strong acids like hydrochloric acid (HCl) or nitric acid (HNO_3). In this process, the mixture is heated to a high temperature, refluxed, and agitated until a gel is formed. LDHs produced through the sol–gel process are less crystalline but more thermally stable when compared to those created using co-precipitation [12, 45]. In the electrochemical deposition technique, NO_3^- is electrically reduced to OH^- at the working electrode, raising the local pH and resulting in the precipitation of the LDH films. This method is effective in depositing LDH films of any desired thickness, morphology, and density on a metal substrate with perfect adhesion. In self-assembly, the exfoliation of precursors is used in the first step. LDHs can be further laminated into 2D positively charged nanosheets, which can be transiently assembled with various negatively charged agents such as polymers, metal complexes, carbon-based materials, and metal nanoparticles (NPs). This results in the fabrication of nanocarbons with appropriate architectures. It should be noted that strong electrostatic forces between the positive and negative charges of LDH sheets make it difficult for their exfoliation. However, some chemical materials, such as formamide and short-chain alcohols, can be introduced into the interlayers of LDHs to improve the interlayer distance and resolve this problem [12, 46]. LDH synthesis can occur through microwaves (MWs) when reaction aging takes place in an MW environment. MW irradiation provides a faster aging process that typically takes around 15–60 min. In MW synthesis, both co-precipitation and urea hydrolysis can be utilized along with MW aging, which is an alternative to reflux aging [47, 48]. The resulting MW aging process leads to homogeneous particle sizes that are smaller compared to those formed by reflux aging, resulting in a slightly higher surface area. A longer duration of MW irradiation also produces well-crystallized materials with a uniform size in comparison to conventional co-precipitation [47, 49]. The improvement in crystallinity depends on the properties of the trivalent metal ions. In general, the porosity, thermal stability, and surface area of synthetic materials increase as the duration of the MW exposure extends. If the aging time is shortened, the formation of impurities that arise when using long aging techniques is prevented [49]. It is important to note that certain parameters such as pH, temperature, and reaction time strongly influence the crystallinity, morphology, and surface area of LDHs. Therefore, it is essential to choose the correct synthesis conditions according to specific situations [44]. LDHs can also be formed using other techniques, such as mechanochemical methods, ultrasonic-assisted extraction, in situ film growth, and so on. The synthesis of LDH-based catalysts still requires much effort to be facile, eco-friendly, and highly controllable.

7.2.4 PROPERTIES

7.2.4.1 Anion Exchange

This is a commonly used method for preparing LDHs with any desired anions in the interlamellar spaces. In practice, LDH precursor materials or pre-prepared LDHs are mixed in an aqueous suspension with an excess of the salt of the anion to be intercalated [47]. The anion exchange that occurs in the solution can be illustrated as:

$$\left[M^{2+} - M^{3+} - A \right] + B \rightarrow \left[M^{2+} - M^{3+} - B \right] + A \tag{7.1}$$

where A and B represent various anions.

The ion exchange capacity of the anion also depends on the weak electrostatic interactions between the positively charged LDH layers and the exchanging anions interspersed in the precursor materials [40]. The anion charge density is the main parameter that determines ion exchange processes, which influence the substitution trend. The charge density of the anion is directly linked to the electrostatic attraction with LDH layers. The denser the anion charge, the higher the electrostatic attraction with LDH layers and the greater the structural stability of LDHs [12, 37]. The combined affinities of different anions and the positively charged LDH layers are arranged as shown as follows:

$$I^- < NO_3^- < Br^- < Cl^- < F^- < OH^- < MoO_4^{2-} < SO_4^{2-} < CrO_4^{2-} < HPO_4^{2-} < CO_3^{2-}$$

Given this classification, Cl^- and NO_3^- are commonly used in this technique because of their weaker interactions compared to carbonates. This method can be applied to the preparation of LDHs with any inorganic or organic anions. However, when organic ions are present, the interaction between carbon chains is affected, and the analysis cannot be conducted based solely on charge density [47]. If co-precipitation is not possible, the anion exchange method can be used. When divalent or trivalent metallic cations or anions are incorporated in an alkaline solution, ion exchange is also exploited as a production route. This is when a direct reaction between the metal ions and the intercalated anions is more favored than intercalation. During this process, the aforementioned anions are exchanged for those in the interlamellar regions of the precursor LDHs, yielding the desired product. All anion exchange preparation processes must be carried out under an inert atmosphere [47]. Figure 7.2 presents a schematic representation of the anion exchange/capture method of LDHs.

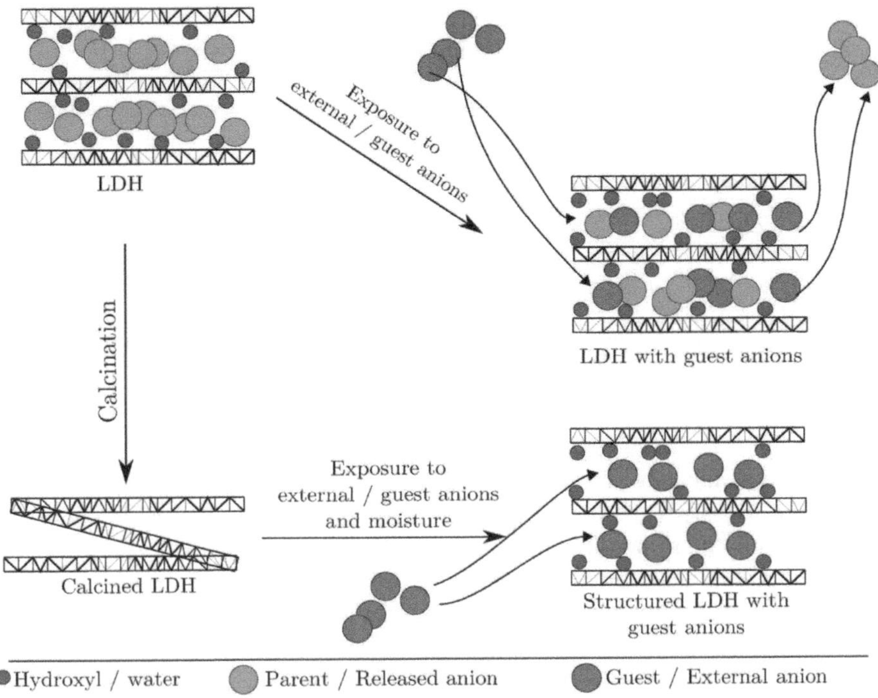

FIGURE 7.2 Schematic representation of the LDH structure and ion exchange/capture mechanism.

Source: Reproduced from Ref. [50].

7.2.4.2 Thermal Stability

The thermal stability of LDHs is an outstanding property that shows a characteristic decomposition that depends on the cations and anions in their structure. Generally, the decomposition process is characterized by three endothermic transitions. The first step represents the loss of water of hydration and adsorbed water molecules, which occurs at around 200 °C [44]. The second transition interval, occurring between 200 °C and 500 °C, results in the loss of some hydroxyl groups and intercalated anions, leading to the formation of an oxide–hydroxide mixture. The third endothermic transition step, which occurs above 500 °C, leads to the decomposition of the remaining hydroxyl groups. This causes the collapse of the lamellar structures, and the amorphous oxides become predominant. At temperatures above 600 °C, LDHs calcine, and their structural breakdown leads to the loss of the reconstitution property of the interlayers [47].

7.2.4.3 Reconstruction/Rehydration Method (Memory Effect)

Miyata (1980) was the first to explain the process of reconstructing the LDH structure through calcined LDH hydration [51]. This process involves two main steps: calcination and rehydration. The first step is to calcine LDHs at a temperature range

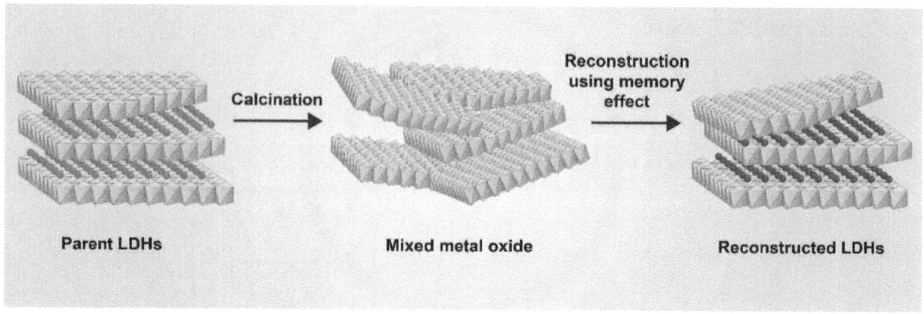

FIGURE 7.3   Simplified representation of the reconstruction process of the memory effect.
Source: Reproduced with modification from Ref. [52] with permission from Springer.

of 450–600 °C to produce well-dispersed mixed metal oxides (MMOs). The second step is to rehydrate the newly produced MMOs by dispersing them in a solution containing the targeted anionic species. This rebuilds the LDHs while integrating the anions into their interlayer regions [12, 52]. Once MMOs are formed, they are dispersed within the solution, which has the anions, resulting in the reformation of LDHs while integrating the anions into the interlayer region. Figure 7.3 provides a representation of the reconstruction process and the memory effect.

7.3   POTENTIAL APPLICATIONS OF LDH-BASED PHOTOCATALYSTS

LDHs, as natural anionic clays, have potent catalytic capabilities and highly organized 2D structures. Thanks to their outstanding features, such as large surface area, strong anion exchange capacity, and adaptable interior space, they have been widely employed as catalysts or catalyst precursors in the photocatalytic process. It is well known that the surface characteristics of the composite and the adsorption capacity can both be improved by modifying LDH components with other materials [53]. Furthermore, LDH materials containing transition metals in their crystalline structures have attracted much attention among scholars due to their greater redox chemistry and structural properties [54–56]. From this perspective, the applications of LDHs and modified LDH-based photocatalysts for pharmaceutical degradation are discussed.

7.3.1   PHOTOCATALYTIC PROCESS

LDHs have structural and electronic properties that make them ideal candidates for developing efficient photocatalysts in the UV–visible (UV–vis) region. They are also efficient photocatalysts because their chemical composition and microstructure can be tuned and then influence their bandgap energies [12, 39]. In recent times, researchers have shown interest in using heterostructures of LDHs for the photo-oxidation of pharmaceutical pollutants. These structures are easy to make and facilitate the transport of photo-excited electrons while preventing electron–hole recombination [12]. Despite these advantages, the photocatalytic efficiency of pristine LDH catalysts is not sufficient due to their short life, fast electron–hole formation, particle aggregation, and lower surface area. In this approach, combining LDHs with semiconductor materials [57, 58], metal-free polymers [59], metals [60, 61], metal oxides, and carbon-based NMs [55, 62] could improve their photocatalytic activities. The formation of exfoliated layers with stronger oxidation capabilities and higher photoactivity can additionally aid in strengthening structural characteristics as well as active sites [12, 63]. The highly hydroxylated layered structure of LDHs is beneficial for photocatalytic pharmaceutical degradation, as it improves the adsorption of pollutants and charge separation through the presence of metal cations, etc. [38].

Metal-modified LDH composites have been further employed to enhance their photocatalytic efficiency. These composites have a surface plasmon resonance (SPR) capability that helps in enhancing their visible-light absorption efficiency and their capacity to act as electron sinks, thereby preventing charge recombination [61, 64]. Shabib et al. [64] observed remarkable photocatalytic efficiency (95.2%) for the degradation of 20 mg/L metoclopramide (MCP) by 0.4 g/L palladium/magnetite $(Pd-Fe_3O_4)$/NiFe LDH composite within 80 min. The incorporation of the Pd particles on LDHs also reduced the recombination rate, and the SPR effect enhanced the UV light absorption and thus enhanced the overall photocatalytic performance. In another study, Kaur et al. [61] conducted a study to improve the photocatalytic performance of wide bandgap Mg–Al LDHs under visible irradiation. They achieved this by depositing various plasmonic metals such as Cu, Ag, and Au, which resulted in metal-loaded composites with high optical response in the visible region, better charge separation, higher surface area, and better degradation of tetracycline (TC) under light-emitting diode (LED) irradiation. It was further reported that the Au-loaded LDH composites demonstrated the highest degradation of TC (99%) within 140 min. A possible mechanism was also proposed for the notable increase in performance caused by the formation of the Schottky barriers and the SPR of plasmonic NPs. Additionally, the total organic carbon (TOC) results showed acceptable mineralization of approximately 84%. Silver-based LDHs have been explored in the field of photocatalysis because of their excellent photosensitive properties. For instance, Fan et al. [65] fabricated 0D/2D Ag_3PO_4/NiAl LDH Z-scheme photocatalysts, which removed 86.9% of tetracycline hydrochloride (TCH) within 60 min. It was believed that the Z-scheme heterojunction can realize the effective spatial separation of the photoinduced electron–hole pairs and consequently enhance the photocatalytic activities. The performance of the LDH-based photocatalysts may be enhanced by the use of transition-metal oxides on LDHs with a greater absorption spectrum ensemble [12]. Ding et al. [66] accordingly reported the preparation of an Ag-decorated Z-scheme CoAl LDH/TiO_2 heterojunction photocatalyst for rapid levofloxacin (LEV) degradation. A 95.2% photocatalytic degradation was thus achieved over 120 min. It was also realized that the internal electric field dominated the Z-scheme charge transfer path between TiO_2 and CoAl LDHs, enabling the effective separation and transfer of the charge carriers while retaining holes and electrons with high redox abilities. In addition, the Ag NPs greatly improved the visible-light uptake capacity of the heterojunction catalyst and generated numerous electrons through the SPR effect. Wang et al. [67] prepared a novel MoO_3/ZnAl LDH composite photocatalyst for the degradation of TC. The photocatalytic degradation performance of MoO_3/ZnAl LDHs (1 g/L) then reached 90.5% removal within 60 min for TC (40 mg/L) at a pH of 9 under visible-light irradiation. The enhanced photocatalysis of composites was due to the migration of the photogenerated charges, which reduced the electron–hole recombination rates. Ran et al. [68] also developed Cu_2O-decorated CuZnAl LDH (Cu_2O/CuZnAl LDHs) heterostructure photocatalysts using facile co-precipitation followed by an in situ reduction approach. They further explored the photocatalytic performance of the as-prepared heterostructure for the removal of tetracycline hydrochloride (TCH). The findings revealed that the visible-light degradation efficiency of TCH catalyzed by Cu_2O/CuZnAl LDHs reached 94.1% within 120 min. The higher TC degradation efficiency observed with the heterostructure was also attributed to the construction of a heterostructure that could inhibit the recombination of relatively useful electrons and holes. Furthermore, the synergistic effects increased the visible-light absorption and the population of adsorption sites, contributing significantly to TC removal. Moreover, Shen et al. [69] synthesized a ternary Ag/ZnO/ZnAl LDH composite photocatalyst via the hydrothermal method, which was employed for bisphenol A (BPA) removal under visible-light irradiation. The degradation rate of the Ag/ZnO/ZnAl LDH was about 4.8-, 2.2-, and 12.0-fold higher than that of ZnO, Ag/ZnO, and ZnAl LDHs, respectively. The improved photocatalytic performance of the ternary composite can be attributed to improvements in optoelectronic properties, such as enhanced sensitivity to visible-light and facilitated charge transfer, leading to a significant increase in ‘OH radical formation. Carbon-based LDHs may enhance fast charge separation due to their distinctive electrical features [70]. In Gholami et al. [70], biochar (BC)-incorporated Zn-Co LDHs (Zn-Co-LDHs@BC) were thus assessed for the degradation of gemifloxacin (GMF) under UV light irradiation. The higher GMF degradation efficiency (92.7%) obtained with Zn-Co-LDHs@BC was accordingly attributed to the increased separation of the photogenerated charge carriers using cobalt hydroxide [$Co(OH)_2$] and BC's ability to prevent the agglomeration of LDH nanostructures. Moreover, the results of the chemical oxygen demand (COD) analysis showed approximately 80% removal efficiency within 200 min, indicating the acceptable mineralization of GMF. Gao et al. [71] also used Ce-doped NiAl LDHs with a reduced graphene oxide (NiAlCe LDHs/rGO) composite photocatalyst to decompose ciprofloxacin (CIP) under visible-light irradiation, wherein the degradation efficiency was 94% after 180 min of irradiation. The presence of Ce and rGO could further accelerate the separation of the photogenerated charges, resulting in an excellent photocatalytic performance for CIP degradation. Figure 7.4 shows the photocatalytic mechanism for the degradation of CIP, catalyzed by NiAlCe LDHs/rGO under visible-light irradiation.

It is also reported that the oxygen vacancies (OVs) on semiconductors modified with LDHs can facilitate light absorption and promote the generation of radicals through the activation of photogenerated carriers. In this regard, Zhang and coworkers [58] used layered lanthanum oxychloride (LaOCl) as an oxyhalide containing OVs to construct a LaOCl/NiAl LDH Z-scheme heterojunction for the photodegradation of norfloxacin (NOR). It was found that the prepared photocatalyst exhibited high efficiency and great stability. Under optimum conditions, 90% of NOR removal was achieved by LaOCl/NiAl LDH, and the reaction rate constant

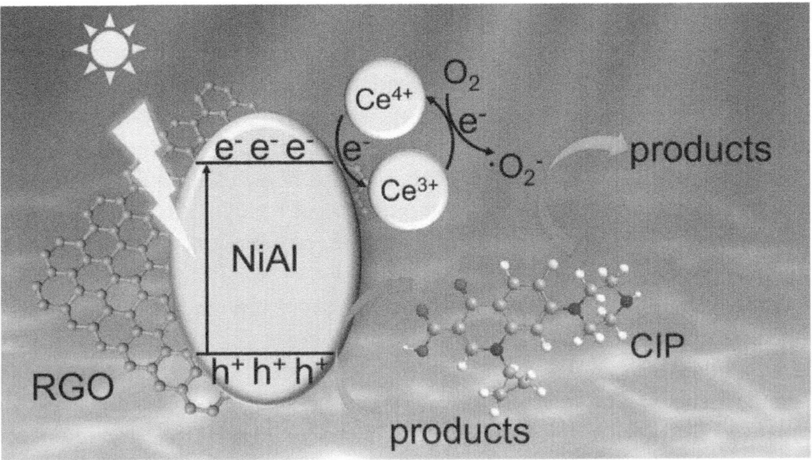

FIGURE 7.4   Proposed photocatalytic mechanism for the degradation of CIP catalyzed by NiAlCe LDHs/rGO under visible-light irradiation.

Source: Reproduced from Ref. [71] with permission from Elsevier.

was twice that of LaOCl and NiAl LDHs, which is attributed to the presence of OVs and the formation of Z-scheme heterojunctions. LDHs can be coupled with metal-free polymer semiconductor materials, such as g-C$_3$N$_4$, to augment photocatalytic performance, which has higher structural stability and light absorption [72]. For instance, Chen et al. [59] synthesized Z-scheme molecularly imprinted photocatalysts with CoZn LDH heterostructures supported by porous g-C$_3$N$_4$ nanorods (MIP-CoZn-LDHs@C$_3$N$_4$). The photodegradation rate of MIP-CoZn-LDHs@C$_3$N$_4$ towards a TC target could further reach 79.8% under visible-light in just 60 min because of the high active cavities and efficient light collection. In another work, Zheng et al. [73] prepared oxygen vacancy-rich g-C$_3$N$_4$/ZnAl LDH composites (g-C$_3$N$_4$/ZnAl-LDH-OVs) with S-scheme heterojunctions for the efficient photocatalytic degradation and mineralization of TC under visible-light irradiation. The g-C$_3$N$_4$/ZnAl-LDH-OVs accordingly achieved outstanding photocatalytic degradation (95%) of TC after 60 min irradiation with visible-light, which was 2.3 and 3.5 times higher than that of pure g-C$_3$N$_4$ and ZnAl-LDH-OVs, respectively. A direct S-scheme mechanism was then proposed to describe the enhanced photocatalytic property of the g-C$_3$N$_4$/ZnAl-LDH-OV heterostructure. Gandamalla et al. [74] also synthesized ZnAl LDH/g-C$_3$N$_4$, tapping a simple MW irradiation method for CIP degradation under visible-light. The ZnAl LDH/g-C$_3$N$_4$ composite accordingly showed the highest photodegradation (84.10%) compared with bare g-C$_3$N$_4$ (64.2%) and ZnAl LDH (61.9%) photocatalysts within 150 min. It was then suggested that the composite material formed between two semiconductor materials with negatively charged g-C$_3$N$_4$ and positively charged metal ions of ZnAl LDHs might have improved the photocatalytic activity. Furthermore, Sahoo et al. [75] fabricated a S,P-co-doped g-C$_3$N$_4$/ZnCr LDH 2D/2D heterostructure by in situ co-precipitation. The 2D/2D alignment of the positively charged ZnCr LDH layers and the negatively charged S,P-co-doped g-C$_3$N$_4$ nanosheets was also established by the strong 2D/2D interface system. It was then established that the prepared S,P-g-C$_3$N$_4$/ZnCr LDH heterostructure had significantly improved photocatalytic performance for CIP degradation (95% in 90 min) under visible-light irradiation, which was mainly attributed to the synergistic interaction between the two oppositely charged species. Moreover, incorporating a non-metal could lead to defects, and the excess electrons in the heteroatoms (S and P) delocalized around the S,P-g-C$_3$N$_4$ matrix could suppress charge carrier recombination, thereby boosting the photocatalytic performance.

Recently, the modification of LDHs with MXene compounds has become a promising technique for photocatalysis. In this vein, Wang et al. [76] developed a 3D marigold-like CoAl LDH/Ti$_3$C$_2$ hybrid photocatalyst by in situ growth of CoAl LDHs on Ti$_3$C$_2$ nanosheets and on-site self-assembly during the hydrothermal process for the degradation of several antibiotics. The optimized CoAl LDH/Ti$_3$C$_2$ exhibits degradation efficiencies of 85, 56, and 70% for tetracycline hydrochloride (TCH), chloramphenicol (CP), and terramycin (TM), corresponding to the 3.1, 3, and 2.1 times that of pure CoAl LDHs. The high removal efficiency of several antibiotics is attributed to the synergistic effect of adsorption and photocatalytic degradation (Figure 7.5).

In another work, Ma et al. [77] designed a 2D/2D OV-embedded NiFe LDH/Ti$_3$C$_2$T$_x$ MXene heterostructured photocatalyst through a hydrothermal method for the elimination of norfloxacin (NOR) from water. They discovered that using hybrid photocatalysts can remove 98% of NOR in 4 h, with the kinetic constant 3.8 times more than that of bare NiFe LDH nanosheets. The OVs in the NiFe LDHs facilitate the separation of photogenerated charge carriers, where the electrons transfer from

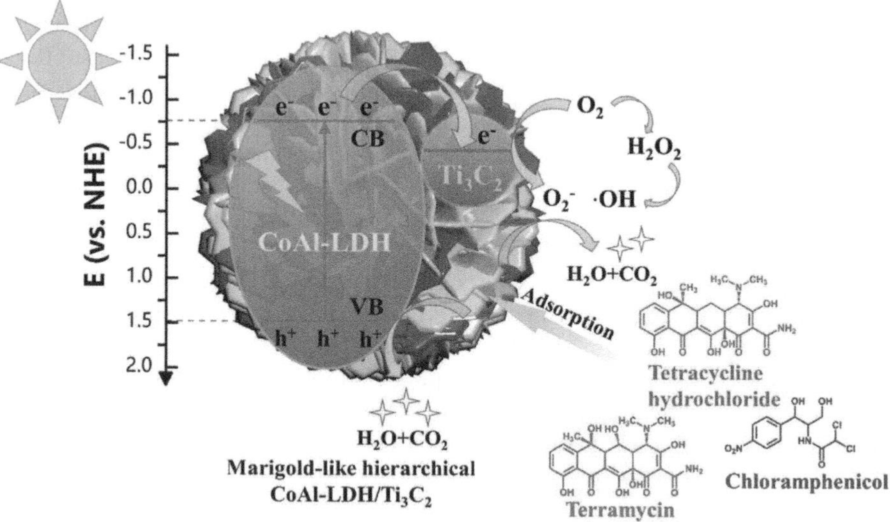

FIGURE 7.5 Schematic diagram of the synergistic effect of adsorption and photocatalytic degradation of CoAl LDH/Ti$_3$C$_2$ to remove the antibiotics.

Source: Reproduced from Ref. [76] with permission from Elsevier.

the NiFe LDH to $Ti_3C_2T_x$ MXene, forming a Schottky barrier at the interfaces of the two components. Grzegórska et al. [78] synthesized a hybrid Zn/Ti LDH photocatalyst coupled with Ti_3C_2, which was used in the photocatalytic degradation of acetaminophen (APAP) and ibuprofen (IBP). The optimized composite exhibited superior activity towards the degradation of APAP (100% within 40 min) and IBP (99.7% within 60 min), which was related to the presence of Ti_3C_2 in the composite structure facilitating electron transfer, thereby inhibiting the electron–hole recombination process. Taken together, LDH-based photocatalysts can be among the first-rate catalysts for purifying pharmaceutical pollutants. Such photocatalysts comparably have enough advantages, so the pharmaceutical wastewater can be decomposed by LDHs, which remarkably helps improve the treatment efficiency of pharmaceutical pollutants.

7.3.2 Sonophotocatalytic Process

Recently, the combinations of several AOPs have been used, mostly to remove harmful chemicals from wastewater. Merging various AOPs can thus elevate degradation efficiency, increase removal rates, and lower treatment costs [79, 80]. Sonication accordingly relies on the cavitation phenomenon, which includes the nucleation, growth, and collapse of solution microbubbles [80–82]. The sonophotocatalytic process, as the combination of sonication and photocatalytic degradation, has thus attracted considerable attention as an efficient integrated technique to decompose and remove pharmaceuticals [83–86]. In this regard, Khataee et al. [85] prepared a NiFe LDH/rGO composite using a hydrothermal method for the sonophotocatalytic degradation of moxifloxacin (MOX). After 60 min, 90.4% degradation was achieved using 1.0 g/L catalyst, 20 mg/L MOX, and 150 W ultrasonic power at the pH of 8 (natural). The nanocomposites also assumed responsibility for the synergistic effect shown in the degradation of MOX during sonophotocatalysis by serving as additional cavitation nuclei and lowering the system threshold energy for a radical generation. Furthermore, the strong interaction between rGO and NiFe/LDH boosted the charge transfer and the pollutant adsorption on the catalyst, thereby increasing the degradation efficiency. For catalytic applications, LDH layers with uniform distributions of metal cations and surface hydroxyl groups provide higher chemical stability. In this line, Abazari et al. [87] prepared g-C_3N_4@NiTi LDH nanocomposites using the hydrothermal method for the sonophotocatalytic removal of amoxicillin (AMX). Like so, 99.5% AMX degradation was achieved within 75 min under optimum conditions. Additionally, no discernible decline was spotted in the sonophotocatalytic activity of the nanocomposites after five consecutive cycles, which confirmed the high durability of the catalyst. Higher performance along with durability in g-C_3N_4–NiTi LDH nanocomposites was thus attributed to improved charge transfer properties, less aggregation, larger surface area, and continual cleaning under ultrasonic irradiation. Furthermore, the efficiency of LDHs could be boosted by various modifications, including the construction of nanocomposites with carbon-based materials. Sadeghi Rad et al. [88] correspondingly synthesized NiCr LDH–rGO nanocomposites via in situ crystallization, followed by hydrothermal treatment. In this respect, an efficiency of 88% was gained within 90 min for the degradation of the rifadin antibiotic solution. An improvement in degradation was then observed due to the lower bandgap of NiCr LDH–rGO (1.58 eV) and the higher surface area (25.5 m^2/g), providing more active sites for the adsorption of rifadin.

7.3.3 Photo-Fenton Process

The Fenton process is currently the most commonly used advanced oxidation process for wastewater treatment [89]. However, it has some limitations such as a restricted pH range (usually acidic), low utilization rate of hydrogen peroxide (H_2O_2), and secondary contamination caused by sludge containing iron-based or other transition metals [90–92]. To address these limitations, the heterogeneous photo-Fenton (PF) process has gained a lot of attention. This process is eco-friendly and highly efficient and involves exposure to natural sunlight or artificial light sources [93, 94]. Modified LDH composites can be used as heterogeneous catalysts for the PF process to overcome the limitations of narrow pH conditions, difficult recovery, and sludge generation. LDH composites have excellent anion exchange capacity and higher surface area [12, 54, 95]. Under light irradiation, LDHs can be excited to generate photogenerated electron–hole pairs. The photogenerated holes can react with water to produce ˙OH radicals, while the photogenerated electrons can accelerate the redox cycles of transition metals to activate H_2O_2 and produce ˙OH radicals. This results in excellent PF efficiency and pollutant degradation [93, 96]. For instance, Costa-Serge et al. [54] synthesized CuMgFe LDHs that exhibited high catalytic performance and stability in degrading the anticancer drug fluorouracil (5FU) in the FP system. They found that borate played a crucial role in electron transfer by reducing the Fe^{3+} and Cu^{2+} species on the surface of LDHs. Fazli et al. [97] developed a magnetic separable $Fe_{2.5}Co_{0.3}Zn_{0.2}O_4$/CuCr LDH composite for the photo-Fenton-like degradation of caffeine (CAF). The photo-Fenton-like degradation process with the presence of the 0.5 g/L nanocomposites and 5 mM H_2O_2 at the pH of 8 demonstrated a 100% degradation of CAF under simulated solar irradiation. The high catalytic performance of nanocomposites was due to the synergistic effect of the photo-Fenton-like process, photoinduced electron transfer on the catalyst surface, and the efficient reduction of Fe^{3+} and Cu^{2+} during the reaction. As well, Zhu et al. [95] developed highly effective and stable Z-scheme NiFe LDH/$Bi_4O_5I_2$ heterogeneous PF catalysts for TC degradation. The NiFe LDH/$Bi_4O_5I_2$/H_2O_2/visible system showed the highest TC degradation rate (> 95%) within 60 min under optimum conditions (TC = 20 mg/L, catalyst = 1 g/L, H_2O_2 = 23.2 mM, and pH = 3). This was attributed to its highly effective Z-scheme charge transport mechanism and the superior decomposition performance of H_2O_2. Radical trapping experiments verified that the ˙OH and $O_2^{˙-}$ radicals significantly contributed to the degradation process. Wu et al. [98] created 2D/2D FeNi LDH/bimetal MOF nanosheets (FeNi LDH/BMNSs) using a semi-sacrificial template. These nanosheets were used as catalysts to degrade antibiotics and were found to be highly effective. The FeNi LDH/BMNSs/H_2O_2/visible system had a TC-HCl removal rate of 95.76% after 60 min. It was discovered that the 2D/2D heterojunctions and the mixed-valence Lewis acidic centers (Fe(III)/Ni(II)) in FeNi LDH/BMNSs were responsible for the fast catalytic process. They accelerated the generation of active species such as ˙OH and $O_2^{˙-}$ radicals. The PF system performed well for the TC-HCl removal (62.50–78.23%) in different composition effluents, including river, lake, and tap water.

Yang et al. [99] identified some issues with FeNi LDHs, such as poor electrical conductivity, high recombination rate of photogenerated carriers, low carrier mobility, and aggregation. To address these limitations, they combined FeNi LDHs with Ti_3C_2 to construct the FeNi LDH/Ti_3C_2 PF system. The 0.2 g/L FeNi LDH/Ti_3C_2 and 20 mM H_2O_2 at the initial pH of 6.08 showed excellent performance (94.7%) in the 20 mg/L TC degradation within 90 min. This was due to the acceleration of the Fe and Ni active site cycles under visible-light, which improved the decomposition efficiency of H_2O_2 and the generation of the reactive radicals. Shao et al. [100] conducted a study on the removal of oxytetracycline (OTC) using bismuth tungstate (Bi_2WO_6)/CoAl LDH nanohybrids with an S-scheme heterojunction. The study found that the prepared Bi_2WO_6/CoAl LDHs significantly improved the photo-Fenton-like catalytic performance of OTC removal through visible-light and H_2O_2 mediation. Under optimal conditions (Catalyst = 100 mg, OTC = 10 mg/L, H_2O_2 = 50 mM, pH = 5), the study achieved 98.47% OTC removal. The high-efficiency photo-Fenton-like catalytic activity of Bi_2WO_6/CoAl LDHs was attributed to the synergistic effect of Fenton-like cobalt ions (Co^{3+}) in CoAl LDHs, an internal electric field, and a heterojunction between Bi_2WO_6 and CoAl LDHs through an S-scheme mechanism. The study also used density functional theory (DFT) calculation and characterization results to support the findings. Furthermore, Kong et al. [101] fabricated a three-dimensional (3D) flower-shaped plasmon Ag/Na-doped defective g-C_3N_4/NiFe LDH (Ag/NaCNN/NiFe LDH) Z-scheme heterojunction using hydrothermal and calcination methods. The study reported that the photocatalyst had a high performance, degrading up to 99% of bisphenol A (BPA) after 135 min in the photothermal PF system. This high performance was attributed to the SPR effect of Ag, Na-doped defects, flower-shaped structure, and the formation of a Z-scheme heterojunction, facilitating spatial charge separation. A schematic diagram of the photocatalytic mechanism of the Ag/NaCNN/NiFe LDH Z-scheme heterojunction under visible-light irradiation was then drawn. The photothermal PF catalytic mechanism of Ag/NaCNN/NiFe LDH for BPA degradation is presented in Figure 7.6.

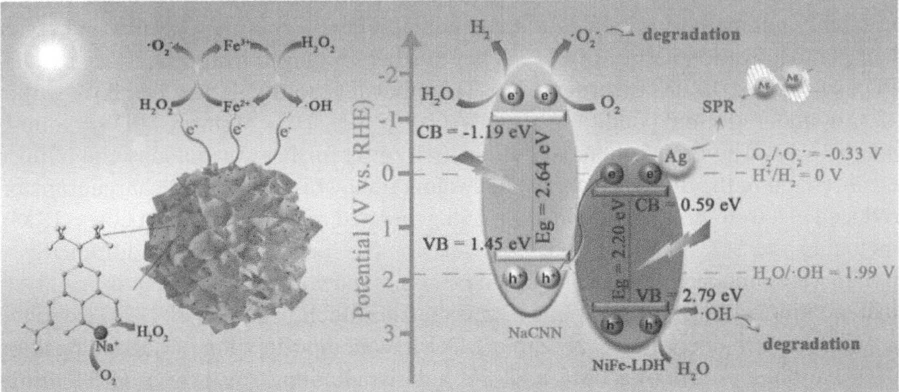

FIGURE 7.6 Schematic diagram of the photocatalytic mechanism of the Ag/NaCNN/NiFe LDH Z-scheme heterojunction under visible-light irradiation.

Source: Reproduced from Ref. [101] with permission from Elsevier.

According to reports, well-designed core–shell nanostructure LDHs can facilitate the rapid transfer of carriers and even improve catalytic activity in the PF system. In this respect, Fang et al. [102] created bifunctional rod-shaped phosphorus-doped CdS@NiFe LDHs (P-CdS@NiFe LDHs) as Z-scheme heterojunctions for BPA degradation. The P-CdS@NiFe LDHs (30 mg) showed high efficiency (98%) in degrading BPA (1 mg/L) with the addition of 30% H_2O_2 within 100 min. It is certain that the synergistic effect of P-doping and Z-scheme heterojunction extended the photoresponse, thereby promoting charge separation.

7.3.4 Photo-Persulfate Process

Over recent decades, photocatalytic technologies have gained popularity due to their low energy consumption and great potential for catalytic performance in energy conversion and environmental remediation [44]. For catalyst-based persulfate (including peroxymonosulfate [PMS] and peroxydisulfate [PDS]) activation, the addition of UV–vis irradiation can thus accelerate the transfer of electrons and boost the generation of radicals in the system [103, 104]. A photocatalytic reaction can further generate reactive species and enhance the charges that activate PMS/PDS to produce the $SO_4^{\bullet-}$ radicals in the solution [44, 105]. In this vein, Zhang et al. [104] prepared a Bi_2O_3/CuNiFe LDH composite using sonication and found that PDS and lomefloxacin (LOM) were adsorbed onto the active catalyst sites after incorporating Bi_2O_3 onto CuNiFe LDHs. Furthermore, the photogenerated electrons were transferred to the LDH layer of Bi_2O_3 and reduced Fe^{3+}, Ni^{3+}, and Cu^{2+}, thus hindering their recombination. The reduced species further activated PDS and produced the $SO_4^{\bullet-}$ radicals, while the holes reacted with water to produce the $^{\bullet}OH$ ones. Under optimal conditions, 84.6% of LOM could be decomposed over 40 min with 0.4 g/L of Bi_2O_3/CuNiFe LDHs and 0.74 mM of PDS under simulated sunlight. The proposed degradation mechanism of LOM by Bi_2O_3/CuNiFe LDHs/PDS/light system is depicted in Figure 7.7.

Li et al. [106] developed an In_2O_3@ZnCr LDH Z-scheme photocatalyst by a hydrothermal method, which exhibited visible-light absorption properties and excellent photoactivation ability of PDS. They further employed the In_2O_3@ZnCr LDH/PDS system for the TC treatment. The TC removal accordingly reached 83% within 45 min under optimal conditions (Catalyst = 0.5 g/L, TC = 50 mg/L, PDS = 8 mM, and pH = 9). The grand photocatalytic activity was further attributed to the built-in electric field in the direct Z-scheme heterojunction, which improved charge transfer, reduced electron–hole recombination, and generated reactive oxygen species (ROS), including the $SO_4^{\bullet-}$, $^{\bullet}OH$, and $O_2^{\bullet-}$ radicals for the removal of TC. Additionally, a polymeric semiconductor g-C_3N_4 could be combined with LDHs to acquire catalysts that are more effective. In this line, Darvishi Cheshmeh Soltani et al. [107] exploited a binary heterojunction g-C_3N_4/ZnFe LDH nanocomposites for the decomposition of TC antibiotics utilizing PMS and UV light irradiation (6 W UVC). The findings indicated that 92.4% of TC was removed within 30 min by the g-C_3N_4/ZnFe LDH/PMS/UV system compared with g-C_3N_4/ZnFe LDHs/PMS (57.2%) and g-C_3N_4/ZnFe LDHs/UV (68.4%). They further claimed that the binary heterojunction nanocomposites not only played the role of a heterogeneous activator for PMS to form the $SO_4^{\bullet-}$

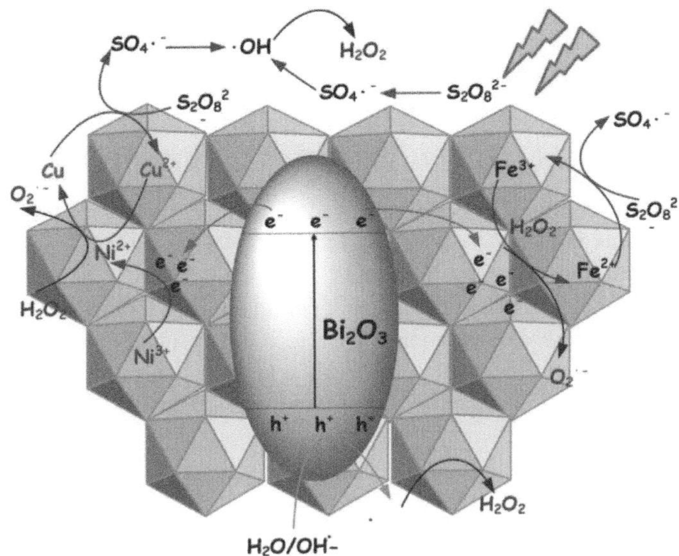

FIGURE 7.7 Proposed mechanism of the Bi$_2$O$_3$/CuNiFe LDHs/PDS/light system.

Source: Reproduced from Ref. [104] with permission from Elsevier.

and ·OH species but also served as a heterogeneous photocatalyst for UV light irradiation. Zeng et al. [108] similarly developed a multi-layered CoAl LDH/g-C$_3$N$_4$ heterostructure via a facile self-assembly process and then assessed the heterogeneous catalytic degradation of sulfadiazine (SDZ) using CoAl LDH/g-C$_3$N$_4$ combined with PMS and visible-light irradiation. Within a 15 min reaction time, 87.1% SDZ (10 µM) was eliminated under the condition of 0.1 g/L CoAl LDH/CN, 0.5 mM PMS, and 6 pH. They also found that the CoAl LDH activated PMS to produce the SO$_4^{•-}$ and ·OH radicals, while g-C$_3$N$_4$ acted as an electron acceptor and a PMS activator. Likewise, Liu et al. [109] synthesized a 2D/2D CoAl LDHs/BiOBr Z-scheme photocatalyst by a simple hydrothermal method for PMS activation. Within 30 min of illumination, the CoAl LDH/BiOBr/PMS/visible system exhibited the highest photocatalytic degradation of the CIP (96%). The presence of PMS could be further activated by Co ions and photogenerated electrons to generate the SO$_4^{•-}$ and ·OH radicals and then facilitate the separation and conversion of charge carriers, thereby enhancing the photocatalytic degradation performance. As reported, LDH calcination (400–600 °C) is an excellent technique for synthesizing uniformly dispersed mixed metal oxides (namely, layered double oxides [LDOs]) with superior catalytic efficiency over conventional methods [7, 110, 111]. From this perspective, LDOs are more chemically stable and have a larger specific area, a lower metal leaching rate, more Lewis base sites, and a unique memory effect in comparison with LDHs. The adsorption capacity and catalytic performance of LDOs is even higher than those of LDHs [111, 112]. In this way, Azolak et al. [111] fabricated a hybrid reusable MnFe–LDO–biochar (BC) catalyst through the co-precipitation–calcination technique for the decomposition of metronidazole (MNZ) under UV light irradiation and found that 92% MNZ

(20 mg/L) was removed within 60 min using 50 mg of MnFe–LDO–BC and 6 mM PDS. Furthermore, MnFe–LDO–BC maintained stable catalytic recycling with an MNZ degradation efficiency of ~88% after three consecutive recycling tests.

7.4 CONCLUSION, FUTURE PROSPECTS, AND CHALLENGES

One of the most pressing environmental concerns today is the presence of pharmaceutical pollutants in our water sources. To address this issue, many researchers are exploring ways to remove these pollutants, particularly from aquatic environments. Water treatment strategies have been developed to purify contaminated water before it is released into these environments. Therefore, it is crucial to develop a comprehensive, profitable, and internationally feasible approach to remove pollutants from water and wastewater. This chapter has revealed that LDHs and modified LDHs have thus far shown more significant potential for the removal of pharmaceuticals from aquatic solutions owing to their stable physicochemical properties, eco-friendliness, and large surface areas, as well as low preparation cost. This chapter first provides an overview of the latest developments in the synthesis, properties, and applications of LDHs and modified LDHs in photo-based water purification processes, including photocatalytic, sonophotocatalytic, photo-Fenton/Fenton-like, and photo-persulfate methods. There is no doubt that LDH-based photocatalysts are indeed one of the most promising candidates in the field of environmental remediation. Although research into LDH-based photocatalysts has made significant developments, the perceptions and evaluations of their reaction mechanisms and large-scale practical applications are still superficial. To fully understand and use LDHs, further studies are needed to assess their structural properties, reaction mechanisms, and potential applications. Moreover, computational chemistry is right for researchers to confirm the structure and composition of LDHs and design catalysts with higher efficiency [113]. Due to the inherent limitations of pristine LDHs, their combination with other innovative materials such as MXenes, metal–organic frameworks (MOFs), and covalent organic frameworks (COFs) may become the dominant trend in the future. Even though heterogeneous catalysts are easier to separate and recover than homogeneous ones, they still pose some challenges. These LDHs may be further incorporated into materials with magnetic properties or immobilized on different substrates, including glass, ceramics, woven cotton fabric, and others [114, 115]. Therefore, no separation step is required. The degradation of various pharmaceuticals by LDHs and their use as modified catalysts in sonophotocatalytic, photo-Fenton, and photo-persulfate processes has not been widely investigated. Thus, much focus should be laid on these processes in further research as well. The cost-effective and long-term durability monitoring of innovative LDH-based catalysts under practical applications are other crucial aspects that need to be taken into account in future research.

REFERENCES

[1] Krishnan RY, Manikandan S, Subbaiya R, et al. Removal of emerging micropollutants originating from Pharmaceuticals and Personal Care Products (PPCPs) in water and wastewater by advanced oxidation processes: A review. Environ Technol Innov. 2021;23:101757.

[2] Scaria J, Anupama KV, Nidheesh PV. Tetracyclines in the environment: An overview on the occurrence, fate, toxicity, detection, removal methods, and sludge management. Sci Total Environ. 2021;771:145291.

[3] Zhou Y, Chen C, Guo K, et al., Kinetics and pathways of the degradation of PPCPs by carbonate radicals in advanced oxidation processes. Water Res. 2020;185:116231.

[4] Li Y, Zhang L, Ding J, et al. Prioritization of pharmaceuticals in water environment in China based on environmental criteria and risk analysis of top-priority pharmaceuticals. J Environ Manage. 2020;253:109732.

[5] Wang X, Yu S, Li Z-H, et al. Fabrication Z-scheme heterojunction of $Ag_2O/ZnWO_4$ with enhanced sonocatalytic performances for meloxicam decomposition: Increasing adsorption and generation of reactive species. Chem Eng J. 2021;405:126922.

[6] Ghanbari F, Giannakis S, Lin K-YA, et al. Acetaminophen degradation by a synergistic peracetic acid/UVC-LED/Fe(II) advanced oxidation process: Kinetic assessment, process feasibility and mechanistic considerations. Chemosphere. 2021;263:128119.

[7] Hassani A, Scaria J, Ghanbari F, et al. Sulfate radicals-based advanced oxidation processes for the degradation of pharmaceuticals and personal care products: A review on relevant activation mechanisms, performance, and perspectives. Environ Res. 2023;217:114789.

[8] Khavari Kashani MR, Kiani R, Hassani A, et al. Electro-peroxone application for ciprofloxacin degradation in aqueous solution using sacrificial iron anode: A new hybrid process. Sep Purif Technol. 2022;292:121026.

[9] Ratnawati R, Enjarlis E, Husnil YA, et al. Degradation of phenol in pharmaceutical wastewater using TiO_2/pumice and O_3/active carbon. Bull Chem React Eng. 2020;15:146–154.

[10] Hassani A, Krishnan S, Scaria J, et al. Z-scheme photocatalysts for visible-light-driven pollutants degradation: A review on recent advancements. Curr Opin Solid State Mater Sci. 2021;25:100941.

[11] Kumar M, Ambika S, Hassani A, et al. Waste to catalyst: Role of agricultural waste in water and wastewater treatment. Sci Total Environ. 2023;858:159762.

[12] Karim AV, Hassani A, Eghbali P, et al. Nanostructured modified Layered Double Hydroxides (LDHs)-based catalysts: A review on synthesis, characterization, and applications in water remediation by advanced oxidation processes. Curr Opin Solid State Mater Sci. 2022;26:100965.

[13] Hassani A, Eghbali P, Mahdipour F, et al. Insights into the synergistic role of photocatalytic activation of peroxymonosulfate by UVA-LED irradiation over $CoFe_2O_4$-rGO nanocomposite towards effective bisphenol A degradation: Performance, mineralization, and activation mechanism. Chem Eng J. 2023;453:139556.

[14] Hassani A, Eghbali P, Kakavandi B, et al. Acetaminophen removal from aqueous solutions through peroxymonosulfate activation by $CoFe_2O_4$/mpg-C_3N_4 nanocomposite: Insight into the performance and degradation kinetics. Environ Technol Innov. 2020;20:101127.

[15] He L-l, Li X-Y, Bai J-Y, et al. A novel $ZnWO_4/MgWO_4$ n-n heterojunction with enhanced sonocatalytic performance for the removal of methylene blue: Characterizations and sonocatalytic mechanism. Surf Interfaces. 2022;31:101980.

[16] Giannakis S, Lin K-YA, Ghanbari F. A review of the recent advances on the treatment of industrial wastewaters by Sulfate Radical-based Advanced Oxidation Processes (SR-AOPs). Chem Eng J. 2021;406:127083.

[17] Yaghoot-Nezhad A, Wacławek S, Madihi-Bidgoli S, et al. Heterogeneous photocatalytic activation of electrogenerated chlorine for the production of reactive oxygen and chlorine species: A new approach for bisphenol A degradation in saline wastewater. J Hazard Mater. 2023;445:130626.

[18] Sun P, Meng T, Wang Z, et al. Degradation of organic micropollutants in UV/NH$_2$Cl advanced oxidation process. Environ Sci Technol. 2019;53:9024–9033.

[19] Pourshirband N, Nezamzadeh-Ejhieh A, A Z-scheme AgI/BiOI binary nanophotocatalyst for the eriochrome black T photodegradation: A scavenging agents study. Mater Res Bull. 2022;148:111689.

[20] Olusegun SJ, Souza TGF, Souza GDO, et al. Iron-based materials for the adsorption and photocatalytic degradation of pharmaceutical drugs: A comprehensive review of the mechanism pathway. J Water Process Eng. 2023;51:103457.

[21] Yu J, Kiwi J, Wang T, et al. Evidence for a dual mechanism in the TiO$_2$/Cu$_x$O photocatalyst during the degradation of sulfamethazine under solar or visible light: Critical issues. J Photochem Photobiol A: Chem. 2019;375:270–279.

[22] Pourshirband N, Nezamzadeh-Ejhieh A. The boosted activity of AgI/BiOI nanocatalyst: A RSM study towards eriochrome black T photodegradation. Environ Sci Pollut Res. 2022;29:45276–45291.

[23] Guo Q, Zhou C, Ma Z, et al. Fundamentals of TiO$_2$ photocatalysis: Concepts, mechanisms, and challenges. Adv Mater. 2019;31:1901997.

[24] Mirzaei A, Chen Z, Haghighat F, et al. Removal of pharmaceuticals and endocrine disrupting compounds from water by zinc oxide-based photocatalytic degradation: A review. Sustain Cities Soc. 2016;27:407–418.

[25] Rasouli K, Alamdari A, Sabbaghi S. Ultrasonic-assisted synthesis of α-Fe$_2$O$_3$@TiO$_2$ photocatalyst: Optimization of effective factors in the fabrication of photocatalyst and removal of non-biodegradable cefixime via response surface methodology-central composite design. Sep Purif Technol. 2023;307:122799.

[26] Marzouqi FA, Al-Balushi NA, Kuvarega AT, et al. Thermal and hydrothermal synthesis of WO$_3$ nanostructure and its optical and photocatalytic properties for the degradation of cephalexin and nizatidine in aqueous solution. Mater Sci Eng B. 2021;264:114991.

[27] Das S, Ahn Y-H. Synthesis and application of CdS nanorods for LED-based photocatalytic degradation of tetracycline antibiotic. Chemosphere. 2022;291:132870.

[28] Shang Q, Chi W, Zhang P, et al. Optimization of Bi$_2$O$_3$/TS-1 preparation and photocatalytic reaction conditions for low concentration erythromycin wastewater treatment based on artificial neural network. Process Saf Environ Prot. 2022;157:297–305.

[29] Nguyen HT, Doan HV, Nguyen TT-B, et al. Nanoarchitectonics of Ag-modified g-C$_3$N$_4$@halloysite nanotubes by a green method for enhanced photocatalytic efficiency. Adv Powder Technol. 2022;33:103862.

[30] Orimolade BO, Idris AO, Feleni U, et al. Recent advances in degradation of pharmaceuticals using Bi$_2$WO$_6$ mediated photocatalysis—A comprehensive review. Environ Pollut. 2021;289:117891.

[31] Garg A, Basu S, Shetti NP, et al. 2D materials and its heterostructured photocatalysts: Synthesis, properties, functionalization and applications in environmental remediation. J Environ Chem Eng. 2021;9:106408.

[32] Bhattacharjee B, Ahmaruzzaman M, Djellabi R, et al. Advances in 2D MXenes-based materials for water purification and disinfection: Synthesis approaches and photocatalytic mechanistic pathways. J Environ Manage. 2022;324:116387.

[33] Javaid A, Latif S, Imran M, et al. MXene-based hybrid composites as photocatalyst for the mitigation of pharmaceuticals. Chemosphere. 2022;291:133062.

[34] Kuthati Y, Kankala RK, Lee C-H. Layered double hydroxide nanoparticles for biomedical applications: Current status and recent prospects. Appl Clay Sci. 2015;112–113:100–116.

[35] Kankala RK. Nanoarchitectured two-dimensional layered double hydroxides-based nanocomposites for biomedical applications. Adv Drug Del Rev. 2022;186:114270.

[36] Chen B-Q, Xia H-Y, Mende LK, et al. Trends in layered double hydroxides-based advanced nanocomposites: Recent progress and latest advancements. Adv Mater Interfaces. 2022;9:2200373.

[37] Mohapatra L, Parida K. A review on the recent progress, challenges and perspective of layered double hydroxides as promising photocatalysts. J Mater Chem A. 2016;4:10744–10766.

[38] Zhang G, Zhang X, Meng Y, et al. Layered double hydroxides-based photocatalysts and visible-light driven photodegradation of organic pollutants: A review. Chem Eng J. 2020;392:123684.

[39] Nava-Andrade K, Carbajal-Arízaga GG, Obregón S, et al. Layered double hydroxides and related hybrid materials for removal of pharmaceutical pollutants from water. J Environ Manage. 2021;288:112399.

[40] Mishra G, Dash B, Pandey S. Layered double hydroxides: A brief review from fundamentals to application as evolving biomaterials. Appl Clay Sci. 2018;153:172–186.

[41] Zhang Y, Xu H, Lu S. Preparation and application of layered double hydroxide nanosheets. RSC Adv. 2021;11:24254–24281.

[42] Takács D, Varga G, Csapó E, et al. Delamination of layered double hydroxide in oonic liquids under ambient conditions. J Phys Chem Lett. 2022;13:11850–11856.

[43] Wang Q, O'Hare D. Recent advances in the synthesis and application of Layered Double Hydroxide (LDH) nanosheets. Chem Rev. 2012;112:4124–4155.

[44] Ge L, Shao B, Liang Q, et al. Layered double hydroxide based materials applied in persulfate based advanced oxidation processes: Property, mechanism, application and perspectives. J Hazard Mater. 2022;424:127612.

[45] Danks AE, Hall SR, Schnepp Z. The evolution of 'sol—Gel' chemistry as a technique for materials synthesis. Mater Horiz. 2016;3:91–112.

[46] Hibino T. Delamination of layered double hydroxides containing amino acids. Chem Mater. 2004;16:5482–5488.

[47] Silva Neto LD, Silva AF, Freire JT, et al. Fundamentals of layered double hydroxides and environmental applications. In: Giannakoudakis D, Meili L, Anastopoulos I, editors. Advanced Materials for Sustainable Environmental Remediation, Elsevier; 2022, p. 301–323.

[48] Rahman A, Pullabhotla RVS. Layered double hydroxide catalysts preparation, characterization and applications for process development: An environmentally green approach. Bull Chem React Eng. 2022;17:163–193.

[49] Banerjee S, Gautam PK. Chapter 8—Layered double hydroxides nanomaterials for water remediation. In: Gautam RK, Chattopadhyaya MC, editors. Nanomaterials for Wastewater Remediation, Butterworth-Heinemann; 2016, p. 161–188.

[50] Mir ZM, Bastos A, Höche D, et al. Recent advances on the application of layered double hydroxides in concrete—A Review. Materials. 2020;13:1426.

[51] Miyata S. Physico-chemical properties of synthetic hydrotalcites in relation to composition. Clays Clay Miner. 1980;28:50–56.

[52] Tichit D, Coq B. Catalysis by hydrotalcites and related materials. Cattech. 2003;7:206–217.

[53] Mohamed F, Abukhadra MR, Shaban M. Removal of safranin dye from water using polypyrrole nanofiber/Zn-Fe layered double hydroxide nanocomposite (Ppy NF/Zn-Fe LDH) of enhanced adsorption and photocatalytic properties. Sci Total Environ. 2018;640–641:352–363.

[54] Costa-Serge NDM, Lima Gonçalves RG, Ramirez-Ubillus MA, et al. Effect of the interlamellar anion on CuMgFe-LDH in solar photo-fenton and fenton-like degradation of the anticancer drug 5-fluorouracil. Appl Catal B: Environ. 2022;315:121537.

[55] Yekan Motlagh P, Khataee A, Sadeghi Rad T, et al. Fabrication of ZnFe-layered double hydroxides with graphene oxide for efficient visible light photocatalytic performance. J Taiwan Inst Chem Eng. 2019;101:186–203.

[56] Liang J, Wei Y, Yao Y, et al. Constructing high-efficiency photocatalyst for degrading ciprofloxacin: Three-dimensional visible light driven graphene based NiAlFe LDH. J Colloid Interface Sci. 2019;540:237–246.

[57] Guo J, Sun H, Yuan X, et al. Photocatalytic degradation of persistent organic pollutants by Co-Cl bond reinforced CoAl-LDH/$Bi_{12}O_{17}Cl_2$ photocatalyst: Mechanism and application prospect evaluation. Water Res. 2022;219:118558.

[58] Zhang W, Meng Y, Liu Y, et al. Boosted photocatalytic degradation of norfloxacin on LaOCl/LDH: Synergistic effect of Z-scheme heterojunction and O vacancies. J Environ Chem Eng. 2022;10:107812.

[59] Chen X, Xu X, Jia X, et al. Surface and interface engineering of Z-scheme 1D/2D imprinted CoZn-LDH/C_3N_4 nanorods for boosting selective visible-light photocatalytic activity. Adv Powder Technol. 2022;33:103531.

[60] Shu J, Wang K, Sharma VK, et al. Efficient micropollutants degradation by ferrate(VI)-Ti/Zn LDH composite under visible light: Activation of ferrate(VI) and self-formation of Fe(III)-LDH heterojunction. Chem Eng J. 2023;456:141127.

[61] Kaur H, Singh S, Pal B. Effect of plasmonic metal (Cu, Ag, and Au) loading over the physicochemical and photocatalytic properties of Mg-Al LDH towards degradation of tetracycline under LED light. Appl Surf Sci. 2023;609:155455.

[62] Li M, Li P, Zhang L, et al. Facile fabrication of ZnO decorated ZnFe-layered double hydroxides @biochar nanocomposites for synergistic photodegradation of tetracycline under visible light. Chem Eng J. 2022;434:134772.

[63] Abazari R, Morsali A, Dubal DP. An advanced composite with ultrafast photocatalytic performance for the degradation of antibiotics by natural sunlight without oxidizing the source over TMU-5@Ni—Ti LDH: Mechanistic insight and toxicity assessment. Inorg Chem Front. 2020;7:2287–2304.

[64] Shabib F, Fazaeli R, Aliyan H, et al. Hierarchical mesoporous plasmonic Pd-Fe_3O_4/NiFe-LDH composites: Characterization, and kinetic study of a photodegradation catalyst for aqueous metoclopramide. Environ Technol Innov. 2022;27:102515.

[65] Fan T, Yang Y, Li P, et al. 0D/2D Ag_3PO_4/Nickel-Aluminum layered double hydroxide Z-scheme photocatalyst for efficient antibiotic degradation. Colloids Surf Physicochem Eng Aspects. 2021;628:127251.

[66] Ding C, Guo J, Gan W, et al. Ag nanoparticles decorated Z-scheme CoAl-LDH/TiO_2 heterojunction photocatalyst for expeditious levofloxacin degradation and Cr(VI) reduction. Sep Purif Technol. 2022;297:121480.

[67] Wang J, Lei X, Huang C, et al. Fabrication of a novel MoO_3/Zn—Al LDHs composite photocatalyst for efficient degradation of tetracycline under visible light irradiation. J Phys Chem Solids. 2021;148:109698.

[68] Ran X, Sun T, Zhou R, et al. In situ formation of Cu_2O decorated CuZnAl-layered double hydroxide heterostructured photocatalysts for enhancing the degradation of tetracycline under visible light. New J Chem. 2023;47:2914–2923.

[69] Shen J, Shi A, Wu M, et al. Efficient degradation of bisphenol A over facilely optimized ternary Ag/ZnO/ZnAl–LDH composite with enhanced photocatalytic performance under visible light irradiation. Solid State Sci. 2022;132:106992.

[70] Gholami P, Khataee A, Soltani RDC, et al. Photocatalytic degradation of gemifloxacin antibiotic using Zn-Co-LDH@biochar nanocomposite. J Hazard Mater. 2020; 382:121070.

[71] Gao Z, Liang J, Yao J, et al. Synthesis of Ce-doped NiAl LDH/RGO composite as an efficient photocatalyst for photocatalytic degradation of ciprofloxacin. J Environ Chem Eng. 2021;9:105405.

[72] Li M, Chen M, Lee SLJ, et al. Facile fabrication of a 2D/2D CoFe-LDH/g-C$_3$N$_4$ nanocomposite with enhanced photocatalytic tetracycline degradation. Environ Sci Pollut Res. 2023;30:4709–4720.

[73] Zheng J, Fan C, Li X, et al. Effective mineralization and detoxification of tetracycline hydrochloride enabled by oxygen vacancies in g-C$_3$N$_4$/LDH composites. Sep Purif Technol. 2023;305:122554.

[74] Gandamalla A, Manchala S, Verma A, et al. Microwave-assisted synthesis of ZnAl-LDH/g-C$_3$N$_4$ composite for degradation of antibiotic ciprofloxacin under visible-light illumination. Chemosphere. 2021;283:131182.

[75] Sahoo DP, Das KK, Patnaik S, et al. Double charge carrier mechanism through 2D/2D interface-assisted ultrafast water reduction and antibiotic degradation over architectural S, P co-doped g-C$_3$N$_4$/ZnCr LDH photocatalyst. Inorg Chem Front. 2020;7:3695–3717.

[76] Wang T, Yang Y, Deng Q, et al. In-situ construction of 3D marigold-like CoAl-LDH/Ti$_3$C$_2$ heterosystem collaborating with 2D/2D interface for efficient photodegradation of multiple antibiotics. Appl Surf Sci. 2021;569:151084.

[77] Ma Y, Xu D, Chen W, et al. Oxygen-vacancy-embedded 2D/2D NiFe-LDH/MXene Schottky heterojunction for boosted photodegradation of norfloxacin. Appl Surf Sci. 2022;572:151432.

[78] Grzegórska A, Wysocka I, Głuchowski P, et al. Novel composite of Zn/Ti-layered double hydroxide coupled with MXene for the efficient photocatalytic degradation of pharmaceuticals. Chemosphere. 2022;308:136191.

[79] Anandan S, Kumar Ponnusamy V, Ashokkumar M. A review on hybrid techniques for the degradation of organic pollutants in aqueous environment. Ultrason Sonochem. 2020;67:105130.

[80] Barik AJ, Gogate PR. Hybrid treatment strategies for 2,4,6-trichlorophenol degradation based on combination of hydrodynamic cavitation and AOPs. Ultrason Sonochem. 2018;40:383–394.

[81] Eghbali P, Hassani A, Sündü B, et al. Strontium titanate nanocubes assembled on mesoporous graphitic carbon nitride (SrTiO$_3$/mpg-C$_3$N$_4$): Preparation, characterization and catalytic performance. J Mol Liq. 2019;290:111208.

[82] Wang X, He X-S, Li C-Y, et al. Sonocatalytic removal of tetracycline in the presence of S-scheme Cu$_2$O/BiFeO$_3$ heterojunction: Operating parameters, mechanisms, degradation pathways and toxicological evaluation. J Water Process Eng. 2023;51:103345.

[83] Sadeghi Rad T, Khataee A, Sadeghi Rad S, et al. Zinc-chromium layered double hydroxides anchored on carbon nanotube and biochar for ultrasound-assisted photocatalysis of rifampicin. Ultrason Sonochem. 2022;82:105875.

[84] Sadeghi Rad T, Khataee A, Arefi-Oskoui S, et al. Graphene-based ZnCr layered double hydroxide nanocomposites as bactericidal agents with high sonophotocatalytic performances for degradation of rifampicin. Chemosphere. 2022;286:131740.

[85] Khataee A, Sadeghi Rad T, Nikzat S, et al. Fabrication of NiFe layered double hydroxide/reduced graphene oxide (NiFe-LDH/rGO) nanocomposite with enhanced sonophotocatalytic activity for the degradation of moxifloxacin. Chem Eng J. 2019;375:122102.

[86] Vinesh V, Ashokkumar M, Neppolian B. rGO supported self-assembly of 2D nanosheet of (g-C$_3$N$_4$) into rod-like nano structure and its application in sonophotocatalytic degradation of an antibiotic. Ultrason Sonochem. 2020;68:105218.

[87] Abazari R, Mahjoub AR, Sanati S, et al. Ni—Ti layered double hydroxide@Graphitic carbon nitride nanosheet: A novel nanocomposite with high and ultrafast sonophotocatalytic performance for degradation of antibiotics. Inorg Chem. 2019;58:1834–1849.

[88] Sadeghi Rad T, Yazici ES, Khataee A, et al. Nanoarchitecture of graphene nanosheets decorated with NiCr layered double hydroxide for sonophotocatalytic degradation of refractory antibiotics. Environ Res. 2022;214:113788.

[89] Scaria J, Gopinath A, Nidheesh PV. A versatile strategy to eliminate emerging contaminants from the aqueous environment: Heterogeneous Fenton process. J Clean Prod. 2021;278:124014.

[90] Fang Z, Liu Y, Qi J, et al. Establishing a high-speed electron transfer channel via CuS/MIL-Fe heterojunction catalyst for photo-Fenton degradation of acetaminophen. Appl Catal B: Environ. 2023;320:121979.

[91] Ghanbari F, Hassani A, Wacławek S, et al. Insights into paracetamol degradation in aqueous solutions by ultrasound-assisted heterogeneous electro-Fenton process: Key operating parameters, mineralization and toxicity assessment. Sep Purif Technol. 2021;266:118533.

[92] Krishnan S, Martínez-Huitle CA, Nidheesh PV. An overview of chelate modified electro-Fenton processes. J Environ Chem Eng. 2022;10:107183.

[93] Yang Z-Z, Zhang C, Zeng G-M, et al. Design and engineering of layered double hydroxide based catalysts for water depollution by advanced oxidation processes: A review. J Mater Chem A. 2020;8:4141–4173.

[94] Ahmadi M, Kakavandi B, Jorfi S, et al. Oxidative degradation of aniline and benzotriazole over PAC@$Fe^{II}Fe_2^{III}O_4$: A recyclable catalyst in a heterogeneous photo-fenton-like system. J Photochem Photobiol A: Chem. 2017;336:42–53.

[95] Zhu C, Wang Y, Qiu L, et al. Z-scheme NiFe LDH/$Bi_4O_5I_2$ heterojunction for photo-fenton oxidation of tetracycline. J. Alloys Compd. 2023;944:169124.

[96] Liu L, Li S, An Y, et al. Hybridization of nanodiamond and CuFe-LDH as heterogeneous photoactivator for visible-light driven photo-fenton reaction: Photocatalytic activity and mechanism. Catalysts. 2019;9:118.

[97] Fazli A, Brigante M, Khataee A, et al. $Fe_{2.5}Co_{0.3}Zn_{0.2}O_4$/CuCr-LDH as a visible-light-responsive photocatalyst for the degradation of caffeine, bisphenol A, and simazine in pure water and real wastewater under photo-fenton-like degradation process. Chemosphere. 2022;291:132920.

[98] Wu Y, Li X, Zhao H, et al. 2D/2D FeNi-layered double hydroxide/bimetal-MOFs nanosheets for enhanced photo-fenton degradation of antibiotics: Performance and synergetic degradation mechanism. Chemosphere. 2022;287:132061.

[99] Yang L, Li L, Liu Z, et al. Degradation of tetracycline by FeNi-LDH/Ti_3C_2 photo-fenton system in water: From performance to mechanism. Chemosphere. 2022;294:133736.

[100] Shao B, Liu Z, Tang L, et al. Construction of Bi_2WO_6/CoAl-LDHs S-scheme heterojunction with efficient photo-fenton-like catalytic performance: Experimental and theoretical studies. Chemosphere. 2022;291:133001.

[101] Kong W, Xing Z, Fang B, et al. Plasmon Ag/Na-doped defective graphite carbon nitride/NiFe layered double hydroxides Z-scheme heterojunctions toward optimized photo-thermal-photocatalytic-fenton performance. Appl Catal B: Environ. 2022;304:120969.

[102] Fang B, Xing Z, Guo M, et al. Phosphorus-doping CdS@NiFe layered double hydroxide as Z-Scheme heterojunction for enhanced photocatalytic and photo-fenton degradation performance. Sep Purif Technol. 2021;274:119066.

[103] Deng J, Xiao L, Yuan S, et al. Activation of peroxymonosulfate by CoFeNi Layered Double Hydroxide/Graphene Oxide (LDH/GO) for the degradation of gatifloxacin. Sep Purif Technol. 2021;255:117685.

[104] Zhang H, Nengzi L-C, Wang Z, et al. Construction of Bi_2O_3/CuNiFe LDHs composite and its enhanced photocatalytic degradation of lomefloxacin with persulfate under simulated sunlight. J Hazard Mater. 2020;383:121236.

[105] Madihi-Bidgoli S, Asadnezhad S, Yaghoot-Nezhad A, et al. Azurobine degradation using Fe_2O_3@multi-walled carbon nanotube activated peroxymonosulfate (PMS) under UVA-LED irradiation: Performance, mechanism and environmental application. J Environ Chem Eng. 2021;9:106660.

[106] Li X, Yuan Z, Huang Z, et al. The photodegradation property and mechanism of tetracycline by persulfate radical activated In_2O_3@LDHs Z-scheme heterojunction. Sep Purif Technol. 2022;302:122077.

[107] Darvishi Cheshmeh Soltani R, Abolhasani E, Mashayekhi M, et al. Degradation of tetracycline antibiotic utilizing light driven-activated oxone in the presence of g-C_3N_4/ZnFe LDH binary heterojunction nanocomposite. Chemosphere. 2022;303:135201.

[108] Zeng H, Zhang H, Deng L, et al. Peroxymonosulfate-assisted photocatalytic degradation of sulfadiazine using self-assembled multi-layered CoAl-LDH/g-C_3N_4 heterostructures: Performance, mechanism and eco-toxicity evaluation. J Water Process Eng. 2020;33:101084.

[109] Liu C, Mao S, Shi M, et al. Peroxymonosulfate activation through 2D/2D Z-scheme CoAl-LDH/BiOBr photocatalyst under visible light for ciprofloxacin degradation. J Hazard Mater. 2021;420:126613.

[110] Azalok KA, Oladipo AA, Gazi M. UV-light-induced photocatalytic performance of reusable MnFe-LDO—Biochar for tetracycline removal in water. J Photochem Photobiol A: Chem. 2021;405:112976.

[111] Azalok KA, Oladipo AA, Gazi M. Hybrid MnFe-LDO—Biochar nanopowders for degradation of metronidazole via UV-light-driven photocatalysis: Characterization and mechanism studies. Chemosphere. 2021;268:128844.

[112] Shen J-C, Zeng H-Y, Chen C-R, et al. A facile fabrication of Ag_2O-Ag/ZnAl-oxides with enhanced visible-light photocatalytic performance for tetracycline degradation. Appl Clay Sci. 2020;185:105413.

[113] Roberts J, Song Y, Crocker M, et al. A genetic algorithmic approach to determine the structure of Li—Al layered double hydroxides. J Chem Inf Model. 2020;60:4845–4855.

[114] Yekan Motlagh P, Khataee A, Hassani A, et al. ZnFe-LDH/GO nanocomposite coated on the glass support as a highly efficient catalyst for visible light photodegradation of an emerging pollutant. J Mol Liq. 2020;302:112532.

[115] Li J, Xue J, Yuan M, et al. Enhanced photocatalytic property of visible light responsive cotton fabric functionalized by NiCrFe layered double hydroxide for rapid sterilization and degradation. Appl Surf Sci. 2022;605:154628.

8 Transition Metal Carbide and Nitride (MXene)-Based Photocatalysts for Pharmaceutical Wastewater Treatment

Nahid Amini-Seresht and Mohsen Sheydaei

8.1 INTRODUCTION

Over the past decade, a growing public concern has arisen regarding the occurrence and fate of pharmaceuticals in the aquatic environment. These chemicals can reach water sources through industrial discharges, sewage systems, aquaculture, and livestock farming. Some pharmaceuticals have been proven to induce adverse health effects on aquatic organisms, such as genomic mutation, even at low concentrations [1]. Therefore, it is necessary to remove pharmaceuticals from polluted sources, which can cause dangers to the natural ecosystems.

Traditional wastewater treatment technologies are not very successful in effectively removing pharmaceutical pollutants and are usually time consuming, require more purification steps, and suffer from the resistance of pollutants [2]. Therefore, finding novel and effective methods is of interest. Photocatalysis, as one of the effective advanced oxidation processes (AOPs), is an economical, simple, and environmentally friendly technology that has shown significant potential in the destructive removal of pharmaceutical pollutants from water. Unlike the conventional chemical treatment process, such as the chlorine oxidation of organics, the formation of hazardous products of AOPs is limited [3]. In addition, the AOPs don't generate a secondary waste stream, as is the case, for example, in membrane separation processes. Furthermore, a considerable mineralization efficiency and rate are the advantages of AOPs [4].

Photocatalysis includes a photoactive semiconductor material under appropriate light radiation. This process involves converting green and inexhaustible solar energy into chemical energy on the surface of an appropriate photocatalyst. The photocatalyst structure consists of valence bands (VBs) and conduction bands (CBs), which are separated by band gap energy [5, 6].

DOI: 10.1201/9781003340164-8

Recently, photocatalytic applications of two-dimensional (2D) nanomaterials have attracted much attention. Among them, transition metal carbides and nitrides (MXenes) with distinct layered structures are widely used in photocatalysis due to their remarkable electrical conductivity, unique optical/thermal properties, large surface area, hydrophilicity, controllable surface chemical properties, high chemical stability, and abundant surface functional groups.

MXenes are a class of new materials with a 2D structure and are prepared from MAX phases as main precursors for MXene production. The MAX phase consists of trilayered nitrides, carbides, and carbonitrides of early transition metals. The chemical formula of the MAX phase is $M_{n+1}AX_n$, and MXenes have the general formula (n = 1, 2, and 3) $M_{n+1}X_nT_x$. The M represents the early transition metals (such as Sc, Ti, V, Nb, Mo, Cr, Ta, Zr, Hf, and W), A is mainly a group IIIA or IVA (i.e., groups 13 or 14) element (Al, Si, P, Ga, Ge, As, In, Sn, Pb, S, and Cd), X is carbon or nitrogen, and T_x is the surface terminal functional group such as -OH, -O, -F, or -Cl [7–10]. For example, the Ti_3C_2 MXene can have at least three formulas: $Ti_3C_2(OH)_2$, $Ti_3C_2O_2$, and $Ti_3C_2F_2$. The values of each terminal group strongly depend on the synthesis process [10]. MXenes that are synthesized with an etching method using a lower hydrofluoric acid (HF) concentration have more -O than -F, while MXenes synthesized in the presence of a higher HF concentration leads to more -F than -O development [11]. The presence of functional groups affects the optical, chemical, and electrical properties of these 2D materials [11, 12]. Hydrophilic functional groups (-OH and -O) on the surfaces of MXenes lead to strong interactions of MXenes with different semiconductor photocatalysts and water molecules to facilitate surface charge transfer. On the other hand, they guarantee the considerable electrical conductivity, adsorption performance, and catalytic capacity of MXenes [11, 13, 14]. MXenes usually show a negative charge because their surface terminals are rich in -F, -O, and -OH functional groups, which improve the interaction and binding of metal cations on the surface [15].

Studies have shown that among -OH, -O, and -F, -OH is the least stable terminal group. The low stability of the -OH terminal is due to the replacement of hydrogen atoms with alkali or intermediate metals and its conversion to -O at high temperatures. Experimentally, the stability order of -F > -O > -OH terminal groups was confirmed for the HF technique, while in thermal treatments and the HCl/LiF method, the -O terminal group shows the most stability and the -OH terminal group shows the least stability. However, it is unclear why they differ based on the order of functionalization [12].

The size and shape of semiconductors have considerable effects on the number of reaction sites and their electronic performance in photocatalysis [16]. MXenes are synthesized in different sizes and shapes, such as 2D nanosheets [17], one-dimensional (1D) specialty nanofibers [18], and zero-dimensional (0D) quantum dots [19].

Among the MXenes, 2D nanoparticles have been considerably applied in photocatalysis. This can be attributed to the low thickness and large surface area of 2D MXenes, which lead to the fast migration of photogenerated electrons to the surface of MXenes and between semiconductors. Furthermore, the density of surface atoms in the planar structure is considerable. Moreover, the simplicity of synthesizing the 2D MXenes has also been influential in the extensive use of this catalyst [20].

The MXene nanofibers also have considerable advantages for photocatalysis. By the way, fewer reports have been reported for the synthesis and application of these catalysts in photocatalysis. It is expected to be intensively studied in the future to synthesize nanofiber MXenes for photocatalysis.

8.2   MXENE SYNTHESIS METHODS

The most common method for preparing 2D MXenes is the separation of one or more atomic layers from the MAX phase layered structure. The MXenes are mainly created by etching the A layers of MAX phases [12]. In 2011, for the first time, $Ti_3C_2T_x$ (MXene) was synthesized by selectively etching the Al atomic layers from Ti_3AlC_2 using 50% HF at room temperature by Naguib et al. [21]. Since the preparation of the first MXene, more than 30 types of MXene have been synthesized [22].

In general, there are two methods for the synthesis of 2D layered nanomaterial: 1) the top-down method and 2) the bottom-up method. Top-down methods include selective etching in a mixture of fluoride salts and various acids, alkaline solutions, with hydrothermal and electrochemical etching methods. Bottom-up methods include chemical vapor deposition, the template method, laser deposition, and pulsed plasma [10]. Also, solvothermal techniques, ball milling, ultrasonic synthesis, and pyrolysis methods are other methods of MXene synthesis [23]. Wet chemical etching is only useful for carbon-based MXenes because it cannot remove the A layer from nitride-based MAX phases. In wet chemical etching, a variety of etchants, such as HF, LiF/HF, and $ZnCl_2$, are used to synthesize MXenes. Different MXenes can be obtained by controlling acid concentration, temperature, and reaction time [23, 24]. In general, MXenes can be synthesized in three steps: 1) preparation of the MAX phase precursor, 2) etching of the Al layer, and 3) intercalation and exfoliation [25]. Generally, in MAX phases with increasing atomic number M and increasing n values, M–A bond energies increase and therefore require stronger etching and/or longer etching time to synthesize the MXene [11, 13]. However, harsh etching can lead to defects in the MXene and thus affect the quality of the MXene [11]. In addition, in the presence of less acidic etchant, a higher temperature is required [13]. MXenes are oxidized and decomposed in humid and oxygen-rich environments. Light irradiation also increases the oxidation rate of MXenes. Therefore, to avoid oxidation, it is better to store MXenes in an oxygen-free and dark environment in the refrigerator. The initiation of oxidation of MXenes is mainly from the edges, which gradually leads to the formation of metal oxide nanocrystals [22].

8.2.1   HYDROFLUORIC ACID ETCHING METHOD

The layers of MAX phases are difficult to exfoliate by mechanical methods due to strong M–A metal covalent bonds. However, M–A covalent bonds are relatively weaker than M–X covalent bonds. Therefore, it is possible to selectively remove A layers with different etchants using chemical etching or heating of the M–A bond [10, 22]. MXenes are mainly prepared by wet etching due to the possibility of destruction of the layered structure at high temperatures. HF is one of the most powerful etchants. The first MXene ($Ti_3C_2T_x$) was synthesized by HF etching [10]. Figure 8.1

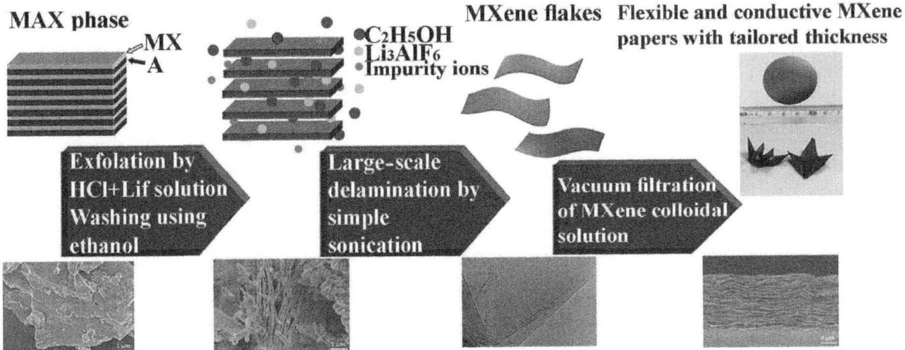

FIGURE 8.1 Preparation of MXenes from MAX phases.

Source: Reproduced from Ref. [26] with permission from Elsevier.

shows the synthesis process of MXenes using HF and a scanning electron microscope micrograph of the prepared sample. The top-down method generally has two steps: etching the MAX phase and layering to separate the plates [10].

After etching the MAX phase with HF, single or multilayer MXene nanosheets are exfoliated and prepared by applying ultrasonication in dimethyl sulfoxide, methanol, or isopropyl alcohol to obtain single or multilayer MXenes [12, 27]. Increasing the corrosive HF concentration and etching time can lead to an increase in defects and a decrease in the lateral size of the prepared MXene. Therefore, the etching conditions should be adjusted according to the desired applications and properties of products. Also, the etching conditions depend on the structure and composition of MAX phases. For example, Ti_3AlC_2 can be etched using a 5% HF, while V- and Nb-based MXenes can only be developed using 50 wt% HF [24]. It has been proven that MXenes with minor defects and large lateral sizes have broad applications in optics, electronics, and electromagnetics, and MXenes with small lateral size are appropriate for catalytic and gas sensing applications. Equation 8.1 shows the production of a MXene from the MAX phase, and Equations 8.2 and 8.3 respectively show the development of -F and -OH functional groups on the surface of a MXene with HF etching [12].

$$Ti_3AlC_2 + 3HF \rightarrow AlF_3 + 3/2H_2 + Ti_3C_2 \qquad (8.1)$$

$$Ti_3C_2 + 2H_2O \rightarrow Ti_3C_2(OH)_2 + H_2 \qquad (8.2)$$

$$Ti_3C_2 + 2HF \rightarrow Ti_3C_2F_2 + H_2 \qquad (8.3)$$

8.2.2 HYDROCHLORIC ACID/LITHIUM FLUORIDE ETCHING METHOD

HF is suitable for etching the MAX phase due to its convenience and low reaction temperature (less than 55 °C) [10, 24]. However, HF is dangerous and highly corrosive. Therefore, in situ HF formation methods are preferred over conventional ones [10]. Mixing fluoride salts such as KF, LiF, NaF, CsF, CaF_2, NH_4HF_2, $NaHF_2$, NH_4F, and KHF_2 with HCl or H_2SO_4 can be a suitable substitute for HF in the process of converting MAX phases to MXenes [10, 11]. For example, using a mixture of 12 M LiF and 9 M HCl with a minimally intensive layer delamination method enabled the exfoliation of MXene layers by simple manual shaking [11]. The developed F⁻ ion reacts with the A atoms of MAX precursors to form AF, H_2, and the desired MXenes [28]. An acid prewash with HCl or H_2SO_4 can also help dissolve salts such as AlF_3 or LiF [10]. The advantages of this method include the use of mild reactants compared to HF, high yield, the elimination of addition steps, high exfoliation efficiency, and obtaining flexible clay-like MXenes [10, 24]. The prepared MXene sheets are resistant to oxidation because they typically have larger lateral dimensions with fewer defects, which are essential properties for some applications that require high electrical conductivity [10].

8.2.3 HYDROTHERMAL METHOD USING ALKALINE SOLUTIONS

In the study of Li et al. [29], a $Ti_3C_2T_x$ MXene was synthesized through the hydrothermal method by removing the Al layer from Ti_3AlC_2 using NaOH. The MXene prepared by this method has -O and -OH surface functional groups. In this method, the reaction is carried out at a very high temperature, and the MXene synthesis is more time consuming than HF chemical methods. The MXene prepared by this method has a larger distance between layers and a higher surface area compared to MXenes prepared by the conventional HF etching method. The fluorine-free hydrothermal process is an effective and environmentally friendly method for producing MXenes. This method leads to the production of MXenes with excellent layering and large interlayer distance, and they can be exfoliated by ultrasonic waves.

8.2.4 LEWIS ACID ETCHING METHOD

The Lewis acid etching method produces various MXenes with -Cl functional groups by the reaction between the MAX phase and molten Lewis salt (Equations 8.4 and 8.5). Li et al. [30], for the first time, synthesized various types of MXenes using a general Lewis acid etching method. Here, to produce chlorine-terminated MXene ($Ti_3C_2Cl_2$ and Ti_2CCl_2), several MAX phases such as Ti_2ZnC, V_2ZnC, Ti_2ZnN, and Ti_3ZnC_2 were mixed with Lewis acidic molten salts, placed in an alumina boat under an argon atmosphere, and heated to 750 °C. This method is a fast, effective, and ideal way to prepare MXenes.

$$Ti_3AlC_2 + 3/2ZnCl_2 \rightarrow Ti_3ZnC_2 + 1/2Zn + AlCl_3 \tag{8.4}$$

$$Ti_3ZnC_2 + ZnCl_2 \rightarrow Ti_3C_2Cl_2 + 2Zn \tag{8.5}$$

8.2.5 ULTRAVIOLET-INDUCED SELECTIVE ETCHING METHOD

Mei et al. [31] used a fluorine-free UV-induced selective etching method to synthesize a Mo_2C MXene from a Mo_2Ga_2C MAX phase (Figure 8.2). In this process, UV irradiation for several hours removes the gallium layers from the MAX phase, and

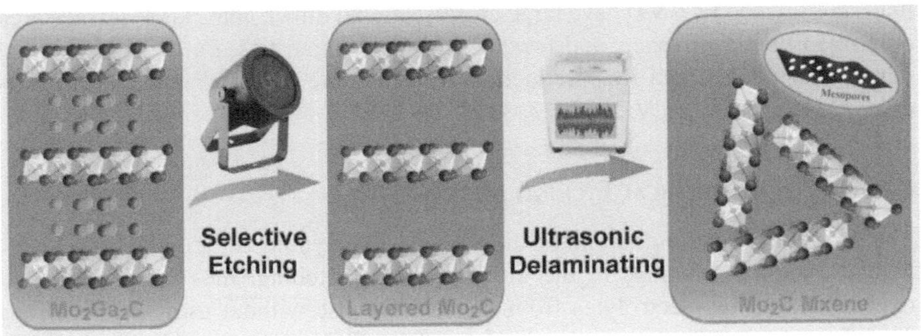

FIGURE 8.2 Synthesis of Mo_2C from Mo_2Ga_2C MAX phase using a UV-induced selective etching method.

Source: Reproduced from Ref. [31] with permission from Elsevier.

a diluted H_3PO_4 solution is used to remove the weak Ga coatings. This method is an excellent method for producing MXenes because it is fast and does not use dangerous and highly corrosive chemicals. By the way, this technique only applies to certain types of MAX phases.

8.2.6 Thermal Reduction Method

Mei et al. [32] synthesized a 2D structure of a sulfur-free Ti_2C MXene from a sulfur-containing Ti_2SC MAX phase using a simple thermal reduction method. In this MAX phase, the sulfur layer is a weak bridge between the titanium and carbon layers. This sulfur layer is removed by simple thermal treatment, and weak Ti_2C MXene layers are created. -OH and -O surface functional groups were observed on the Ti_2C MXene surface due to oxygen absorption. The simple thermal reduction method provides a fast, efficient, and environmentally friendly alternative to chemical etching.

8.2.7 Molten Salt Etching Method

In the molten salt etching method, MXenes are prepared by etching the MAX phase at high temperature. For example, Urbankowski et al. [33] synthesized Ti_4N_3 materials by adding the Ti_4AlN_3 MAX phase to a molten fluoride mixture at 550 °C to etch Al atoms in an argon gas environment. The $Ti_4N_3T_x$ prepared through the molten fluoride salt etching had less crystallinity than another $Ti_3C_2T_x$ etched with HF. This suggests there may still be problems in the synthesis of nitride MXenes via the molten salt etching process [9]. The synthesis of MXenes must be carried out at a suitable temperature; otherwise, a stereoscopic phase is formed [22].

8.2.8 Chemical Vapor Deposition (CVD)

Using CVD technology, it is possible to produce various 2D monolayer materials, including MXenes and graphene, in a controllable environment [22]. The 2D materials created by CVD have large cross-sectional dimensions, large surface area, controlled interlayer space, and fewer defects and impurities. By the way, the high manufacturing cost and low production efficiency are its drawbacks and limit the commercialization of CVD [22, 23, 34].

8.2.9 ELECTROCHEMICAL ETCHING METHOD

Electrochemical etching is one of the fluoride-free synthesis methods of MXenes. Unlike chemical etching, in the electrochemical etching, the A layer is selectively removed by electrolytes from the MAX phase without using fluoride ions. Therefore, no fluoride ions appear on the surface of MXenes. Considering that chlorine ions and alkaline solutions are commonly applied as the electrolyte of the electrochemical etching method, chloride and hydroxyl groups may exist on the surface of developed MXenes. Using this method, the A layer can corrode due to a strong reaction between Cl^- ions and A or between OH^- ions and A [27]. In addition, the Li^+ ions in the electrolyte can increase the distance between the layers by entering the MXene screen, in addition to weakening the interaction between the adjacent layers. MXenes can also be exfoliated using the ultrasonic method without using any harmful organic chemical reagents. In electrochemical etching, by applying a certain voltage (between 0 and 2.5 V) to the MAX phase as an electrode, the A layer is selectively removed. Finally, a MXene is prepared by this method [24]. A gradual increase in voltage removes more of the M layer. As a result, amorphous carbon materials are created [11]. Therefore, controlling the etching potential and time is necessary to synthesize appropriate MXenes. [24]. The electrochemical etching method, as a green and safe synthetic method, has advantages such as low reaction temperature, low energy consumption, and minimal use of corrosive acid. By the way, it is often criticized due to the presence of a carbide-derived carbon layer on MXenes at the same time. Therefore, it is unsuitable for large-scale preparation [24]. Electrochemical etching reactions can be defined as follows in Equations 8.6–8.8 [24]:

$$Ti_3AlC_2 - 3e^- + 3Cl^- \rightarrow Ti_3C_2 + AlCl_3 \tag{8.6}$$

$$Ti_3C_2 + 2OH^- - 2e^- \rightarrow Ti_3C_2(OH)_2 \tag{8.7}$$

$$Ti_3C_2 + 2H_2O \rightarrow Ti_3C_2(OH)_2 + H_2 \tag{8.8}$$

8.2.10 HALOGEN ETCHING METHOD

Halogens can also be used as an etchant in the preparation of MXenes. In the study of Shi et al. [35] for the synthesis of $Ti_3C_2I_x$, I_2 was used to etch the Ti_3AlC_2 MAX phase in anhydrous acetonitrile at 100 °C. The product was later exfoliated by exposure to 1 M HCl solution to remove the AlI_3 produced during the etching process. For the synthesis of $Ti_3C_2T_x$, Jawaid et al. selectively removed Al from the Ti_3AlC_2 MAX phase using organic solvents, including halogen and interhalogen.

8.2.11 OTHER ETCHING METHODS

In addition to the mentioned methods, other methods, such as plasma-enhanced pulsed laser deposition, have also been reported by researchers to prepare MXenes. Another method is the template method, which has significantly higher production efficiency than CVD methods [10]. Also, studies have shown that MXenes can be prepared from the MAX phase by using algae. In the study of Zada et al. [36], organic acids produced from algae were used to engrave Al atoms in the V_2AlC MAX phase and V_2C synthesis. Here, the bioactive compounds could exfoliate and cut the MAX phases. In Table 8.1, a summary of the synthesis methods of MXenes is given. Figure 8.3 shows the different synthesis techniques of MXenes from 2011 to 2023.

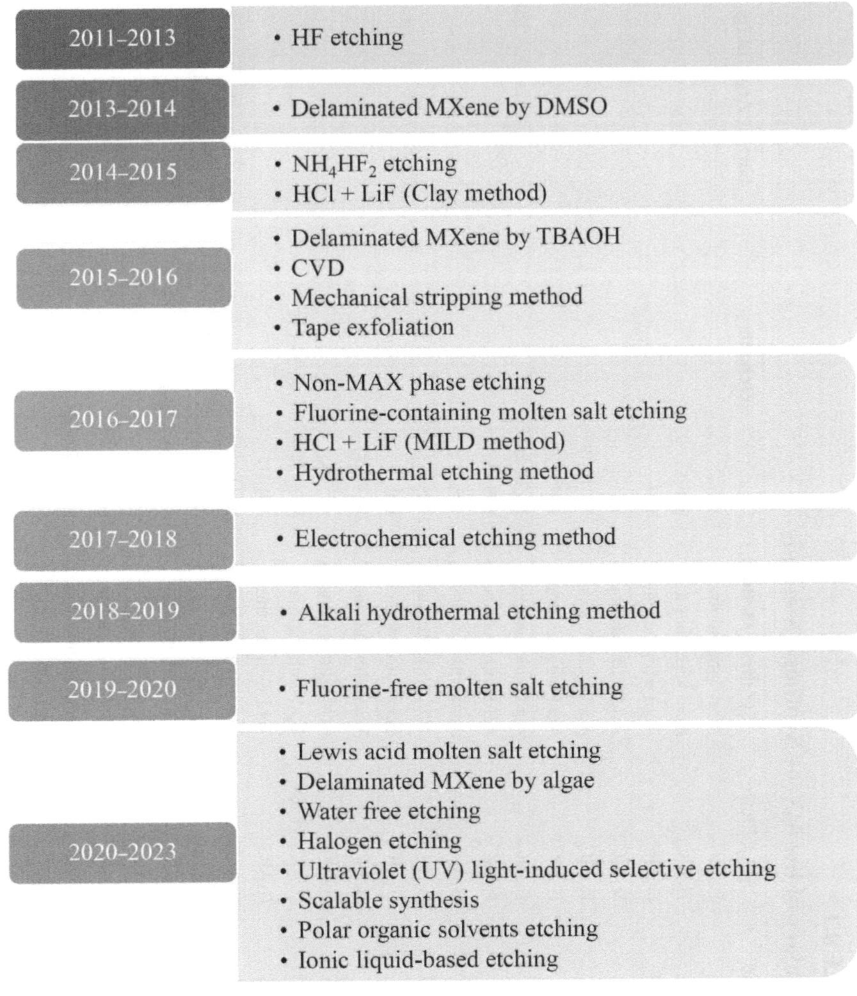

FIGURE 8.3   Synthesis methods of MXenes from 2011–2023.

TABLE 8.1

Summary of the MXene Synthesis Methods

MXene	MAX Phase	Synthesis Method	Etchants	Surface Area (m²/g)	Time	Temp. (°C)	Ref.
Ti_3C_2	Ti_3AlC_2	Acid with fluorine	HF	–	24 h	Room temperature (RT)	[38]
$Ti_3C_2T_x$	Ti_3AlC_2	Acid with fluorine	HF	–	24 h	RT	[39]
$Ti_3C_2T_x$	Ti_3AlC_2	Acid with fluorine	HF	–	24 h	RT	[40]
Ti_3C_2	Ti_3AlC_2	Acid with fluorine	HF	–	24 h	60	[41]
Ti_3C_2	Ti_3AlC_2	Acid with fluorine	HCl/LiF	–	48 h	50	[42]
Ti_3C_2	Ti_3AlC_2	Acid with fluorine	HF	1.17	72 h	RT	[43]
Ti_3C_2	Ti_3AlC_2	Acid with fluorine	HCl/NaF	–	12 h	60	[44]
Ti_3C_2	Ti_3AlC_2	Acid with fluorine	HF	–	72 h	35	[2]
Ti_3C_2	Ti_3AlC_2	Acid with fluorine	HF	–	72 h	35	[45]
$Ti_3C_2T_x$	Ti_3AlC_2	Acid with fluorine	HF	–	24 h	40	[46]
Ti_3C_2	Ti_3AlC_2	Acid with fluorine	HF	–	48 h	40	[47]
Ti_2C_2	Ti_3SiC_2	Acid with fluorine	HF/H_2O_2	–	40 h	RT	[48]
$Ti_3C_2T_x$	Ti_3AlC_2	Acid with fluorine	HF/HCl	–	24 h	35	[49]
Ti_2C, Ti_3C_2	Ti_2AlC, Ti_3AlC_2	Acid with fluorine	$HCl/(NaF, KF, or NH_4F)$	–	48 h, 24 h	60, 50	[50]
$Ti_3C_2T_x$	Ti_3AlC_2	Acid with fluorine	NH_4HF_2	–	2 h	1350	[51]
$Ti_3C_2T_x$	Ti_3AlC_2	Hydrothermal	NaOH	16	–	270	[29]
Ti_3C_2	Ti_3AlC_2	Hydrothermal	$NaBF_4/HCl$	44.6	35 h	180	[52]
$Ti_3C_2T_x$	Ti_3AlC_2	Acid with fluorine	$EMIMBF_4, BMIMPF_6$	–	44 h	90	[53]
Ti_3CT_x	Ti_3AlC	Electrochemical	HCl	–	–	RT	[54]
Ti_3CT_x	Ti_3AlC	Electrochemical	HCl	–	9 h	50	[55]
$Ti_3C_2T_x$	Ti_3AlC_2	Electrochemical	$NH_4Cl/TMAOH$	–	5 h	RT	[56]
V_2CT_x	V_2AlC	Electrochemical	HCl	–	9 h	50	[55]

$Ti_3C_2T_x$	Ti_3SiC_2	Lewis acid	$CuCl_2$/NaCl/KCl	–	24 h	750	[30]
$Ti_3C_2Cl_2$	Ti_3ZnC_2	Lewis acid	$ZnCl_2$	–	5 h	550	[57]
$Ti_3C_2Br_2$	Ti_3AlC_2	Lewis acid	$CdCl_2$/$CdBr_2$	–	-	610	[58]
Ti_3CNT_x	Ti_3AlCN	Lewis acid	$CuCl_2$	–	24 h	700	[30]
Nb_2CT_x	Nb_2AlC	Lewis acid	AgCl	–	24 h	700	[30]
Ta_2CT_x	Ta_2AlC	Lewis acid	AgCl	–	24 h	700	[30]
V_2CT_x	V_2AlC	Lewis acid	$ZnCl_2$	–	5 h	550	[57]
Ti_2C	Ti_2SC	Thermal reduction	Ar/H_2	–	0.5 h	400–900	[32]
Mo_2C	Mo_2Ga_2C	UV induced	H_3PO_4	–	5 h	RT	[31]
$a\text{-}Mo_2C$	–	CVD	CH_4/Mo/Cu	–	3 h	1100	[59]
$Ti_4N_3T_x$	Ti_4AlN_3	Molten salts	LiF/NaF/KF	–	0.3 h	550	[33]
$Ti_3C_2Br_2$	Ti_3AlC_2	Molten salts	Li_2Te/Li_2S/Li_2Se/Li_2O/$NaNH_2$/LiH	–	–	600	[58]
$Ti_3C_2T_x$	Ti_3AlC_2	Other etching methods	Algae	–	1 day	RT	[36]
$Ti_3C_2T_x$	Ti_3AlC_2	Other etching methods	Surface acoustic waves, LiF	–	Millisecond	–	[60]

8.3 PHOTOCATALYTIC DEGRADATION OF PHARMACEUTICAL POLLUTANTS BY MXENES

MXenes have been used as an auxiliary catalyst to increase the ability of photocatalytic degradation of hazardous pollutants such as pharmaceutical pollutants [23, 61]. In addition to MXenes, various catalysts such as TiO_2, g-C_3N_4, and CdS have also been used to degrade pharmaceutical compounds. By the way, the rapid recombination of photogenerated electron/hole (e^-/h^+) pairs limits their practical application in the photocatalytic degradation process [8]. Therefore, in hybrid or heterogeneous structures, MXenes as auxiliary catalysts and effective support can enhance the excited electron transfer, increase the absorption of visible light, prevent the rapid recombination of charge carriers, and lead to the improvement of the catalytic performance of MXene-based photocatalysts [23]. This is due to three features of MXenes:

1) Surface functional groups in MXenes such as -F, -OH, and -O are effective for creating hybrid structures due to their strong interaction with other semiconductors.
2) Modifying the surface chemistry of the MXene affects its band gap alignment.
3) The presence of metal cores in the layered structure of MXene leads to superior metal conductivity and supercharge carrier transport [5, 8].

Among MXenes, the $Ti_3C_2T_x$ with considerable photocatalytic ability has been widely studied for the photocatalytic degradation of organic pollutants, especially pharmaceutical contaminants. The hybridization of MXenes with other materials such as MOFs, metal sulfide, metal oxides, graphene, graphitic carbon nitride, and polymers improves their photocatalytic properties [23]. The mechanism of photocatalytic degradation of pharmaceutical pollutants by MXene-based photocatalysts is shown in Figure 8.4. Zhou et al. proposed a schematic of ciprofloxacin degradation via graphitic carbon nitride/Ti_3C_2 MXene/black phosphorus (CN/MX/BP), which is illustrated in Figure 8.4a. According to this figure, the electrons excited by visible-light irradiation migrate from g-C_3N_4 to the BP nanoplate using the Ti_3C_2 MXene as an electron mediator due to its high conductivity. Accumulated holes in the g-C_3N_4 valence bands can react with dissolved O_2 to produce $^1O_2$ or directly react with ciprofloxacin. In this process, the active radicals were $O_2^{-\bullet}$ and $^1O_2$, which mainly attacked the piperazine ring in ciprofloxacin, and the final products were CO_2, H_2O, and inorganic ions (F^- and NH_4^+) [62].

According to Figure 8.4b, light irradiation on the 2D ternary ZnO/Bi_2WO_6/Ti_3C_2 photocatalyst leads to the effective separation of electrons and holes. MXenes, as good electron carriers, play an essential role in separating charge carriers, and thus increasing photocatalytic activity. Finally, as a result of oxidation–reduction reactions, hydroxyl radicals are produced, which attack ciprofloxacin molecules and destroy them [63].

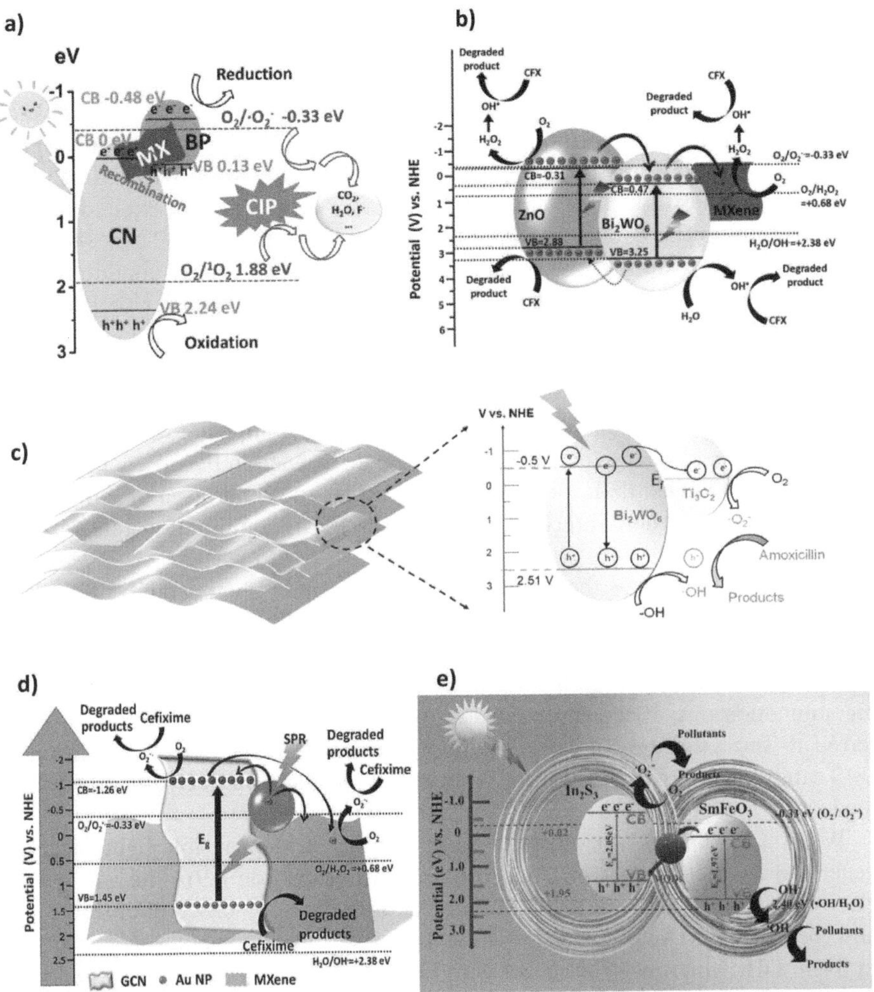

FIGURE 8.4 Mechanism of photocatalytic degradation of pharmaceutical pollutants by MXene-based photocatalysts. (a) Ciprofloxacin via graphitic carbon nitride/Ti_3C_2 MXene/black phosphorus (CN/MX/BP). (b) Ciprofloxacin via 2D ternary nanocomposites of ZnO/Bi_2WO_6/Ti_3C_2 MXene. (c) Amoxicillin via 2D/2D Bi_2WO_6/Ti_3C_2 hybrid. (d) Cefixime via nanomaterial-decorated g-C_3N_4/MXene (AGM). (e) Sulfamethoxazole via In_2S_3/Ti_3C_2 MXene quantum dots/$SmFeO_3$ Z-scheme.

Source: Reproduced from Ref. (a) [62], (b) [63], (c) [2], (d) [64] and (e) [41] with permission from Elsevier.

As shown in Figure 8.4c, in the 2D/2D Bi_2WO_6/Ti_3C_2 hybrid, a lower Fermi level of the Ti_3C_2 nanosheet and the close contact between it and Bi_2WO_6 leads to a fast electron transfer between the two catalysts. In addition, the Ti_3C_2 leads to the effective separation of charge carriers and improves photocatalytic performance due to its excellent photothermal conversion properties. This results in increased light absorption and rapid charge carrier transfer, which essentially increases the production of reactive oxygen species, mainly holes, and $O_2^{-\bullet}$ [2].

The mechanism of a new plasmonic photocatalyst composed of Au plasmonic nanoparticles, graphite carbon nitride, and Ti_3C_2 MXene for cefixime degradation under visible-light irradiation is shown in Figure 8.4d. According to this figure, by visible-light irradiation, the electrons in $g\text{-}C_3N_4$ are excited from the valence band to the conduction band, thus creating e^-/h^+ pairs. Then, the excited electrons move from the $g\text{-}C_3N_4$ conduction band to the highly conductive Ti_3C_2 MXene surface. In addition, gold nanoparticles can also transfer their electrons to Ti_3C_2 MXene. The 2D Ti_3C_2 MXene nanosheets, as excellent electron sinks, are helpful in adsorbing pollutants due to their high surface area. Furthermore, due to the metallic conductivity and abundant functional groups of MXenes, the incorporation of MXenes as co-catalysts in the nanocomposite facilitates the surface separation of charge carriers. Due to the difference in Fermi energy levels of $g\text{-}C_3N_4$ and MXenes, their close contact increases the separation of light-induced charge carriers at their interface. It decreases the recombination rate of charge carriers [64].

Figure 8.4e illustrates the mechanism of visible-light photocatalytic degradation of sulfamethoxazole (SMX) by In_2S_3/MQDs/$SmFeO_3$ (IMS) composites [41]. In this Z-scheme photocatalytic system, the heterogeneous structure of the photocatalyst leads to the separation of photogenerated charge carriers and reduces the e^-/h^+ pair recombination rate. Electrons excited in the conduction band of $SmFeO_3$ are transferred to In_2S_3 through MQDs. The MQDs, as intermediate electronic bridges in IMS catalysts, improve charge transfer between In_2S_3 and $SmFeO_3$ and increase photocatalytic activity.

When a MXene as a photocatalyst absorbs the incident visible light energy, the generated charge carriers migrate to its surface (Equation 8.9). The valence band electrons are excited to the conduction band of the MXene-based photocatalyst and absorbed by the MXene sheet (Equation 8.10). The developed holes reacts with OH^- to form $^\bullet OH$, while electrons react with O_2 to form $O_2^{-\bullet}$ (Equations 8.11 and 8.12). The developed radicals play an essential role in the degradation of pharmaceutical and organic pollutants. As a result, less harmful products, such as H_2O and CO_2, are formed (Equations 8.13 and 8.14) [8, 65].

$$MXene - composite + h\nu \rightarrow h^+ + e^- \tag{8.9}$$
$$e^- + MXene - composite \rightarrow e^- \text{ (MXene trapping sites)} \tag{8.10}$$
$$e^- \text{ (MXene trapping sites)} + O_2 \rightarrow O_2^{-\bullet} \tag{8.11}$$
$$h^+ + H_2O \rightarrow {}^\bullet OH \tag{8.12}$$
$${}^\bullet OH + \text{Pharmaceutical pollutant} \rightarrow \text{degradation products} \tag{8.13}$$
$$O_2^{-\bullet} + \text{Pharmaceutical pollutant} \rightarrow \text{degradation products} \tag{8.14}$$

Atomic modulation [66], multiphase photosystem [67], and morphology modulation [68] are various techniques developed to enhance e^-/h^+ separation to improve the efficiency of photocatalytic degradation. Meanwhile, morphological modulation, as a method to design and fabricate unique nanostructures such as 2D ultrathin films, has attracted much attention. This method leads to a larger surface area, shorter charge transfer distance, and more active sites. Finally, it facilitates photocatalytic reactions [2].

In the study of Wu et al. [43], g-C_3N_4 and TiO_2 attached to graphene layers were prepared using MXenes (Ti_3C_2) through one-step in situ calcination. These were used for the visible-light photocatalytic degradation of tetracycline and ciprofloxacin. The degradation efficiency reached 66.3% for tetracycline after 80 min and 41.8% for Ciprofloxacin after 60 min. The presence of MXenes increases the photocatalytic activity by enhancing the absorption of visible light and the separation of photogenerated carriers. The pollutants are mainly degraded by $O_2^{-\bullet}$ and $\cdot OH$ with high oxidation activity. The synergistic effects and strong interactions between g-C_3N_4, graphene, and TiO_2, increased the rate of radical development and pollutant photocatalytic degradation.

In the study of Zhou et al. [69], a Z-scheme heterojunction of g-C_3N_4/Ti_3C_2 MXene/$MoSe_2$ (CXM) was fabricated. The optimal ratio of this photocatalyst (CXM-0.01) achieved the ability of visible-light photocatalytic degradation of enoxacin. Here, a Ti_3C_2 MXene as an intermediate between g-C_3N_4 and $MoSe_2$ inhibited the developed electron and hole recombination by efficient photo-excited electron transfer. Also, Ti_3C_2 MXenes increased active sites and separated charge carriers to produce more active species such as $O_2^{-\bullet}$, and as a result, improved photocatalytic activity. This photocatalyst also presented good performance in the degradation of ofloxacin (100%), moxifloxacin (100%), levofloxacin (100%), ciprofloxacin (80%), norfloxacin (80%), and gatifloxacin (65%).

In the study of Cai et al. [44], for the first time, a heterogeneous catalyst was synthesized by the electrostatic self-assembly method and used for visible-light photocatalytic degradation of tetracycline hydrochloride. The rate of e^-/h^+ pair recombination in Ag_3PO_4 is high, making it easily reduced to silver elements during the photocatalysis process. Therefore, the use of Ti_3C_2 as an auxiliary catalyst by reducing the rate of recombination of charge carriers improved the photocatalytic activity (68.4%) and also inhibited corrosion of the Ag_3PO_4/Ti_3C_2 composite. The catalyst maintained its stability after eight photocatalysis cycles.

Jang et al. [70] synthesized a magnetic titanium carbide ($Ti_3C_2T_x$) MXene using the co-precipitation synthesis method by ammonium bifluoride as a mild etchant and applied in the photocatalytic degradation of diclofenac under a UV/chlorine process. The concentrations of $\cdot OH$ and active chlorine species in the UV/chlorine process were higher compared to UV, chlorination, and UV/H_2O_2. Magnetic nanoparticles on the surfaces of MXene nanosheets can increase the absorption capacity of the synthesized nanocomposite and enhance the speed of electron transfer in the solution. The prepared MXenes showed considerable stability during seven consecutive cycles of photocatalytic degradation reaction.

In Zhao et al.'s study [71], a $Ti_3C_2T_x$ nanosheet/Cu_2O composite with Schottky heterojunction connection was synthesized by a precipitation method and applied for visible-light photocatalytic degradation of tetracycline. According to the result, the $Ti_3C_2T_x$ nanosheet/Cu_2O composite with 7.0 wt% $Ti_3C_2T_x$ nanosheets showed the highest photocatalytic performance with an efficiency of about 97.6% for tetracycline degradation in 50 min. During the degradation of tetracycline, superoxide radicals and holes were the main active and reactive species. In this work, more effective e^-/h^+ pair separation and faster surface charge transfer in this heterogeneous Schottky junction led to increased photocatalytic activity.

Wu et al. [72] fabricated a Ti_3C_2-Bi/BiOCl heterogeneous photocatalyst by the solvothermal method and investigated its visible-light photocatalytic performance in the degradation of ciprofloxacin antibiotic. During this process, active superoxide radicals and h^+ were responsible for the rapid degradation of the ciprofloxacin. The heterogeneous structure of Ti_3C_2-Bi/BiOCl increased the absorption of visible light and the separation and transfer of photocarriers, thus increasing the photocatalytic activity, and the ciprofloxacin degradation efficiency improved to 89% within 100 min.

In the study of Shen et al. [73], the CeO_2/Ti_3C_2 MXene composite was synthesized using an ultrathin Ti_3C_2 MXene as a co-catalyst through a hydrothermal method. After exposure to sunlight, the CeO_2/Ti_3C_2 MXene with the optimal ratio of the Ti_3C_2 MXene (CeO_2/MX 5%) exhibited the highest photocatalytic performance in tetracycline degradation with 80.2% in 60 min, which was 6.3 times higher than pristine CeO_2 activity. The improved activity of CeO_2/Ti_3C_2 MXene was attributed to Schottky coupling induced by the internal electric field between CeO_2 and the Ti_3C_2 MXene, which drives photo-excited electrons from CeO_2 to the Ti_3C_2 MXene and reduces the rate of combined charge carriers.

In the study of Saravanakumar et al. [41], novel Ti_3C_2 MXene quantum dots (MQDs) modified with In_2S_3/MQDs/$SmFeO_3$ (IMS) were successfully prepared by solvothermal method. The prepared composite was applied for the visible-light photocatalytic degradation of sulfamethoxazole. The degradation rate of sulfamethoxazole using the optimal ternary composite ratio was 98% in 120 min. The significant increase in photoactivity of the IMS composite was attributed to the role of MQDs as charge transfer bridges and, consequently, effective charge separation and transfer in the Z-scheme system. Furthermore, the MQDs accelerate the oxidation–reduction rates of the IMS catalyst surface. Thus, it stimulates the development of reactive species for pollutant degradation.

Table 8.2 shows various MXene-based catalysts as photocatalysts for successful photocatalytic degradation of pharmaceutical pollutants.

8.3.1 Kinetic Studies

The reaction rate is a significant characteristic of the catalytic process and is usually considered representative of the catalyst's role. In the heterogeneous catalytic reaction, the rate depends upon various variables. Therefore, kinetic studies are essential in understanding the rate and the factors influencing the process to exhibit the real photodegradation process. Generally, the photocatalysis rate can be described by a pseudo-first-order kinetic model. Therefore, this model was applied by many researchers in the photocatalytic degradation of pharmaceutical pollutants using MXenes. Using the pseudo-first-order kinetic model, the rate constants were obtained by the fit line in the plot of $-\ln(C_t/C_0)$ versus reaction time t where C_0 and C_t were the pollutant concentration at time 0 and t, respectively [85].

TABLE 8.2

Performance of MXene-Based Catalysts as Photocatalysts for Successful Degradation of Pharmaceutical Pollutants

Photocatalyst	Synthesis Method	Pollutant	Light Source	Intensity of Lamp	Removal Efficiency (%)	Ref.
$Ti_3C_2/Ag/Ag_3VO_4$	In situ partial reduction along with hydrothermal	Tetracycline	300 W Xe lamp with a UV-cut-off filter (15 A, 15 cm far away from the photocatalytic reactor)	~120 mW/cm^2	97% in 20 min	[74]
MXene/TiO_2/g-C_3N_4	One-step in situ calcination	Ciprofloxacin, tetracycline	Visible-light illumination ($\lambda >$ 400 nm) with a 300 W xenon lamp at 14 V and 21 A	300 mW/cm^2	41.8% in 60 min, 66.3% in 80 min	[43]
C–TiO_2/Bi_4NbO_8Cl	Fabricated via molten salt flux process	Ciprofloxacin	300 W xenon lamp equipped with a condensing collector	–	>95%	[42]
Rod-like Nb_2O_5/Nb_2CT_x composites	Hydrothermal	Tetracycline	500 W Xe lamp	–	~91.2% in 180 min	[75]
TiO_2/$Ti_3C_2T_x$	Hydrothermal	Carbamazepine	Solar and UV light	Solar light simulator (100 mW/cm^2, AM 1.5 G filter)	~98.67% in 180 min	[76]
Carbon nitride coupled with Ti_3C_2 MXene-derived amorphous titanium (Ti) peroxo heterojunction	One-step H_2O_2 oxidation	Tetracycline	270 W xenon lamp ($\lambda >$ 420 nm)	–	~86.34% within 60 min	[77]
Magnetic-$Ti_3C_2T_x$	One-step chemical co-precipitation	Diclofenac	20 W Hg lamp (254 nm)	1.35 ± 0.2 mW/cm^2	100% in 30 min	[70]

(Continued)

TABLE 8.2 (*Continued*)

Performance of MXene-Based Catalysts as Photocatalysts for Successful Degradation of Pharmaceutical Pollutants

Photocatalyst	Synthesis Method	Pollutant	Light Source	Intensity of Lamp	Removal Efficiency (%)	Ref.
$CuFe_2O_4/Ti_3C_2$	Sol-hydrothermal	Sulfamethazine	300 W xenon lamp (common filter 56 × 0, cut-off filter, $\lambda > 420$ nm)	–	59.4% in 60 min	[78]
$Bi/BiOCl/Ti_3C_2$	Solvothermal	Ciprofloxacin	300 W Xe lamp with a cut-off filter ($\lambda > 420$ nm)	–	89% in 100 min	[72]
2D/2D Bi_2WO_6/Ti_3C_2 MXene composite	Hydrothermal-assisted in situ growth	Amoxicillin	300 W Xe lamp with an AM 1.5 cut-off filter	–	100% in less than 40 min	[79]
$CdS/Ti_3C_2/TiO_2$	Calcination and hydrothermal	Sulfachloropyridazine	300 W Xe lamp (cut-off $\lambda <$ 400 nm)	300 mW/cm^2	100% in 88 min	[80]
MXene Ti_3C_2/MoS_2	Hydrothermal	Ranitidine	UV-free 25 W LED lamp (using an ultraviolet filter)	40 mW/cm^2	88.4% in 60 min	[81]
$Ti_3C_2/TiO_2/BiOCl$	Hydrothermal	Tetracycline	500 W Xe lamp	–	100% in 120 min	[82]
$Ti_3C_2/g-C_3N_4$	Sonochemical	Levofloxacin	300 W Xe lamp with a cut-off filter ($\lambda > 420$ nm)	–	72% in 30 min	[83]
Oxygen-vacancy-embedded 2D/2D NiF LDH/$Ti_3C_2T_x$ MXene composite	One-step hydrothermal	Norfloxacin	Simulated sunlight	–	98% in 240 min	[84]
$g-C_3N_4/Ti_3C_2$/black phosphorus	Calcination process	Ciprofloxacin	300 W xenon lamp with a cut-off filter ($\lambda > 420$ nm)	–	Over 99% within 60 min	[62]
Au NP/$g-C_3N_4/Ti_3C_2$	Electrostatic assembly route	Cefixime	45 W compact fluorescent lamp (CFL)	–	64.7% in 105 min	[64]

CeO_2/Ti_3C_2 MXene	Hydrothermal	Tetracycline	350 W xenon lamp	–	80.2% within 60 min	[73]
$Ti_3C_2T_x$ nanosheets/Cu_2O	Precipitation	Tetracycline	500 W xenon lamp	–	97.6% in 50 min	[71]
Ag_3PO_4/Ti_3C_2	Electrostatically driven self-assembly	Tetracycline HCl	300 W xenon lamp with a cut-off filter ($\lambda > 420$ nm)	–	68.4%	[44]
g-C_3N_4/Ti_3C_2 MXene/ $MoSe_2$ (CXM)	Calcination treatment	Enoxacin, moxifloxacin, ofloxacin, levofloxacin, norfloxacin, ciprofloxacin, gatifloxacin	300 W xenon lamp with a cut-off filter ($\lambda > 420$ nm)	–	100%,100% 100%, 100%, 80%, 80%, 65%	[69]
In_2S_3/MQDs/$SmFeO_3$ (IMS)	Solvothermal reaction followed by calcination	Sulfamethoxazole	Visible-light irradiation	–	98% in 120 min	[41]
TiO_2@Ti_3C_2/$MgIn_2S_{4-}$15 microflower	Two-step hydrothermal	Ciprofloxacin	Visible-light irradiation	–	92% in 60 min	[40]
Ti_3C_2@TiO_2 Hybrid	Facile reflux method under acidic conditions	Tetracycline	Visible-light irradiation	–	90% in 90 min	[39]
3D/2D TiO_2@Ti_3C_2 MXene/Bi_2S_3 20%	In situ hydrothermal oxidation	Tetracycline	150 W xenon lamp with a cut-off filter ($\lambda > 420$ nm)	–	84.13% in 135 min	[38]

TABLE 8.3

Efficiency and Kinetic Rate Constant Value for Visible-Light Degradation of Different Pharmaceutical Pollutants Using MXene-Based Photocatalysts

Catalyst	Pollutant	Degradation Efficiency (%)/time (min)	Catalytic Rate Constant (1/min)	Ref.
g-C_3N_4/Ti_3C_2 MXene/$MoSe_2$ (0.2 g/L)	Moxifloxacin (20 mg/L)	100/60	0.12	[69]
g-C_3N_4/Ti_3C_2 MXene/$MoSe_2$ (0.2 g/L)	Ofloxacin (20 mg/L)	100/60	0.088	[69]
BiOBr/MXene/g-C_3N_4 (0.2 g/L)	Tetracycline (20 mg/L)	100/30	0.204	[87]
BiOBr/Bi_2MoO_6/Ti_3C_2/MMT_{ex} (0.4 g/L)	Levofloxacin (10 mg/L)	61/60	0.039	[88]
CuO-MXene-OCN (0.2 g/L)	Metoprolol (20 mg/L)	75.6/60	0.0220	[89]
CuO-MXene-OCN (0.2 g/L)	Tetracycline (20 mg/L)	98.2/60	0.0652	[89]
CuO-MXene-OCN (0.2 g/L)	Diclofenac (20 mg/L)	92.6/60	0.0412	[89]
g-C_3N_4/MXene/Ag_3PO_4 (0.6 g/L)	Tetracycline (20 mg/L)	88.4/60	0.03525	[90]

Xuanying et al. (2020) studied the kinetics of tetracycline photocatalytic degradation using $Ti_3C_2T_x$/alkalized g-C_3N_4 and found that the reaction followed pseudo-first-order kinetics [85]. $Ti_3C_2T_x$/alkalized g-C_3N_4 exhibited the kinetic rate constant value of $0.0307\,min^{-1}$, about 1.75-fold that of mechanically mixed $Ti_3C_2T_x$ and alkalized g-C_3N_4 samples ($0.0175\,min^{-1}$). This was due to the catalytic synergistic effect between $Ti_3C_2T_x$ and alkalized g-C_3N_4 in the prepared composite. Additionally, the kinetic study indicated that the $^{\bullet}OH$ and $^{\bullet}O^{2-}$ radicals were the most effective radicals produced during the photocatalytic process and played the main role in tetracycline degradation [86].

Results of the kinetics study for applying various MXene-based catalysts as photocatalysts for the photocatalytic degradation of pharmaceuticals are summarized in Table 8.3. The reported results showed the considerable efficiency and kinetic rate constant value for visible-light degradation of different pharmaceutical pollutants using MXene-based catalysts. This indicates the remarkable ability of these catalysts in visible-light photocatalysis.

8.3.2 Reusability of MXene-Based Photocatalysts

The stability of a catalyst for long-term applications is a considerable challenge to apply in practical applications and is an important factor from an economic point of view. Therefore, the possibility of recovering and reusing a catalyst is considered an important characteristic of the photocatalyst. Accordingly, the stability and reusability of MXene-Based photocatalysts was investigated and reported.

Gao et al. (2023) evaluated BiOBr/MXene/g-C_3N_4 long-term activity through its application in eight successive runs of tetracycline photocatalytic degradation and found a negligible decrease in the degradation efficiency after eight cycles [87].

Zhou et al. (2021) studied the reusability of the g-C_3N_4/Ti_3C_2 MXene/$MoSe_2$ for five cycles with a decline in photocatalyst efficiency [69]. The photocatalyst maintained a high ability of enoxacin degradation after five cycles.

Xu et al. (2022) investigated the visible-light photocatalytic degradation of tetracycline by g-C_3N_4/MXene/Ag_3PO_4 for five cycles and found that the catalyst was not deactivated after five successive recycles [91].

The result of the degradation of pharmaceutical pollutants using MXene-based photocatalysts at different consecutive photocatalysis runs is summarized in Table 8.4.

Furthermore, the characterization of catalysts before and after successive photocatalysis runs was compared in various works. For example, Huang et. al (2021) compared the XRD patterns of a Ag/g-C_3N_4/Ti_3C_2 photocatalyst before and after four successive runs of photocatalytic degradation of tetracycline [96]. The obtained result indicated that the typical peak still appeared and showed almost no obvious change in the crystalline structure of the applied Ag/g-C_3N_4/Ti_3C_2 sample compared to that of the as-synthesized one [96].

Zhou et al. (2021) compared the XPS spectra of fresh g-C_3N_4/Ti_3C_2/black phosphorus samples after reuse five times. There was no apparent change in the XPS spectra of the recycled photocatalyst, indicating that the g-C_3N_4/Ti_3C_2/black phosphorus is stable during the photocatalytic degradation of ciprofloxacin [98].

These results support the high stability and reusability of MXene-based photocatalysts and confirm the considerable potential of these catalysts as effective and reusable materials for photocatalysis.

TABLE 8.4

Results of the Degradation of Pharmaceutical Pollutants Using MXene-Based Photocatalysts in Different Consecutive Photocatalysis Runs

Catalyst	Pollutant	Number of Consecutive Runs	Degradation Efficiency (first run)	Degradation Efficiency (last run)	Ref.
ZnO/TiO_2 nanorod MXene	Ceftriaxone	8	99.4%	93%	[92]
TiO_2/MXene/$MnFe_2O_4$	Carbamazepine	4	100%	100%	[93]
TiO_2/Ti_3C_2	Tetracycline	5	97.6%	94.3%	[94]
g-C_3N_5/Ti_3C_2	Tetracycline	10	97.9%	97.9%	[95]
Ag/g-C_3N_4/Ti_3C_2	Tetracycline	4	92.1%	>85%	[96]
TiO_2@Ti_3C_2 MXene/Bi_2S_3	Tetracycline	3	84.13%	75.96%	[97]
g-C_3N_4/Ti_3C_2/black phosphorus	Ciprofloxacin	5	99.8%	96.9%	[98]

8.4 CONCLUSIONS AND PERSPECTIVES

In summary, research on the synthesis and application of MXene-based photocatalysts for the degradation of pharmaceuticals is increasing considerably. Ti_3C_2 is the most studied MXene in the photocatalytic degradation of pharmaceuticals. Various methods were applied for MXene synthesis, such as HF, HCl/LiF, hydrothermal, Lewis acid, UV-induced, halogen, electrochemical, thermal reduction, and molten salt etching methods.

Numerous works were reported for the successful application of MXene-based photocatalysts in the degradation of pharmaceuticals. This can be attributed to the unique role of MXenes in the MXene-based photocatalysts. MXenes, as auxiliary catalysts and strong supports, can increase the absorption of light for electron photogeneration, enhance the electron transfer of excited semiconductors, suppress the rapid recombination of charge carriers, and consequently lead to the improvement of the photocatalytic performance of MXene-based composites. Furthermore, abundant surface functional groups enhance the ability of photocatalysts in the adsorption of pharmaceuticals. By the way, despite the outstanding performance of 2D MXenes in photocatalysis, much more research is still needed to further identify the scientific and technological potential of these semiconductors in the future. Furthermore, scalable processing and commercialization are required to turn MXenes into real products. In addition, it is expected to be intensively studied in the future for the synthesis of 1D and 0D MXenes for photocatalysis.

REFERENCES

[1] Khataee A, Kıranşan M, Karaca S, et al. Photocatalytic ozonation of metronidazole by synthesized zinc oxide nanoparticles immobilized on montmorillonite. J Taiwan Inst Chem Eng. 2017;74:196–204.

[2] Xu D, Ma Y, Wang J, et al. Interfacial engineering of 2D/2D MXene heterostructures: Face-to-face contact for augmented photodegradation of amoxicillin. Chem Eng J. 2021;426:131246.

[3] Pablos C, Marugán J, van Grieken R, et al. Emerging micropollutant oxidation during disinfection processes using UV-C, UV-C/H_2O_2, UV-A/TiO_2 and UV-A/TiO_2/H_2O_2. Water Res. 2013;47(3):1237–1245.

[4] Ma D, Yi H, Lai C, et al. Critical review of advanced oxidation processes in organic wastewater treatment. Chemosphere. 2021;275:130104.

[5] Li M-F, Liu Y-G, Zeng G-M, et al. Graphene and graphene-based nanocomposites used for antibiotics removal in water treatment: A review. Chemosphere. 2019;226:360–380.

[6] Gusain R, Gupta K, Joshi P, et al. Adsorptive removal and photocatalytic degradation of organic pollutants using metal oxides and their composites: A comprehensive review. Adv Colloid and Interface. 2019;272:102009.

[7] Zhang C, Ma Y, Zhang X, et al. Two-dimensional transition metal carbides and nitrides (MXenes): Synthesis, properties, and electrochemical energy storage applications. Energy Environ Mater. 2020;3(1):29–55.

[8] Tawalbeh M, Mohammed S, Al-Othman A, et al. MXenes and MXene-based materials for removal of pharmaceutical compounds from wastewater: Critical review. Environ Res. 2023:115919.

[9] Siwal SS, Sheoran K, Mishra K, et al. Novel synthesis methods and applications of MXene-Based Nanomaterials (MBNs) for hazardous pollutants degradation: Future perspectives. Chemosphere. 2022;293:133542.

[10] Nikkhah A, Nikkhah H, Nouri A, et al. MXene: From synthesis to environment remediation. Chin J Chem Eng. 2023;61:260–280.

[11] Murali, G., et al. A review on MXene synthesis, stability, and photocatalytic applications. J ACS Nano. 2022;16(9):13370–13429.

[12] Ayodhya DJD, Materials R. A review of recent progress in 2D MXenes: Synthesis, properties, and applications. Diam Relat Mater. 2022;109634.

[13] Zhan X, Si C, Zhou J, et al. MXene and MXene-based composites: Synthesis, properties and environment-related applications. Nanoscale Horiz. 2020;5(2):235–258.

[14] Irfan S, Khan SB, Din MAU, et al. Retrospective on exploring MXene-based nanomaterials: Photocatalytic applications. Molecules. 2023;28(6):2495.

[15] Badr A, Taghizadeh Tabrizi A, Aghajani HJME. Two-dimensional Mxene structures: Comprehensive review on synthesis methods & applications. Metall Eng. 2021;24(3):236–259.

[16] Yin X, Liu D, Zhou L, et al. Structure and dimension effects on the performance of layered triboelectric nanogenerators in contact-separation mode. ACS Nano. 2018;13(1):698–705.

[17] Xiao R, Zhao C, Zou Z, et al. In situ fabrication of 1D CdS nanorod/2D Ti_3C_2 MXene nanosheet Schottky heterojunction toward enhanced photocatalytic hydrogen evolution. Appl Catal B: Environ. 2020;268:118382.

[18] Rana SS, Rahman MT, Salauddin M, et al. Electrospun PVDF-TrFE/MXene nanofiber mat-based triboelectric nanogenerator for smart home appliances. ACS Appl Mater Interfaces. 2021;13(4):4955–4967.

[19] Safaei M, Shishehbore MR. Energy conversion and optical applications of MXene quantum dots. J Mater Sci. 2021;1–37.

[20] Zhang K, Li D, Cao H, et al. Insights into different dimensional MXenes for photocatalysis. Chem Eng J. 2021;424:130340.

[21] Naguib M, Kurtoglu M, Presser V, et al. Two-dimensional nanocrystals produced by exfoliation of Ti_3AlC_2. Adv Mater. 2011;23(37):4248–4253.

[22] Fu B, Sun J, Wang C, et al. MXenes: Synthesis, optical properties, and applications in ultrafast photonics. Small. 2021;17(11):2006054.

[23] Iravani S, Varma RSJM. MXene-based photocatalysts in degradation of organic and pharmaceutical pollutants. Molecules. 2022;27(20):6939.

[24] Wei Y, Zhang P, Soomro RA, et al. Advances in the synthesis of 2D MXenes. Adv Mater. 2021;33(39):2103148.

[25] Sohan A, Banoth P, Aleksandrova M, et al. Review on MXene synthesis, properties, and recent research exploring electrode architecture for supercapacitor applications. Int J Energy Res. 2021;45(14):19746–19771.

[26] Zhang T, Pan L, Tang H, et al. Synthesis of two-dimensional Ti_3C_2Tx MXene using HCl+LiF etchant: Enhanced exfoliation and delamination. J Alloys Compd. 2017;695:818–826.

[27] Guan X, Yang Z, Zhou M, et al. 2D MXene nanomaterials: Synthesis, mechanism, and multifunctional applications in microwave absorption. Small Structures. 2022;3(10):2200102.

[28] Kumar JA, Prakash P, Krithiga T, et al. Methods of synthesis, characteristics, and environmental applications of MXene: A comprehensive review. Chemosphere. 2022;286:131607.

[29] Li T, Yao L, Liu Q, et al. Fluorine-free synthesis of high-purity Ti_3C_2Tx (T= OH, O) via alkali treatment. Angew Chem Int Ed. 2018;57(21):6115–6119.

[30] Li Y, Shao H, Lin Z, et al. A general lewis acidic etching route for preparing MXenes with enhanced electrochemical performance in non-aqueous electrolyte. Nat Mater. 2020;19(8):894–899.

[31] Mei J, Ayoko GA, Hu C, et al. Two-dimensional fluorine-free mesoporous Mo_2C MXene via UV-induced selective etching of Mo_2Ga_2C for energy storage. 2020;25:e00156.

[32] Mei J, Ayoko GA, Hu C, et al. Thermal reduction of sulfur-containing MAX phase for MXene production. Sustain Mater Technol. 2020;395:125111.

[33] Urbankowski P, Anasori B, Makaryan T, et al. Synthesis of two-dimensional titanium nitride Ti_4N_3 (MXene). Nanoscale. 2016;8(22):11385–11391.

[34] Solangi NH, Karri RR, Mazari SA, et al. MXene as emerging material for photocatalytic degradation of environmental pollutants. Coord Chem Rev. 2023;477:214965.

[35] Shi H, Zhang P, Liu Z, et al. Ambient-stable two-dimensional titanium carbide (MXene) enabled by iodine etching. Angew Chem Int Ed. 2021;60(16):8689–8693.

[36] Zada S, Dai W, Kai Z, et al. Algae extraction controllable delamination of vanadium carbide nanosheets with enhanced near-infrared photothermal performance. Angew Chem Int Ed. 2020;59(16):6601–6606.

[37] Parihar A, Singhal A, Kumar N, et al. Next-generation intelligent MXene-based electrochemical aptasensors for point-of-care cancer diagnostics. Nano-Micro Lett. 2022;14(1):100.

[38] Hosseini SF, Dorraji MSS. Rasoulifard MHJCPBE: Boosting photo-charge transfer in 3D/2D $TiO_2@Ti_3C_2$ MXene/Bi_2S_3 schottky/Z-scheme heterojunction for photocatalytic antibiotic degradation and H_2 evolution. Compos B Eng. 2023;110820.

[39] Biswal L, Mishra BP, Das S, et al. Nanoarchitecture of a $Ti_3C_2@TiO_2$ hybrid for photocatalytic antibiotic degradation and hydrogen evolution: Stability, kinetics, and mechanistic insights. Inorg Chem. 2023;62(19):7584–7597.

[40] Biswal L, Acharya L, Mishra BP, et al. Interfacial solid-state mediator-based Z-scheme heterojunction $TiO_2@ Ti_3C_2$/$MgIn_2S_4$ microflower for efficient photocatalytic pharmaceutical micropollutant degradation and hydrogen generation: Stability, kinetics, and mechanistic insights. ACS Appl Energy Mater. 2023;6(3):2081–2096.

[41] Saravanakumar K, Yun K, Maheskumar V, et al. Construction of novel In_2S_3/Ti_3C_2 MXene quantum dots/$SmFeO_3$ Z-scheme heterojunctions for efficient photocatalytic removal of sulfamethoxazole and 4-chlorophenol: Degradation pathways and mechanism insights. Chem Eng J. 2023;451:138933.

[42] Jiang D, Sun X, Wu X, et al. MXene-Ti_3C_2 assisted one-step synthesis of carbon-supported TiO_2/Bi_4NbO_8Cl heterostructures for enhanced photocatalytic water decontamination. Nanophotonics. 2020;9(7):2077–2088.

[43] Wu Z, Liang Y, Yuan X, et al. MXene Ti_3C_2 derived Z—Scheme photocatalyst of graphene layers anchored TiO_2/g–C_3N_4 for visible light photocatalytic degradation of refractory organic pollutants. Chem Eng J. 2020;394:124921.

[44] Cai T, Wang L, Liu Y, et al. Ag_3PO_4/Ti_3C_2 MXene interface materials as a schottky catalyst with enhanced photocatalytic activities and anti-photocorrosion performance. Appl Catal B: Environ. 2018;239:545–554.

[45] Chen Z, Xu X, Ding Z, et al. Ti_3C_2 MXenes-derived $NaTi_2(PO_4)_3$/MXene nanohybrid for fast and efficient hybrid capacitive deionization performance. Chem Eng J. 2021;407:127148.

[46] Li J, Li Z, Liu X, et al. Interfacial engineering of Bi_2S_3/$Ti_3C_2T_x$ MXene based on work function for rapid photo-excited bacteria-killing. Nat Commun. 2021;12(1):1224.

[47] Deng H, Li Z-J, Wang L, et al. Nanolayered Ti_3C_2 and $SrTiO_3$ composites for photocatalytic reduction and removal of uranium (VI). ACS Appl Nano Mater. 2019;2(4):2283–2294.

[48] Alhabeb M, Maleski K, Mathis TS, et al. Selective etching of silicon from Ti_3SiC_2 (MAX) to obtain 2D titanium carbide (MXene). Angew Chem. 2018;130(19):5542–5546.

[49] Huang X, Wu PJAFM. A facile, high-yield, and freeze-and-thaw-assisted approach to fabricate MXene with plentiful wrinkles and its application in on-chip micro-supercapacitors. Adv Funct Mater. 2020;30(12):1910048.

[50] Liu F, Zhou A, Chen J, et al. Preparation of Ti_3C_2 and Ti_2C MXenes by fluoride salts etching and methane adsorptive properties. Appl Surf Sci. 2017;416:781–789.

[51] Karlsson LH, Birch J, Halim J, et al. Atomically resolved structural and chemical investigation of single MXene sheets. Nano Lett. 2015;15(8):4955–4960.

[52] Peng C, Wei P, Chen X, et al. A hydrothermal etching route to synthesis of 2D MXene (Ti_3C_2, Nb_2C): Enhanced exfoliation and improved adsorption performance. Ceram Int. 2018;44(15):18886–18893.

[53] Husmann S, Budak Ö, Shim H, et al. Ionic liquid-based synthesis of MXene. Chem Comm. 2020;56(75):11082–11085.

[54] Sun W, Shah S, Chen Y, et al. Electrochemical etching of Ti_2AlC to Ti_2CTx (MXene) in low-concentration hydrochloric acid solution. J Mater Chem A. 2017;5(41):21663–21668.

[55] Pang S-Y, Wong Y-T, Yuan S, et al. Universal strategy for HF-free facile and rapid synthesis of two-dimensional MXenes as multifunctional energy materials. J Am Chem Soc. 2019;141(24):9610–9616.

[56] Yang S, Zhang P, Wang F, et al. Fluoride-free synthesis of two-dimensional titanium carbide (MXene) using a binary aqueous system. Angew Chem. 2018;130(47):15717–15721.

[57] Li M, Lu J, Luo K, et al. Element replacement approach by reaction with Lewis acidic molten salts to synthesize nanolaminated MAX phases and MXenes. J Am Chem Soc. 2019;141(11):4730–4737.

[58] Kamysbayev V, Filatov AS, Hu H, et al. Covalent surface modifications and superconductivity of two-dimensional metal carbide MXenes. Science. 2020;369(6506):979–983.

[59] Geng D, Zhao X, Li L, et al. Controlled growth of ultrathin Mo_2C superconducting crystals on liquid Cu surface. 2D Mater. 2016;4(1):011012.

[60] Ghazaly AE, Ahmed H, Rezk AR, et al. Ultrafast, one-step, salt-solution-based acoustic synthesis of Ti_3C_2 MXene. ACS Nano. 2021;15(3):4287–4293.

[61] Khatami M, Iravani SJCOIC. MXenes and MXene-based materials for the removal of water pollutants: Challenges and opportunities. Comments Inorg Chem. 2021;41(4):213–248.

[62] Zhou Y, Yu M, Liang H, et al. Novel dual-effective Z-scheme heterojunction with g-C_3N_4, Ti_3C_2 MXene and black phosphorus for improving visible light-induced degradation of ciprofloxacin. Appl Catal B: Environ. 2021;291:120105.

[63] Sharma V, Kumar A, Kumar A, et al. Enhanced photocatalytic activity of two dimensional ternary nanocomposites of ZnO–Bi_2WO_6–Ti_3C_2 MXene under natural sunlight irradiation. Chemosphere. 2022;287:132119.

[64] Kumar A, Majithia P, Choudhary P, et al. MXene coupled graphitic carbon nitride nanosheets based plasmonic photocatalysts for removal of pharmaceutical pollutant. Chemosphere. 2022;308:136297.

[65] Irfan S, Khan SB, Din MAU, et al. Retrospective on exploring MXene-based nanomaterials: Photocatalytic applications. Molecules. 2023;28(6):2495.

[66] Huang Y, Wang Y, Tang C, et al. Atomic modulation and structure design of carbons for bifunctional electrocatalysis in metal—Air batteries. Adv Mater. 2019;31(13):1803800.

[67] Du P, Carneiro JT, Moulijn JA, et al. A novel photocatalytic monolith reactor for multiphase heterogeneous photocatalysis. Appl Catal A: Gen. 2008;334(1–2):119–128.

[68] Zhao Q, Ju D, Deng X, et al. Morphology-modulation of SnO_2 hierarchical architectures by Zn doping for glycol gas sensing and photocatalytic applications. Sci Rep. 2015;5(1):7874.

[69] Zhou Y, Yu M, Zhan R, et al. Ti_3C_2 MXene-induced interface electron separation in g-C_3N_4/Ti_3C_2 MXene/$MoSe_2$ Z-scheme heterojunction for enhancing visible light-irradiated enoxacin degradation. Sep Purif Technol. 2021;275:119194.

[70] Jang J, Shahzad A, Woo SH, et al. Magnetic Ti_3C_2Tx (Mxene) for diclofenac degradation via the ultraviolet/chlorine advanced oxidation process. Environ Res. 2020;182:108990.

[71] Zhao Q, Wang J, Li Z, et al. Two-dimensional Ti_3C_2TX-nanosheets/Cu_2O composite as a high-performance photocatalyst for decomposition of tetracycline. Carbon Resour Convers. 2021;4:197–204.

[72] Wu S, Su Y, Zhu Y, et al. In-situ growing Bi/BiOCl microspheres on Ti_3C_2 nanosheets for upgrading visible-light-driven photocatalytic activity. Appl Surf Sci. 2020;520:146339.

[73] Shen J, Shen J, Zhang W, et al. Built-in electric field induced CeO_2/Ti_3C_2-MXene schottky-junction for coupled photocatalytic tetracyline degradation and CO_2 reduction. Ceram Int. 2019;45(18, Part A):24146–24153.

[74] Fang H, Pan Y, Yan H, et al. Facile preparation of Yb^{3+}/Tm^{3+} co-doped Ti_3C_2/Ag/Ag_3VO_4 composite with an efficient charge separation for boosting visible-light photocatalytic activity. Appl Surf Sci. 2020;527:146909.

[75] Cui C, Guo R, Ren E, et al. Facile hydrothermal synthesis of rod-like Nb_2O_5/Nb_2CTx composites for visible-light driven photocatalytic degradation of organic pollutants. Environ Res. 2021;193:110587.

[76] Shahzad A, Rasool K, Nawaz M, et al. Heterostructural TiO_2/Ti_3C_2Tx (MXene) for photocatalytic degradation of antiepileptic drug carbamazepine. Chem Eng J. 2018;349:748–755.

[77] Tu W, Liu Y, Chen M, et al. Carbon nitride coupled with Ti3C2-Mxene derived amorphous Ti-peroxo heterojunction for photocatalytic degradation of rhodamine B and tetracycline. Colloids Surf A: Physicochem Eng Asp. 2022;640:128448.

[78] Cao Y, Fang Y, Lei X, et al. Fabrication of novel $CuFe_2O_4$/MXene hierarchical heterostructures for enhanced photocatalytic degradation of sulfonamides under visible light. J Hazard Mater. 2020;387:122021.

[79] Xu D, Ma Y, Wang J, et al. Interfacial engineering of 2D/2D MXene heterostructures: Face-to-face contact for augmented photodegradation of amoxicillin. Chem Eng J. 2021;426:131246.

[80] Liu Q, Tan X, Wang S, et al. MXene as a non-metal charge mediator in 2D layered CdS@Ti_3C_2@TiO_2 composites with superior Z-scheme visible light-driven photocatalytic activity. Environ Sci. 2019;6(10):3158–3169.

[81] Zou X, Zhao X, Zhang J, et al. Photocatalytic degradation of ranitidine and reduction of nitrosamine dimethylamine formation potential over MXene–Ti_3C_2/MoS_2 under visible light irradiation. J Hazard Mater. 2021;413:125424.

[82] Liu H, Yang C, Jin X, et al. One-pot hydrothermal synthesis of MXene Ti_3C_2/TiO_2/BiOCl ternary heterojunctions with improved separation of photoactivated carries and photocatalytic behavior toward elimination of contaminants. Colloids Surf A. 2020;603:125239.

[83] Liu W, Sun M, Ding Z, et al. Ti_3C_2 MXene embellished g-C_3N_4 nanosheets for improving photocatalytic redox capacity. J Alloys Compd. 2021;877:160223.

[84] Ma Y, Xu D, Chen W, et al. Oxygen-vacancy-embedded 2D/2D NiFe-LDH/MXene Schottky heterojunction for boosted photodegradation of norfloxacin. App Surf Sci. 2022;572:151432.

[85] Bekele, T. Synthesis and characterization of C_eO_2/Ag_3PO_4 p-n heterojunction photocatalyst: Its photocatalytic activity for the degradation of Alizarin Yellow Dye, *Journal of Nanomaterials*, 2023; 2023, 7140181.

[86] Yi X, Yuan J, Tang H, et al. Embedding few-layer $Ti_3C_2T_x$ into alkalized g-C_3N_4 nanosheets for efficient photocatalytic degradation. J Colloid Interface Sci. 2020;571:297–306.

[87] Gao K, Hou L-A, An X, et al. BiOBr/MXene/gC$_3$N$_4$ Z-scheme heterostructure photocatalysts mediated by oxygen vacancies and MXene quantum dots for tetracycline degradation: Process, mechanism and toxicity analysis. Appl Catal B: Environ. 2023;323:122150.

[88] Liu K, Wang L, Fu T, et al. Oxygen-functionalized Ti$_3$C$_2$ MXene/exfoliated montmorillonite supported S-scheme BiOBr/Bi$_2$MoO$_6$ heterostructures for efficient photocatalytic quinolone antibiotics degradation. Chem Eng J. 2023;457:141271.

[89] Dai S, Wang Y, Xiao L, et al. 2D/2D/2D CuO-MXene-OCN heterojunction with enhanced photocatalytic removal of pharmaceuticals and personal care products: Characterization, efficiency and mechanism. J Alloys and Compd. 2022;919:165873.

[90] Xu Q, Wang P, Wang Z, et al. Aerosol self-assembly synthesis of g-C$_3$N$_4$/MXene/Ag$_3$PO$_4$ heterostructure for enhanced photocatalytic degradation of tetracycline hydrochloride. Colloids Surf A. 2022;648:129392.

[91] Xu Q, Wang P, Wang Z, et al. Aerosol self-assembly synthesis of g-C$_3$N$_4$/MXene/Ag$_3$PO$_4$ heterostructure for enhanced photocatalytic degradation of tetracycline hydrochloride. Colloids Surf A: Physicochem Eng Asp. 2022;648:129392.

[92] Abbas KK, AbdulkadhimAl-Ghaban AMH, Rdewi EH. Synthesis of a novel ZnO/TiO$_2$-nanorod MXene heterostructured nanophotocatalyst for the removal pharmaceutical ceftriaxone sodium from aqueous solution under simulated sunlight. J Environ Chem Eng. 2022;10(4):108111.

[93] Grzegórska A, Ofoegbu JC, Cervera-Gabalda L, et al. Magnetically recyclable TiO$_2$/MXene/MnFe$_2$O$_4$ photocatalyst for enhanced peroxymonosulphate-assisted photocatalytic degradation of carbamazepine and ibuprofen under simulated solar light. J Environ Chem Eng. 2023;11(5):110660.

[94] Qiang W, Qu X, Chen C, et al. Ti$_3$C$_2$ MXene derived (001)TiO$_2$/Ti$_3$C$_2$ heterojunctions for enhanced visible-light photocatalytic degradation of tetracycline. Mater Today Commun. 2022;33:104216.

[95] Guan Y, Cao Y, Ma S, et al. Metal-free g-C$_3$N$_5$ photocatalyst coupling MXenes Ti$_3$C$_2$ for tetracycline degradation: Insight for electron transfer mechanism, degradation mechanism and photothermal effect. J Alloys Compd. 2023;951:169864.

[96] Huang K, Li C, Wang L, et al. Layered Ti$_3$C$_2$ MXene and silver co-modified g-C$_3$N$_4$ with enhanced visible light-driven photocatalytic activity. Chem Eng J. 2021;425:131493.

[97] Hosseini SF, Seyed Dorraji MS, Rasoulifard MH. Boosting photo-charge transfer in 3D/2D TiO$_2$@Ti$_3$C$_2$ MXene/Bi$_2$S$_3$ Schottky/Z-scheme heterojunction for photocatalytic antibiotic degradation and H2 evolution. Compos B Eng. 2023;262:110820.

[98] Zhou Y, Yu M, Liang H, et al. Novel dual-effective Z-scheme heterojunction with g-C$_3$N$_4$, Ti$_3$C$_2$ MXene and black phosphorus for improving visible light-induced degradation of ciprofloxacin. Appl Catal B: Environ. 2021;291:120105.

9 Metal–Organic Framework (MOF)-Based Photocatalysts for Pharmaceutical Wastewater Treatment

Venkateshwaran Gopal and Ambika Selvaraj

9.1 INTRODUCTION

MOFs are a new class of porous material consisting of a metal core coupled with organic bridging ligands. This unique linkage leads to the production of a closed ring-like crystalline structure with unique surface textures, as shown in Figure 9.1. These MOFs have gained significant importance in the last two decades, mainly due to their tailorable properties with higher surface areas and tuneable internal structures [1, 2, 3]. These unique properties allow MOFs to find applications in various fields such as energy conversion [4], catalysis [5, 6], gas storage [7], light harvesting, and drug delivery [8].

MOFs have also gained recognition in wastewater treatment because of their potential to work as photocatalysts. This is possible by incorporating a redox-sensitive metal with a light-harvesting organic ligand, leading to the formation of metal–organic cluster [9, 10, 11]. The ligands in MOFs usually act as the primary receptors of light, which upon irradiation, generate electrons (e^-). This photoexcited e^- then moves from the highest occupied molecular orbital (HOMO) to the lowest unoccupied molecular orbital (LUMO) of MOFs, and subsequently e^- moves to the surface of metal nodes [11]. These e^- trapped at metal node and, consequently, the holes (h^+) generated at the valence band of the MOF, are responsible for the production of reactive oxygen species (ROS) such as $^{\bullet}OH$ and $O_2^{\bullet-}$. These three species (h^+, $^{\bullet}OH$, and $O_2^{\bullet-}$) play a significant role in the photodegradation of any organic pollutants. So far, researchers have successfully synthesized and tested various MOF combinations for the degradation of organic pollutants such as dyes, pesticides, pharmaceuticals, and personal care products [10, 11, 12].

The success of MOFs towards organic pollutant degradation is attributed mainly to their efficient e^-–h^+ pair separation and remarkable pore structure. The former e^-–h^+ pair separation is due to the semiconductor property of the tailored

DOI: 10.1201/9781003340164-9

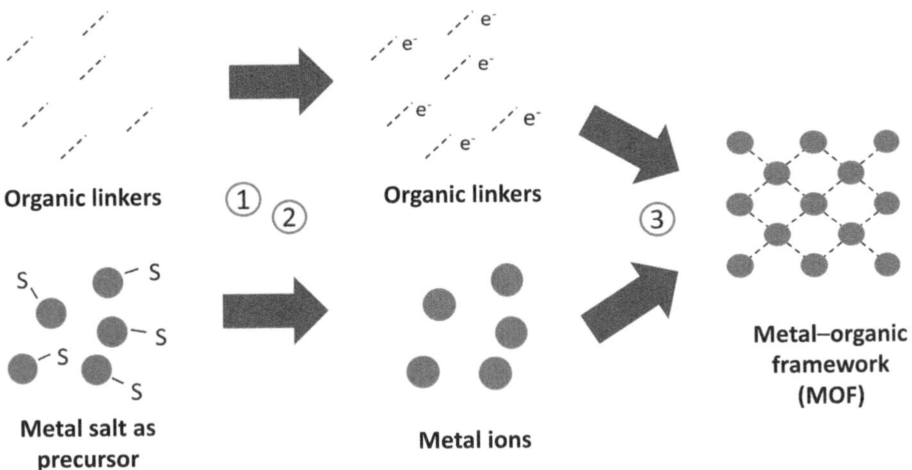

FIGURE 9.1 General structural components of MOF.

MOFs, which upon light excitation leads to the generation of e⁻h⁺ pairs and subsequent photoreactions. The latter property results in the direct adsorption of pollutants onto the surface of the MOFs through π–π interaction and surface complexation [10], thus allowing for the concentration of pollutants on the MOF surface and subsequent degradation by the reactive active species generated through photocatalysis.

In order to obtain an effective performance, the synthesized MOFs should be stable in water over a wide range of pH values and a longer detention time. The stability of thus produced MOFs is mainly dependent on the metal–ligand coordination and bond strength [10, 12]. Hence, the choice of metal–ligand combination should be selected very carefully. Pearson's hard and soft acids and bases (HSAB) is the principle most followed for choosing a suitable combination. According to this principle, hard metal acids can form a stable complex with hard base ligand compounds and vice versa [10]. Based on this principle, a series of representative MOF complexes have been designed, as shown in Figure 9.2. This includes (a) the MIL (materials from Institute Lavoisier) series: MOFs made up of high-valent-metal cations and carboxylate-based ligands [13]; (b) UiO-66 (University of Oslo): consisting of a zirconium (Zr) metal node and terephthalic acid ligands [14]; (c) ZIF (zeolitic imidazolate frameworks): made up of a weak acid and a weak base [10, 15]. Hence, appropriate metal–organic hybrids should be studied for their stability before going for real-time applications. As of now, MOFs made up of various transition metals (such as Fe, Cu, Co, and Ni), lanthanide series, alkaline earth metals, and organic ligands (consisting of C, H, O, N, and S) have been successfully explored by researchers [1, 16].

This chapter currently discusses various MOFs utilized for the degradation of pharmaceutical compounds from water. Different synthesis methods followed, along with the modifications employed to increase MOFs' performance, are explained in

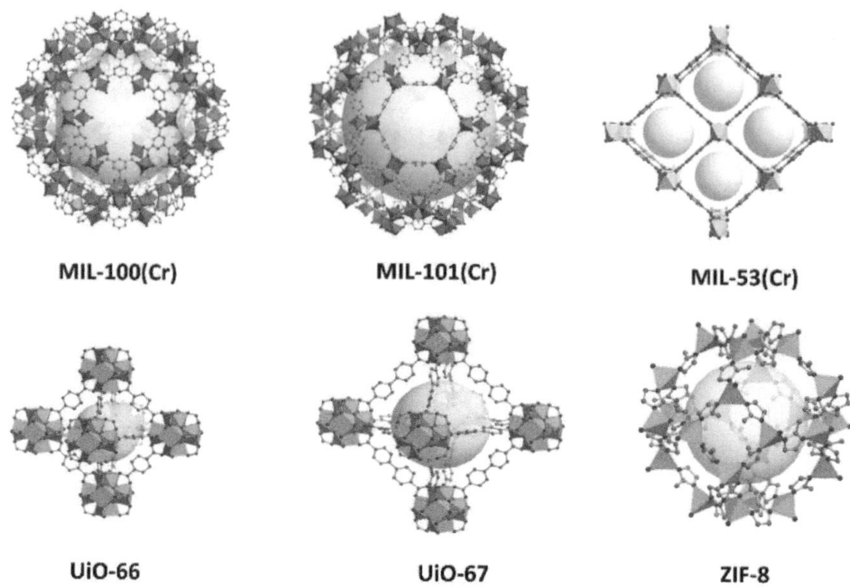

FIGURE 9.2 Some commonly reported MOF complexes.

subsequent sub-sections. Finally, the drawbacks associated with MOFs and their future opportunities are discussed in the latter part of this chapter.

9.2 SYNTHESIS OF MOFs

Generally, the synthesis of MOFs occurs in a liquid phase, where a metal core and organic ligand are mixed in the presence of a suitable solvent. The obtained product is then filtered and dried at high temperatures to produce the desired MOF [17]. The most commonly used methods for the synthesis of MOFs include the solvothermal/hydrothermal method, microwave-assisted synthesis, sonochemical method, electrochemical method, and mechanochemical method (Table 9.1) [1, 12]. The type of preparation method employed and optimum reaction conditions are crucial factors in synthesizing MOFs with the desired structure.

The synthesis of MOFs through a solvothermal process requires highly soluble organic solvents such as ethanol, methanol, acetone, acetonitrile, and dimethylformamide. The hydrothermal method is more eco-friendly than the solvothermal method, as it replaces the organic solvent with water [18]. However, the quality of MOFs obtained through this conventional method was found to be limited by various factors such as type of metal salt precursor, pH, reactant concentration, temperature, and reaction time [18]. The solvothermal/hydrothermal method is a simple process that gives a high yield with precise control over shape and size. But it requires a longer retention time than other techniques [6, 19].

TABLE 9.1

Various Synthesis Routes Adopted for MOF Preparation

Synthesis Method	MOF	Metal Salt	Ligand	Solvent	Reaction Conditions	Surface Area (m^2/g)	Ref.
Solvothermal	MIL-53(Fe)	$FeCl_3 \cdot 6H_2O$	H_2BTC	DMF	170°C, 24 hr	1890	[29]
	UiO-66(Zr)	$ZrCl_4$	H_2BTC	DMF	120°C, 24 hr	389	[30]
	MIL-125ML	TTIP	H_2BDC/NH_2BDC	DMF	150°C, 48 hr	1001	[31]
	MIL-53(Fe)	$FeCl_3 \cdot 6H_2O$	H_2BDC	DMF	393 K, 3 d	184	[32]
	Bi-BTC	$Bi(NO_3)_3 \cdot 5H_2O$	H_3BTC	DMF and MeOH	120°C, 24 hr	148	[33]
Hydrothermal	MIL-101(Cr)	$Cr(NO_3)_3 \cdot 9H_2O$	H_2BDC	H_2O	220°C, 8.5 hr	2518	[34]
	MIL-100(Fe)	$FeCl_3 \cdot 6H_2O$	H_3BTC	H_2O	150°C, 6 d	1766	[35]
	MIL-101(Fe)	$FeCl_3 \cdot 6H_2O$	H_2BDC	H_2O	110°C, 20 hr	253	[36]
	MIL-100(Fe)	$Fe(NO_3)_3 \cdot 9H_2O$	H_3BTC	H_2O	160°C, 12 hr	1203	[36]
Microwave-assisted synthesis	MIL-100(Fe)	$FeCl_3 \cdot 6H_2O$	H_3BTC	H_2O	403 K, 10 min	1245	[37]
	UiO-66-NH_2	$ZrOCl_2 \cdot 8H_2O$	NH_2BDC	DMF	120–180°C, 7 min	583–712	[38]
	NH_2-MIL-125	TTIP	NH_2BDC	DMF	200°C, 15 min	1030	[39]
	MOF-808	$ZrOCl_2 \cdot 8H_2O$	H_3BTC	DMF	Medium heat, 5 min	545	[40]
	MOF-177	$Zn(NO_3)_2 \cdot 6H_2O$	H_3BTB	NMP	378 K, 35 min	4197	[41]
	Fe_3O_4@ MIL-100(Fe)	Fe_3O_4 microspheres	H_3BTC	H_2O	150°C, 30 min	1244.62	[35]
	MIL-53(Fe)	$FeCl_3 \cdot 6H_2O$	H_2BDC	DMF	150°C, 30 min	124	[42]
Sonochemical	$Zn_6(IDC)_4(OH)_2(Hprz)_2$	$Zn(NO_3)_2 \cdot 6H_2O$	prz	H_2O	40 kHz, 90 min	889	[43]
	S-MOF-525	$ZrOCl_2 \cdot 8H_2O$	Benzoic acid	DMF	20 kHz, 3 hr	2557	[44]
	ZIF-8	$Zn(NO_3)_2 \cdot 6H_2O$	MeIM	DMF	20 kHz, 1 hr	1249	[45]
	AM-Co1	$Co(NO_3)_2 \cdot 6H_2O$	bpfb and bpta	DMF	–	748	[46]
	Cu-BDC-NH_2	$Cu(NO_3)_2 \cdot 3H_2O$	H_2BDC-NH_2	DMF	100 W, 10 min	234.7	[47]
	MOF-177	$Zn(NO_3)_2 \cdot 6H_2O$	H_3BTB	NMP	500 W, 20 kHz, 40 min	4898	[41]

(Continued)

TABLE 9.1 (*Continued*)

Various Synthesis Routes Adopted for MOF Preparation

Synthesis Method	MOF	Metal Salt	Ligand	Solvent	Reaction Conditions	Surface Area (m²/g)	Ref.
Mechanochemical	ZIF-8	Zinc acetate	MeIM	–	2 hr	1881	[48]
	ZIF-8	ZnO	MeIM	–	25 Hz, 45 min	1256	[45]
	Cu-BTC@GO	$Cu(OAc)_2 \cdot H_2O$	H_3BTC	–	40 Hz, 30 min	1362.7	[49]
	HKUST-1	$Cu(CH_3COO)_2 \cdot H_2O$	H_3BTC	–	50 Hz, 25 min	692–1944	[50]
	MOF-14	$Cu(CH_3COO)_2 \cdot H_2O$	H_3BTB	–	50 Hz, 25 min	1502	[50]
Electrochemical	$Zn(1,3\text{-bdc})_{0.5}bzi,$	Zn plate	H_2BDC, Hbzim	DMF/EtOH	Electrolyte: $NaNO_3$, 40–80 mA, 2 hr	14.404	[51]
	HKUST-1	Cu plate	H_3BTC	MTBS and EtOH/H_2O	2.7 V, 20–30 min	–	[52]
	MIL-100(Fe)	Fe plate	H_3BTC	EtOH/H_2O	2–20 mA/cm²	2200	[53]
	Fe_2O_3@MIL-88a	Fe plate	Fumaric acid	EtOH/H_2O	Electrolyte: KCl, 15 V, 30 min	128.5	[54]

DMF = N,N-dimethylformamide; MeOH = methanol; H_2BTC = 1,4-benzenedicarboxylic acid; H_3BTC = trimesic acid; TTIP = titanium isopropoxide; H_2BDC = terephthalic acid; NH_2BDC = 2-aminoterephthalic acid; prz = piperazine; bpfb = N,N-(1,4-phenylene) diisonicotinamide; bpta = N,N-bis(4-pyridinyl)-terephthalamide; H_3BTB = 4,4',4''-benzene-1,3,5-triyl-tribenzoic acid; MeIM = 2-methylimidazole; Hbzim = benzimidazole

In order to overcome the extended time requirement for crystallization, microwave-assisted synthesis was developed. By employing microwave synthesis, Vakili et al. [20] observed a 1000-fold decrease in the retention time as compared to conventional thermal heating. The time reduction was associated with directly heating the reaction medium with microwaves. Internally, these microwaves lead to the rotation of dipoles in the molecules of the reaction vessel, leading to the overall heating of the solution in a short time without any temperature gradient. In contrast, conventional thermal heating usually delivers heat through conduction, convection, and radiation [21]. Along with a lesser time requirement, the synthesis of MOFs through microwave synthesis also possesses reasonable control over particle size and morphology [20, 22]. However, this method requires careful monitoring of reaction conditions to prevent unwanted side reactions or thermal damage to MOF crystals.

In sonochemical synthesis, the reaction is brought by the phenomenon of cavitation. In this method, ultrasound waves are passed through the reaction chamber, resulting in the formation of small bubbles. These bubbles are further subjected to alternate compression and expansion by the ultrasound waves, forming local hotspots. The collapse of these bubbles results in the production of very high temperatures and pressures, thus assisting in the formation of nucleation sites and the crystallization of MOFs [23, 24].

Mechanochemical synthesis has recently attracted the attention of many researchers due to its ability to synthesize MOFs with large surface areas and uniform particle sizes. Unlike the previously mentioned methods, here, the MOF synthesis involves simple agitation and collision between substances, thus requiring no additional organic solvent [25]. Here, the agitation provided by milling leads to the cleavage of intramolecular bonds, which is followed by the chemical transformation of reactants into the desired MOF [26]. This method offers numerous benefits, such as a faster reaction rate, lower energy requirement, and potential for large-scale applications. However, to ensure the production of MOF materials with good quality and purity, it is crucial to carefully manage the milling conditions, including milling duration, milling speed, and the type of milling equipment employed. Also, the solid–solid interactions in mechanochemical synthesis lead to the production of MOFs with irregular crystallinity. For instance, Leng et al. [27] compared the degree of crystallinity of MIL-100(Cr) synthesized by both hydrothermal and mechanochemical methods. The MOF synthesized by the hydrothermal process was observed to possess a perfect octahedral crystal, whereas the latter method produced an MOF with lower crystallinity.

The electrochemical synthesis of MOFs involves the use of electric potential to drive the formation reactions. This method is gaining more importance due to its ability to produce MOFs continuously. In this method, the metal precursor is in the form of an anode electrode compared to the conventional salt precursor. As the reaction proceeds, the metal is released into the solution through continuous anode dissolution, which reacts with the organic ligand in the reaction chamber and forms the MOF [28]. Figure 9.3 shows the operational parameters of different synthesis methods.

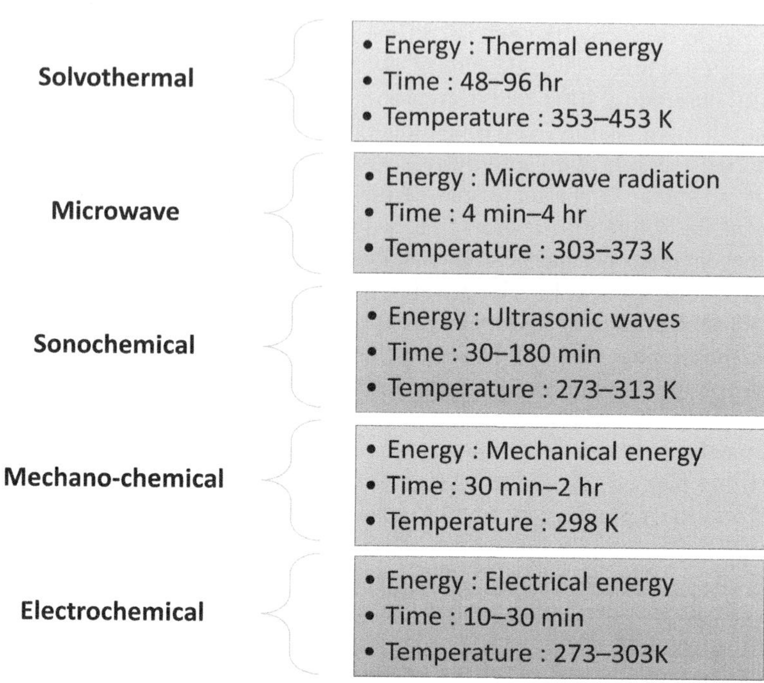

FIGURE 9.3 Different synthesis methods used for MOF preparation and their operational parameters.

9.3 MODIFIED MOFs FOR PHARMACEUTICAL WASTEWATER TREATMENT

Although pristine MOFs have attracted much attention due to their tailorable properties, there are still many drawbacks to using them as potential candidates for photocatalysts. So, in order to increase the overall performance, some modifications in the existing structure have been performed by researchers. These modifications are collectively aimed towards (a) increasing the light absorption spectrum of prepared MOFs to the visible region, (b) achieving the efficient separation of photogenerated e^-h^+ pairs, and (c) enhancing the stability of MOFs in water. Some commonly reported modifications in various studies include the functionalization of MOFs, doping with metal/metal oxide nanoparticles, and the introduction of heterojunctions by incorporating semiconductors, which are discussed as follows.

9.3.1 FUNCTIONALIZATION OF MOFs

One of the simplest ways to functionalize MOFs is to use an organic linker with a desired functional group as a building block. These organic bridging ligands or linkers are the primary receptors of incoming light irradiation [9, 18]. Hence, an appropriate ligand with a suitable bandgap is necessary to achieve efficient photodegradation. One major drawback of pristine MOFs is their efficient application only in the UV region. But UV represents only 4% of incoming solar irradiation. Thus, the modification of ligands with suitable functional groups may shift the photo-response of MOFs from UV to the visible spectrum. The use of N, O-donor, and N-containing ligands has been successfully reported for synthesizing functionalized MOFs. This is mainly attributed to the presence of a lone pair of e^- in N-containing ligands, which upon light irradiation, is directly transferred to the metal nodes [11, 12]. This was proven using an amine-functionalized MOF (NH_2-UiO-66) as a photocatalyst for tetracycline (TC) degradation. Under solar light irradiation, the author observed about 71.3% of TC degradation after 280 min [55]. The use of amide ($-NH_2$)-functionalized ligands has resulted in rise of HOMO level, thus resulting in an overall decrease in the bandgap of MOFs. This rise in HOMO level was attributed to the presence of a chromophore block with a conjugated π bond system, which extends the absorption spectrum of NH_2-UiO-66 to the visible-light region.

9.3.2 METAL DEPOSITION

Metal deposition on the surface of MOFs has seemingly attracted the attention of many researchers. This is due to the introduction of new energy levels, resulting in metal–metal charge transfer [1]. To say more clearly, these newly introduced energy levels act as trapping sites for photogenerated e^-, resulting in suppression of e^-h^+ pair recombination. This suppression in turn increases the overall efficiency of photocatalytic treatment. Recently, the introduction of nanoscale materials has gained quite popular attention in the field of water treatment. In contrast to their bulk metal predecessor, these nanoparticles (NPs) have distinct optical, electrical, and thermal properties [3, 56]. Incorporating such versatile metal nanoparticles into the structure of MOFs has seemingly improved photocatalytic activity.

Studies have shown that the deposition of noble metal NPs (such as Au, Ag, Pd, and Pt) on the surfaces of MOFs can extend the light absorption spectrum to the visible region. This phenomenon was ascribed to the local surface plasmon resonance effect created by electron oscillation in metal NPs [57]. Also, the metal NPs act as a reservoir and collect e^- from MOFs upon light irradiation. This transfer of e^- from MOF to NPs, by overcoming the Schottky barrier, prevents its recombination with h^+. This suppression of e^-h^+ recombination was evident in the study conducted by Liang et al. [3], where the author tested the efficiency of Pd NPs loaded MIL-100(Fe) for the degradation of theophylline and ibuprofen under visible-light irradiation. The author observed that the 1% Pd@ MIL-100(Fe) performed better than MIL-100(Fe) due to the efficient suppression of photoexcited e^- by Pd NPs, which was confirmed by the photoluminescence (PL) spectrum study conducted at a wavelength of 340 nm. The PL intensity of Pd@ MIL-100(Fe) was lower as compared to MIL-100(Fe), implying the longer detention time of charge carriers in Pt@ MIL-100(Fe). Similarly, Muelas-Ramos et al. [56] obtained a visible-light-active MOF by loading Ag, Pt, and Pd on MIL-125. The enhancement of visible-light absorption and reduction in the recombination rate of e^-h^+ was explained by UV–vis data and a reduction in the PL spectra of metal-loaded MIL-125 (Figure 9.4) [56].

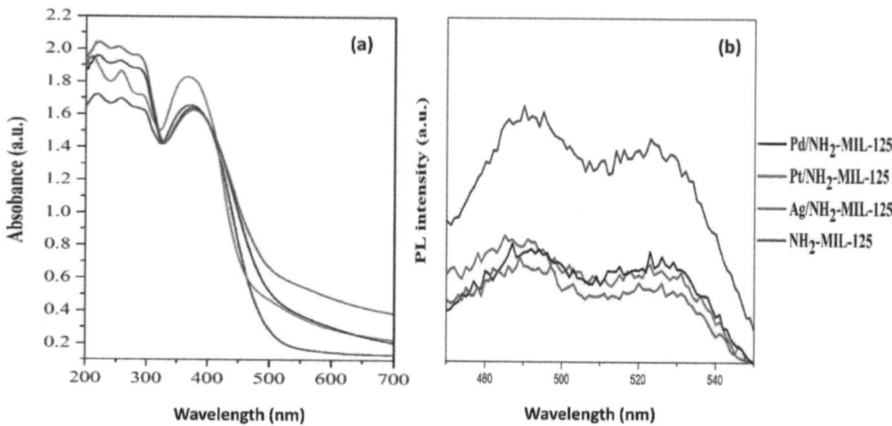

FIGURE 9.4 (a) UV-Vis absorption spectrum, (b) PL spectra of noble metal loaded MOF [56].

Similar results were obtained by Huang et al. [2], where the author prepared a hybrid compound (Ag/AgCl@MIL-88A(Fe)) for ibuprofen degradation. Almost complete removal of ibuprofen was observed after 3 hr reaction time. Here, Ag/AgCl NPs enhance the light-harvesting ability of MIL-88A(Fe) due to the surface plasmon resonance effect. As displayed in Figure 9.5a, the photoexcited h⁺ in metallic Ag NPs traps the excited e⁻ from MIL-88A(Fe), thus enhancing the separation of e⁻h⁺ in MIL-88A(Fe). As a result, the photoinduced h⁺ on MIL-88A(Fe) effectively participated in the oxidation of ibuprofen. Furthermore, there was negligible leaching of Ag (0.4%wt of total Ag) and Fe (0.25%wt of total Fe), even after four cycles (Figure 9.5b), thus showing the exceptional stability of the prepared composite MOF. Analogous to this, Fan et al. [58] observed the degradation of acetaminophen by Ag/AgCl@ZIF-8, where about 98.5% degradation achieved in 60 min, which was 77.9% higher than the pristine ZIF-8. The formation of a Ag/AgCl@ZIF-8 composite was confirmed by the XRD results showing distinct peaks. Here, the Ag/AgCl improved the light-harvesting ability of ZIF-8 due to the plasmon resonance effect of Ag⁰, which was evident from the UV–vis spectra of ZIF-8, Ag/AgCl, and Ag/AgCl@ZIF-8, as shown in Figure 9.6 [58].

Apart from the introduction of new energy levels, the binary metal frameworks can also modify the adsorption capacity of the MOF [32, 59]. The synergistic effect of adsorption and photocatalytic behaviour may result in superior performance even at low pollutant concentrations. Recently, Chatterjee et al. [59] prepared a novel bimetallic MOF comprising Cu and Fe (MIL-53(Fe-Cu)) via solvothermal route. The prepared catalyst exhibited an adsorption capacity of about 190.4 mg/g for ciprofloxacin. As a result, the maximum degradation efficiency of about 74.48% and 57.88% was observed under UV and visible-light irradiation.

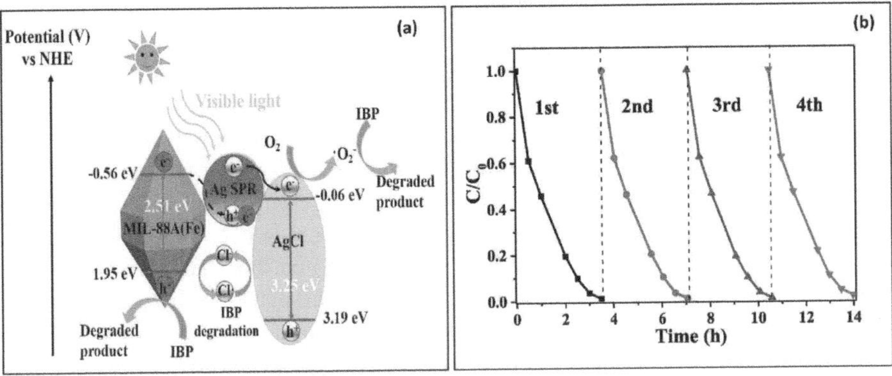

FIGURE 9.5 (a) Schematic of e⁻h⁺ separation mechanism in Ag/AgCl@MIL-88A(Fe), (b) Stability performance of Ag/AgCl@MIL-88A(Fe) [2].

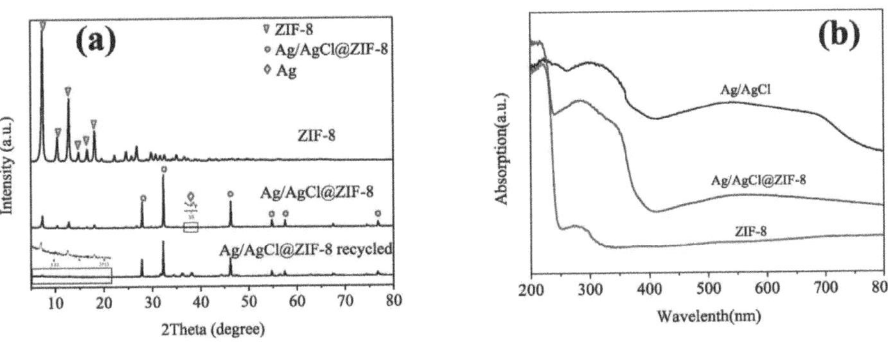

FIGURE 9.6 (a) XRD pattern of Ag/AgCl@ZIF-8 and pristine ZIF-8, (b) UV–vis spectra of Ag/AgCl, Ag/AgCl@ZIF-8 and ZIF-8 [58].

9.3.3 Semiconductor Incorporation

The development of heterojunctions by semiconductor incorporation is another of the most commonly used methods for improving the photocatalytic activity of MOFs. These heterojunctions prevent the recombination of e⁻h⁺ pairs, resulting in the increased production of ROS. In this strategy, the MOF porosity network may facilitate the diffusion of semiconductor materials, resulting in more active centres. Researchers have successfully utilized various semiconductors such as TiO_2, ZnO_2, $BiVO_4$, and In_2S_3 to improve the charge transfer efficiency of MOFs [9, 16].

Apart from the compounds mentioned earlier, heterojunctions can be formed by anchoring the MOFs with graphene-based materials. These graphene compounds, namely graphene oxide (GO) and reduced graphene oxide (rGO), have an abundant π–π conjugated system, allowing for rapid charge transfer and attenuated recombination of e⁻h⁺ pairs [6]. For example, Yang et al. [6] reported 93% degradation of amoxicillin (AMX) by using MIL-68(In)-NH$_2$/(GrO) after 120 min of visible-light irradiation.

Aside from the common metal oxide semiconductors, some metal-free organic semiconductors have also been successfully studied. For instance, Panneri et al. [60] synthesized a g-C$_3$N$_4$/ZIF-8 novel photocatalyst for tetracycline degradation under natural sunlight. The author observed that the prepared MOF presented a good adsorption ability (420 mg/g), thus offering sufficient contact between the pollutant and catalyst surface. The increased surface area resulted in over 96% degradation of tetracycline in 1 hr reaction time [60]. It should be noted that the ZIF-8 MOF has a very good porous network, but it lacks efficient light-harvesting capacity and has poor charge separation. Alternatively, g-C$_3$N$_4$ is a narrow-band-gap (2.6–2.7 eV) material with remarkable chemical and physical stability [61]. Therefore, the drawbacks were removed by adding g-C$_3$N$_4$ to ZIF-8, creating a novel compound with ideal synergy. This hypothesis was verified by the N$_2$ adsorption–desorption isotherm, where the pristine g-C$_3$N$_4$ and ZIF-8 showed type IIb (adsorption on a sheet-like structure) and type I isotherms (microporous adsorption), and the composite g-C$_3$N$_4$/ZIF-8 exhibited a combination of both characteristics (sheet adsorption at high P/P$_0$ values and microporous adsorption at relatively low P/P$_0$ values), as shown in Figure 9.7.

In continuation, many researchers have tried utilizing ternary and hybrid compounds to increase the degradation output of MOFs. For instance, Jiang et al. [62] used Pd-phosphotungstic acid (PTA)-MIL-100(Fe) for the photodegradation of theophylline and ibuprofen under visible-light irradiation. They observed that the

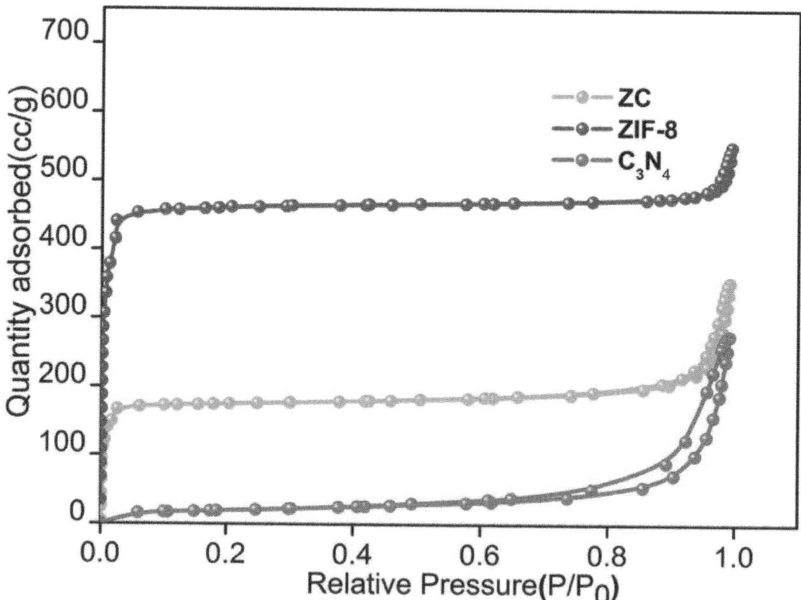

FIGURE 9.7　N$_2$ adsorption–desorption isotherms of C$_3$N$_4$, ZIF-8, and ZC (C$_3$N$_4$/ZIF-8) composite [60].

Pd-PTA-MIL-100(Fe) showed about 99.5% removal of theophylline. The superior performance was attributed to the perfect synergy between the three compounds (i.e., the enhancement of electrical conduction by Pd, fast e^- transport by PTA, and visible-light adsorption by MIL-100(Fe)) and well-matched bond structure. Similarly, Askari et al. [63] prepared a ternary heterojunction using $CuWO_4$-Bi_2S_3-ZIF-67 and tested its efficiency towards the degradation of antibiotics (metronidazole and cephalexin) under visible light. The authors observed a nearly 5.5-fold increase in reaction rate compared to that of pristine ZIF-67 [63]. Recently, a composite catalyst made up of ZIF-8, GO, Fe_3O_4, and semiconductor (ZnO) showed excellent performance both in terms of pollutant degradation as well as stability in water. The prepared composite was able to rapidly degrade the target pollutants (sulphamethazine, metronidazole, norfloxacin, and acetaminophen) within 1 hr of solar irradiation [64]. Also, the catalyst was able to perform well even after ten cycles, thus showing its stability. From the UV–vis diffuse reflectance spectra, the authors observed that the incorporation of GO enhanced the light absorption of Fe_3O_4 and ZnO.

It should be noted that numerous experts have undertaken a great deal of research using various metal–organic unit combinations. However, the primary goal of these efforts is to create a photocatalyst that is not only functional but also stable for use in real-time applications. Table 9.2 summarizes various MOFs that researchers have utilized for the remediation of pharmaceutical pollutants.

9.4 MECHANISM OF PHOTOCATALYTIC DEGRADATION OF PHARMACEUTICALS BY MOFs

With respect to the growing interest in photocatalytic properties of MOFs, it is essential to understand the underlying degradation mechanism. There are three significant steps involved in the photocatalysis of pharmaceuticals by MOFs: (a) photo-absorption, (3) photogenerated e^-h^+ pair separation/transport, and (c) redox reaction on the surface of the photocatalyst [1]. This photo-absorption refers to the absorption of incident light by MOF units. The organic bridging linker/ligand acts as a primary receptor of light. This is an important step, as the necessary photoreactions are initiated only upon light irradiation with an energy greater than the bandgap of the MOFs. In many studies, researchers have used UV–vis diffuse reflectance spectra for determining the photo-absorption capacity of photocatalysts.

In the second step, the photoexcited e^- moves from the HOMO to LUMO of MOFs and is subsequently transported to the metal node. This phenomenon is represented as a ligand-to-metal charge transfer mechanism (LMCT) [11, 36]. This separation of charge carriers is considered the most crucial step, as its recombination may stop the photoreactions in MOFs. Hence, the charge-separation efficiency of an MOF is an important parameter to be considered before its application for treatment. PL spectra and electrochemical impedance spectroscopy (EIS) are the most common studies used to analyze the charge-separation efficiency of a prepared photocatalyst. The PL spectra usually measure the photocurrent response, and the higher the intensity of PL spectrum, the higher the recombination of e^-h^+ pairs. Hence, a lower PL intensity refers to the efficient suppression of e^-h^+ pairs recombination.

TABLE 9.2

MOF-Based Photocatalysts for the Degradation of Pharmaceutical Pollutants

MOF	Synthesis Method	Target Compound	Parameters	Degradation Efficiency (%)	Reference
MIL-53(Fe)/H_2O_2	Solvothermal	CA	500 W xenon lamp, MOF = 5 mg, CBZ = 40 mg/L, H_2O_2 = 20 mM	98.2%, 240 min	[32]
MIL-53(Fe)/H_2O_2	Solvothermal	CBZ	500 W xenon lamp, MOF = 5 mg, CBZ= 40 mg/L, H_2O_2 = 20 mM	90.1%, 240 min	[32]
Pd@MIL-100(Fe) with H_2O_2	Solvothermal	TP	Visible lamp, MOF = 5 mg, TP = 20 mg/L, H_2O_2 = 40 μL	99.5%, 150 min	[3]
Pd@MIL-100(Fe) with H_2O_2	Solvothermal	IP	Visible lamp, MOF = 5 mg, IP = 20 mg/L, H_2O_2 = 40 μL	99.5%, 150 min	[3]
Pt@ NH_2-MIL-125(Ti)	Solvothermal	ACE	Xenon lamp, MOF = 250 mg/L, ACE = 5 mg/L	100%, 180 min	[56]
Pd@ NH_2-MIL-125(Ti)	Solvothermal	ACE	Xenon lamp, MOF = 250 mg/L, ACE = 5 mg/L	100%, 180 min	[56]
Ag@ NH_2-MIL-125(Ti)	Solvothermal	ACE	Xenon lamp, MOF = 250 mg/L, ACE = 5 mg/L	100%, 180 min	[56]
MIL-88 A (Fe)@Zn In_2S_4	Solvothermal	SMZ	500 W xenon lamp, MOF = 20 mg, SMZ = 20 mg/L	99.6%, 60 min	[19]
AgI/UiO-66	Solvothermal	SMZ	300W xenon lamp, MOF = 25 mg, SMZ = 5 mg/L	99.6%, 20 min	[65]
In_2S_3@ MIL-125(Ti)	Solvothermal	TC	300 W xenon lamp, MOF = 30 mg, TC = 46 mg/L	63.3%, 60 min	[66]
MIL-68(In)-NH_2/ GrO	Solvothermal	AMX	300 W xenon lamp, MOF = 120 mg, AMX = 20 mg/L	93%, 120 min	[6]
g-C_3N_4/ MIL-68(In)-NH_2	Solvothermal	IP	300 W xenon lamp, MOF = 30 mg, IP = 20 mg/L	93%, 120 min	[61]
Bi_2WO_6/ NH_2-UiO-66	Hydrothermal	CFX	100 W xenon lamp, MOF = 25 mg, CFX = 100 mg/L	83.1%, 60 min	[18]
$CuWO_4$/Bi_2S_3/ ZIF-67	Hydrothermal	MDZ	LED, MOF = 0.3 g/L, MDZ = 20 mg/L	95.6%, 80 min	[63]

TABLE 9.2 (*Continued*)
MOF-Based Photocatalysts for the Degradation of Pharmaceutical Pollutants

MOF	Synthesis Method	Target Compound	Parameters	Degradation Efficiency (%)	Reference
$CuWO_4/Bi_2S_3/$ ZIF-67	Hydrothermal	CPX	LED, MOF = 0.3 g/L, CPX = 20 mg/L	90.1%, 80 min	[63]
MIL-101(Fe)/ TiO_2	Solvothermal	TC	300 W xenon lamp, MOF = 1 g/L, TC = 20 mg/L	92.76%, 10 min	[67]
NH_2-MIL-125(Ti)/BiOBr	Solvothermal	TC	Xenon lamp, MOF = 10 mg, TC = 25 mg/L	88%, 90 min	[9]
Ag/AgCl@ MIL-88 A(Fe)	Solvothermal	IP	500 W Xenon lamp, MOF = 20 mg, IP = 10 mg/L	100%, 210 min	[2]
Pd-PTA-MIL-100(Fe)	Solvothermal	TP	Xenon lamp, MOF = 5 mg, TP = 20 mg/L	88.2%, 150 min	[62]
Pd-PTA-MIL-100(Fe)	Solvothermal	IP	Xenon lamp, MOF = 5 mg, IP = 20 mg/L	99.5%, 150 min	[62]
Basolite F300 with H_2O_2	–	LIN	1500 W xenon lamp, MOF = 100 mg/L, LIN = 10 mg/L, H_2O_2 = 100 mg/L	95%, 60 min	[68]
ZnO/Fe_3O_4-GO/ ZIF	Solvothermal	NFX	MOF = 500 mg/L, NFX = 10 mg/L	99.3%, 45 min	[64]
ZnO/Fe_3O_4-GO/ ZIF	Solvothermal	ACE	MOF = 500 mg/L, ACE = 10 mg/L	99.1%, 45 min	[64]
ZnO/Fe_3O_4-GO/ ZIF	Solvothermal	SMZ	MOF = 500 mg/L, SMZ = 10 mg/L	97.7%, 45 min	[64]

CA = clofibric acid; CBZ = carbamazepine; TP = theophylline; IP = ibuprofen; ACE = acetaminophen; SMZ = sulphamethoxazole; TC = tetracycline; AMX = amoxicillin; CFX = ciprofloxacin; MDZ = metronidazole; CPX = cephalexin; LIN = lincomycin; NFX = norfloxacin

This charge transfer (excitation of e^-) leads to the generation of h^+ with high oxidizing capacity in the HOMO of MOFs [11]. The h^+ can either directly oxidize the organic pollutant or react with water molecules to generate $^\bullet OH$. Similarly, the photoexcited e^- can react with oxygen to form $O_2^{\bullet-}$. In the last step, these reactive species, such as $^\bullet OH$, $O_2^{\bullet-}$, e^-, and h^+, undergo redox reactions leading to the mineralization of parent compounds into simpler units such as CO_2 and H_2O. However, the predominant charge species responsible for degradation may vary depending on the bandgap of the photocatalyst and pollutant type. Hence, researchers have conducted scavenger studies to access the primary species of interest. Some of the commonly used scavengers for $^\bullet OH$, $O_2^{\bullet-}$, e^-, and h^+ are t-butanol, benzoquinone, carbon tetrachloride, and isopropanol. For instance, Yang et al. [6] found that $O_2^{\bullet-}$ and h^+ are the active species responsible for the degradation of AMX by MIL-68(In)-NH_2/

GrO. Another study reported that $^{•}OH$ and $O_2^{•-}$ species are primarily responsible for SMZ degradation by using AgI/UIO-66 as an MOF photocatalyst [65]. The most commonly reported degradation pathway for pharmaceuticals is as follows: The first step is the hydroxylation of the double bond and aromatic rings in target compounds. This is followed by the cleavage of intermediates to produce compounds with single aromatic rings and simple aliphatic dicarboxylic acids. The final step involves further degrading daughter compounds into CO_2 and H_2O [69, 70].

Currently, there has been increased research focussed on the use of adsorptive photocatalysts as an MOF. These catalysts possess both adsorptive and photocatalytic behaviours. The adsorption of target pollutants on the surface of the photocatalyst occurs due to the presence of active sites. Figure 9.8 shows some of the commonly reported adsorptive mechanisms through which pollutant sorption occurs, including π–π interactions, hydrogen bonding, and electrostatic interactions [71, 72]. Since most of the photocatalytic reactions occur on the surface of the catalyst, this pre-adsorption increases the overall rate of degradation and reduces the reactive time. This was evident from the study conducted by Wang et al. [66], where the author used the novel core–shell In_2S_3@MIL-125(Ti) for tetracycline (TC) degradation. The larger pore diameter, good adsorption capacity (157.2 mg/g), and mesoporous structure of the prepared MOF overcame the blockage of pores by TC, thus providing more contact between the catalyst and TC [66].

Apart from the pathways mentioned earlier, there have been a few studies where the authors tried to improve the degradation efficiency of MOFs by the addition of oxidizing agents such as H_2O_2 and persulphate. For instance, the addition of H_2O_2 to Fe-containing MOFs has resulted in a photo-Fenton-like reaction, resulting in the increased production of $^{•}OH$ radicals [62, 68]. Equations 9.1–9.5 describe the degradation mechanism of Pd-PTA-MIL-100(Fe) in the presence of H_2O_2. Initially, upon light irradiation, e^- and h^+ are generated in Pd-PTA-MIL-100(Fe). The photoexcited e^- may react with the oxidizing species such as H_2O_2 and produce $^{•}OH$. Subsequently, h^+ reacts with H_2O to form $^{•}OH$. Further, Fe species and H_2O_2 result in a Fenton-like reaction, producing more $^{•}OH$ [62]. Altogether, they increase the production of $^{•}OH$, thus enhancing the degradation efficiency of pharmaceuticals in water.

$$Pd@MIL-100(Fe) \xrightarrow{hv} Pd@MIL-100(Fe)(h^+ + e^-) \tag{9.1}$$

$$e^- + H_2O_2 \longrightarrow OH + OH^- \tag{9.2}$$

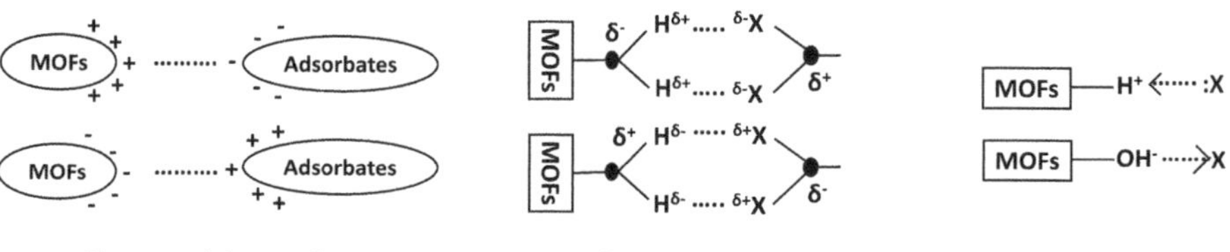

FIGURE 9.8 Various adsorptive mechanisms for pharmaceutical pollutant removal [72].

$$h^+ + H_2O \longrightarrow OH + H^+ \tag{9.3}$$

$$\equiv Fe^{3+} + H_2O_2 \longrightarrow \equiv Fe^{2+} + HOO^\bullet + H^+ \tag{9.4}$$

$$\equiv Fe^{2+} + H_2O_2 \longrightarrow \equiv Fe^{3+} + OH + OH^- \tag{9.5}$$

9.5 CHALLENGES AND FUTURE PERSPECTIVES

The choice of an MOF as a photocatalyst mainly depends on its photocatalytic activity and stability in the environment. The photocatalytic activity of an MOF is associated with its visible-light absorption and efficient suppression of e^-h^+ pair recombination. As explained in the previous sections, numerous studies have reported the synthesis of MOFs with good photocatalytic properties. However, the stability of these prepared MOFs acts as a hinge for their applications in the water environment.

The instability arises from the dissolution of metal ions or organic ligands in water under harsh environmental conditions (acidic or alkaline). Also, the nucleophilic components of water, such as hydroxides, amines, and alkoxides, frequently outcompete the ligand and destroy MOF complexes, as shown in Equations 9.6 and 9.7 [10]. These leached compounds or the by-products formed are sometimes more toxic than the pollutant of concern. Also, so far, no toxicity studies on these leached compounds have been reported. Hence, this issue should be addressed before going for commercialization.

$$M - L + H_2O \rightarrow M - OH_2 \ldots L \tag{9.6}$$

$$M - L + H_2O \rightarrow M - (OH) + (L - H) \tag{9.7}$$

The reusability of an MOF is another essential criterion that should be taken into account. Reusability is closely related to the stability of the MOF. If an MOF becomes unstable, it loses its chemical structure during the first cycle; as a result, the subsequent use of the unstable MOF for the next cycle becomes inefficient. Reusability is also affected by the deactivation of active sites by the reactants or products, resulting in reduced efficiency. Even though studies have reported the use of MOFs with good stability and reusability, the commercial implementation of these MOFs is still unclear. Recently, there has been an increased usage of hybrid MOFs that have synergistic effects for processes such as adsorption and photocatalysis. These hybrids have been found to possess good stability and recyclability. Hence, further research should be conducted in this direction to extend their application to real-world scenarios. Apart from the challenges mentioned earlier, there are certain factors that limit the application of MOF for real-case scenarios, which are listed as follows:

- Synthesis method: The type of synthesis method used for obtaining an MOF is also a crucial step. For instance, most of the studies reported used the solvothermal method. But it has several disadvantages, such as longer detention time and the requirement of organic solvents such as DMF. The use of toxic or expensive solvents limits the production of MOFs for large-scale applications. Therefore, a sustainable and environmentally benign synthesis approach with a high output rate is required.

- Porosity: Although MOFs are known for their highly porous structure, they have some drawbacks. For instance, large-sized pollutants may find it hard to access the pores of MOFs, thus limiting their application to specific scenarios.
- Fouling: Since pharmaceutical wastewater contains sophisticated organic chemicals, there is a possibility for the potential build-up of contaminants on the MOF surface. Over time, this may lead to fouling of the MOF, resulting in a reduction in adsorption and degradation ability.
- Selectivity: The selectivity of the MOF is yet another important factor that determines the efficiency of a treatment system. Some MOFs might not be sufficiently selective to target particular pollutants in pharmaceutical effluent, which might result in inadequate treatment or the removal of beneficial substances. Thus, before deciding on a choice of MOF, its selectivity must be studied thoroughly.
- Characterization: The complex structures of MOFs makes them difficult to characterize, thus requiring the use of sophisticated techniques.
- Reproducibility: Different research groups have followed variable synthetic methods or reaction conditions for preparing MOFs. This has resulted in the production of MOFs with variable textural properties, which affects the reproducibility of results. This can restrict the possibility for the broad usage of MOFs in many applications and makes it challenging to compare the findings from various investigations.

9.6 CONCLUSIONS

The utilization of MOFs as photocatalysts for the remediation of pharmaceutical pollutants shows significant potential in tackling the rising challenge of water pollution. This is mainly associated with the unique properties of MOFs, such as their high surface areas, tuneable structures, and outstanding stability, which makes them appealing candidates for pharmaceutical wastewater treatment. Moreover, the tunability of MOFs enables the modification of their structures and compositions to improve photocatalytic efficiency, selectivity, and stability. So far, various studies have been carried out where the authors have successfully adjusted the pore diameters, surface functional group, and metal centres of MOFs to optimize their performance for selective pharmaceutical pollutants, assuring optimal removal from wastewater. However, certain obstacles remain in the actual application of MOFs as photocatalysts, including scale-up, long-term stability, and cost-effective synthesis. Hence, these issues must be addressed to ensure the successful incorporation of MOFs into existing water treatment systems. In conclusion, the use of MOFs as photocatalysts offers enormous promise for the treatment of pharmaceutical pollutants in wastewater. Further research and development are needed to improve their performance, overcome technical problems, and pave the road for broader use as an effective and sustainable solution in water treatment operations.

REFERENCES

1. Mumtaz N, Javaid A, Imran M, et al. Nanoengineered metal-organic framework for adsorptive and photocatalytic mitigation of pharmaceuticals and pesticide from wastewater. Environ Pollut. 2022;308:119690.
2. Huang W, Jing C, Zhang X, et al. Integration of plasmonic effect into spindle-shaped MIL-88A (Fe): Steering charge flow for enhanced visible-light photocatalytic degradation of ibuprofen. Chem Eng J. 2018;349:603–612.
3. Liang R, Luo S, Jing F, et al. A simple strategy for fabrication of Pd@ MIL-100 (Fe) nanocomposite as a visible-light-driven photocatalyst for the treatment of Pharmaceuticals and Personal Care Products (PPCPs). Appl Catal B. 2015;176:240–248.
4. Wei T, Zhang M, Wu P, et al. POM-based metal-organic framework/reduced graphene oxide nanocomposites with hybrid behaviour of battery-supercapacitor for superior lithium storage. Nano Energy. 2017;34:205–214.
5. Jiao L, Wang Y, Jiang HL, et al. Metal—Organic frameworks as platforms for catalytic applications. Adv Mater. 2018;30(37):1703663.
6. Yang C, You X, Cheng J, et al. A novel visible-light-driven In-based MOF/graphene oxide composite photocatalyst with enhanced photocatalytic activity toward the degradation of amoxicillin. Appl Catal B. 2017;200:673–680.
7. Ding M, Flaig RW, Jiang HL, et al. Carbon capture and conversion using metal—Organic frameworks and MOF-based materials. Chem Soc Rev. 2019;48(10):2783–2828.
8. Karmakar A, Prabakaran V, Zhao D, et al. A review of Metal-Organic Frameworks (MOFs) as energy-efficient desiccants for adsorption driven heat-transformation applications. Appl Energy. 2020;269:115070.
9. Wang Y, Feng S, Wu W, et al. Ionic liquid-assisted solvothermal construction of NH_2-MIL-125 (Ti)/BiOBr heterojunction for removing tetracycline under visible light. Opt Mater. 2022;123:111817.
10. Wen Y, Zhang P, Sharma VK, et al. Metal-organic frameworks for environmental applications. Cell Reports Phy Sci. 2021;2(2):100348.
11. Zhang X, Wang J, Dong XX, et al. Functionalized metal-organic frameworks for photocatalytic degradation of organic pollutants in environment. Chemosphere. 2020;242:125144.
12. Bedia J, Muelas-Ramos V, Peñas-Garzón M, et al. (2019). A review on the synthesis and characterization of metal organic frameworks for photocatalytic water purification. Catalysts. 2019;9(1):52.
13. Yuan S, Feng L, Wang K, et al. Stable metal—Organic frameworks: Design, synthesis, and applications. Adv Mater. 2018;30(37):1704303.
14. Cavka JH, Jakobsen S, Olsbye U, et al. A new zirconium inorganic building brick forming metal organic frameworks with exceptional stability. J Am Chem Soc. 2008;130(42):13850–13851.
15. Wang B, Côté AP, Furukawa H, et al. Colossal cages in zeolitic imidazolate frameworks as selective carbon dioxide reservoirs. Nature. 2008;453(7192):207–211.
16. Gautam S, Agrawal H, Thakur M, et al. Metal oxides and metal organic frameworks for the photocatalytic degradation: A review. J Env Chem Eng. 2020;8(3):103726.
17. Dai J, Xiao X, Duan S, et al. Synthesis of novel microporous nanocomposites of ZIF-8 on multiwalled carbon nanotubes for adsorptive removing benzoic acid from water. Chem Eng J. 2018;331:64–74.
18. Karuppannan R, Mohan S, Do To: Amine-functionalized metal—Organic framework integrated bismuth tungstate (Bi_2WO_6/NH_2-UiO-66) composite for the enhanced solar-driven photocatalytic degradation of ciprofloxacin molecules. New J Chem. 2021;45(48):22650–22660.

19. Yuan R, Qiu J, Yue C, et al. (2020). Self-assembled hierarchical and bifunctional MIL-88A (Fe)@ $ZnIn_2S_4$ heterostructure as a reusable sunlight-driven photocatalyst for highly efficient water purification. Chem Eng J. 2020;401:126020.

20. Vakili R, Xu S, Al-Janabi N, et al. Microwave-assisted synthesis of zirconium-based Metal Organic Frameworks (MOFs): Optimization and gas adsorption. Microporous Mesoporous Mater. 2018;260:45–53.

21. Zorainy MY, Alalm MG, Kaliaguine S, et al. Revisiting the MIL-101 metal—Organic framework: Design, synthesis, modifications, advances, and recent applications. J Mater Chem A. 2021;9(39):22159–22217.

22. Liang W, D'Alessandro DM. Microwave-assisted solvothermal synthesis of zirconium oxide based metal—Organic frameworks. Chem Comm. 2013;49(35):3706–3708.

23. Vaitsis C, Sourkouni G, Argirusis C. Sonochemical synthesis of MOFs. In: Metal-Organic Frameworks for Biomedical Applications, Woodhead Publishing; 2020. p. 223–244.

24. Khan NA, Jhung SH. Synthesis of Metal-Organic Frameworks (MOFs) with microwave or ultrasound: Rapid reaction, phase-selectivity, and size reduction. Coord Chem Rev. 2015;285:11–23.

25. Klimakow M, Klobes P, Thünemann AF, et al. Mechanochemical synthesis of metal–Organic frameworks: A fast and facile approach toward quantitative yields and high specific surface areas. Chem Mater. 2010;22(18):5216–5221.

26. Stolle A, Ranu B. Ball milling towards green synthesis: Applications, projects, challenges. Royal Soc Chem. 2014;2:34–55.

27. Leng K, Sun Y, Li X, et al. Rapid synthesis of metal—Organic frameworks MIL-101 (Cr) without the addition of solvent and hydrofluoric acid. Cryst Growth Des. 2016;16(3):1168–1171.

28. Varsha MV, Nageswaran G. Direct electrochemical synthesis of metal organic frameworks. J Electrochem Soc. 2020;167(15):155527.

29. Jiang D, Zhu Y, Chen M, et al. Modified crystal structure and improved photocatalytic activity of MIL-53 via inorganic acid modulator. Appl Catal B. 2019;255:117746.

30. Dong W, Wang D, Wang H, et al. Facile synthesis of In_2S_3/UiO-66 composite with enhanced adsorption performance and photocatalytic activity for the removal of tetracycline under visible light irradiation. J Colloid Interface Sci. 2019;535:444–457.

31. Salimi M, Esrafili A, Jafari AJ, et al. Photocatalytic degradation of cefixime with MIL-125 (Ti)-mixed linker decorated by g-C_3N_4 under solar driven light irradiation. Colloids Surf A: Physicochem Eng Asp. 2019;582:123874.

32. Gao Y, Yu G, Liu K, et al. Integrated adsorption and visible-light photodegradation of aqueous clofibric acid and carbamazepine by a Fe-based metal-organic framework. Chem Eng J. 2017;330:157–165.

33. Nguyen VH, Van Tan L, Lee T, et al. Solvothermal synthesis and photocatalytic activity of metal-organic framework materials based on bismuth and trimesic acid. Sustain Chem Pharm. 2021;20:100385.

34. Huo Q, Qi X, Li J, et al. Preparation of a direct Z-scheme α-Fe_2O_3/MIL-101 (Cr) hybrid for degradation of carbamazepine under visible light irradiation. Appl Catal B. 2019;255:117751.

35. Li S, Cui J, Wu X, et al. Rapid in situ microwave synthesis of Fe_3O_4@ MIL-100 (Fe) for aqueous diclofenac sodium removal through integrated adsorption and photodegradation. J Hazard Mater. 2019;373:408–416.

36. Wang D, Jia F, Wang H, et al. Simultaneously efficient adsorption and photocatalytic degradation of tetracycline by Fe-based MOFs. J Colloid Interface Sci. 2018;519:273–284.

37. Sánchez NC, Palomino GT, Cabello CP. Sulfonic-functionalized MIL-100-Fe MOF for the removal of diclofenac from water. Microporous Mesoporous Mater. 2023;348:112398.

38. Solís RR, Peñas-Garzón M, Belver C, et al. Highly stable UiO-66-NH$_2$ by the microwave-assisted synthesis for solar photocatalytic water treatment. J Environ Chem Eng. 2022;10(2):107122.

39. Solís RR., Gómez-Avilés A, Belver C, et al. Microwave-assisted synthesis of NH$_2$-MIL-125 (Ti) for the solar photocatalytic degradation of aqueous emerging pollutants in batch and continuous tests. J Environ Chem Eng. 2021;9(5):106230.

40. Wang F, Xue R, Ma Y, et al. Study on the performance of a MOF-808-based photocatalyst prepared by a microwave-assisted method for the degradation of antibiotics. RSC Adv. 2021;11(52):32955–32964.

41. Jung DW, Yang DA, Kim J, et al. Facile synthesis of MOF-177 by a sonochemical method using 1-methyl-2-pyrrolidinone as a solvent. Dalton Trans. 2010;39(11):2883–2887.

42. Chakhtouna H, Benzeid H, Zari N, et al. Microwave-assisted synthesis of MIL—53 (Fe)/biochar composite from date palm for ciprofloxacin and ofloxacin antibiotics removal. Sep Purif Technol. 2023;308:122850.

43. Abazari R, Mahjoub AR, Shariati J. Synthesis of a nanostructured pillar MOF with high adsorption capacity towards antibiotics pollutants from aqueous solution. J Hazard Mater. 2019;366:439–451.

44. Yu K, Lee YR, Seo JY, et al. Sonochemical synthesis of Zr-based porphyrinic MOF-525 and MOF-545: Enhancement in catalytic and adsorption properties. Microporous Mesoporous Mater. 2021;316:110985.

45. Lee YR, Jang MS, Cho HY, et al. ZIF-8: A comparison of synthesis methods. Chem Eng J. 2015;271:276–280.

46. Bagheri S, Pazoki F, Heydari A. Ultrasonic synthesis and characterization of 2D and 3DMetal—Organic frameworks and their application in the oxidative amidation reaction. ACS Omega. 2020;5(34):21412–21419.

47. Dastbaz A, Karimi-Sabet J, Moosavian MA. Sonochemical synthesis of novel decorated graphene nanosheets with amine functional Cu-terephthalate MOF for hydrogen adsorption: Effect of ultrasound and graphene content. Int J Hydrog Energy. 2019;44(48):26444–26458.

48. Taheri M, Enge TG, Tsuzuki T. Water stability of cobalt doped ZIF-8: A quantitative study using optical analyses. Mater Today Chem. 2020;16:100231.

49. Li Y, Miao J, Sun X, et al. Mechanochemical synthesis of Cu-BTC@ GO with enhanced water stability and toluene adsorption capacity. Chem Eng J. 2016;298:191–197.

50. Klimakow M, Klobes P, Rademann K, et al. Characterization of mechanochemically synthesized MOFs. Microporous Mesoporous Mater. 2012;154:113–118.

51. de Lima Neto OJ, de Oliveira Frós AC, Barros BS, et al. Rapid and efficient electrochemical synthesis of a zinc-based nano-MOF for Ibuprofen adsorption. New J Chem. 2019;43(14):5518–5524.

52. Van Assche TR, Desmet G, Ameloot R, et al. Electrochemical synthesis of thin HKUST-1 layers on copper mesh. Microporous Mesoporous Mater. 2012;158:209–213.

53. Campagnol N, Van Assche T, Boudewijns T, et al. High pressure, high temperature electrochemical synthesis of metal—Organic frameworks: Films of MIL-100 (Fe) and HKUST-1 in different morphologies. J Mater Chem A. 2013;1(19):5827–5830.

54. Kuzharov AA, Gritsai MA, Butova VV, et al. One-step electrochemical synthesis of γ-Fe$_2$O$_3$@ MIL-88a magnetic composite for heterogeneous Fenton-like catalysis. Ceram Int. 2022;48(23):34864–34876.

55. Kaur M, Mehta SK, Kansal SK. Construction of multifunctional NH$_2$-UiO-66 metal-organic framework: Sensing and photocatalytic degradation of ketorolac tromethamine and tetracycline in aqueous medium. Environ Sci Pollut Res. 2022;1–21.

56. Muelas-Ramos V, Belver C, Rodriguez JJ, et al. Synthesis of noble metal-decorated NH$_2$-MIL-125 titanium MOF for the photocatalytic degradation of acetaminophen under solar irradiation. Sep Purif Technol. 2021;272:118896.

57. Wang M, Tang Y, Jin Y. Modulating catalytic performance of metal—Organic framework composites by localized surface plasmon resonance. ACS Catal. 2019;9(12):11502–11514.

58. Fan G, Zheng X, Luo J, et al. Rapid synthesis of Ag/AgCl@ ZIF-8 as a highly efficient photocatalyst for degradation of acetaminophen under visible light. Chem Eng J. 2018;351:782–790.

59. Chatterjee A, Jana AK, Basu JK. A binary MOF of iron and copper for treating ciprofloxacin-contaminated wastewater by an integrated technique of adsorption and photocatalytic degradation. New J Chem. 2021;45(37):17196–17210.

60. Panneri S, Thomas M, Ganguly, P et al. C_3N_4 anchored ZIF 8 composites: Photoregenerable, high capacity sorbents as adsorptive photocatalysts for the effective removal of tetracycline from water. Catal Sci Technol. 2017;7(10):2118–2128.

61. Cao W, Yuan Y, Yang C, et al. In-situ fabrication of g-C_3N_4/MIL-68 (In)-NH_2 heterojunction composites with enhanced visible-light photocatalytic activity for degradation of ibuprofen. Chem Eng J. 2020;391:123608.

62. Liang R, Huang R, Ying S, et al. Facile in situ growth of highly dispersed palladium on phosphotungstic-acid-encapsulated MIL-100 (Fe) for the degradation of pharmaceuticals and personal care products under visible light. Nano Res. 2018;11(2):1109–1123.

63. Askari N, Beheshti M, Mowla D, et al. Fabrication of $CuWO_4$/Bi_2S_3/ZIF67 MOF: A novel double Z-scheme ternary heterostructure for boosting visible-light photodegradation of antibiotics. Chemosphere. 2020;251:126453.

64. Chen L, Peng J, Wang F, et al. ZnO nanorods/Fe_3O_4-graphene oxide/metal-organic framework nanocomposite: Recyclable and robust photocatalyst for degradation of pharmaceutical pollutants. Environ Sci Pollut Res. 2021;28(17):21799–21811.

65. Wang C, Xue Y, Wang P, et al. Effects of water environmental factors on the photocatalytic degradation of sulfamethoxazole by AgI/UiO-66 composite under visible light irradiation. J Alloys Compd. 2018;748:314–322.

66. Wang H, Yuan X, Wu Y, et al. In situ synthesis of In_2S_3@ MIL-125 (Ti) core—shell microparticle for the removal of tetracycline from wastewater by integrated adsorption and visible-light-driven photocatalysis. Appl Catal B. 2016;186:19–29.

67. He L, Dong Y, Zheng Y, et al. A novel magnetic MIL-101 (Fe)/TiO_2 composite for photo degradation of tetracycline under solar light. J Hazard Mater. 2019;361:85–94.

68. Kontogiannis A, Evgenidou E, Nannou C, et al. MOF-based photocatalytic degradation of the antibiotic lincomycin enhanced by hydrogen peroxide and persulfate: Kinetics, elucidation of transformation products and toxicity assessment. J Environm Chem Eng. 2022;10(4):108112.

69. Mestre AS, Carvalho AP. Photocatalytic degradation of pharmaceuticals carbamazepine, diclofenac, and sulfamethoxazole by semiconductor and carbon materials: A review. Molecules. 2019;24(20):3702.

70. Yang B, Wei T, Xiao K, et al. Effective mineralization of anti-epilepsy drug carbamazepine in aqueous solution by simultaneously electro-generated H_2O_2/O_3 process. Electrochim Acta. 2018;290:203–210.

71. Jin E, Lee S, Kang E, et al. Metal-organic frameworks as advanced adsorbents for pharmaceutical and personal care products. Coord Chem Rev. 2020;425:213526.

72. Hasan Z, Jhung SH. Removal of hazardous organics from water using Metal-Organic Frameworks (MOFs): Plausible mechanisms for selective adsorptions. J Hazard Mater. 2016;283:329–339.

10 Polymer-Based Photocatalysts for Pharmaceutical Wastewater Treatment

Nurul Izzati Izhar, Zul Adlan Mohd Hir, Nurul Syarima Nadia Sazman, Nur Ramadhan Mohamad Azaludin, Shaari Daud, and Hartini Ahmad Rafaie

10.1 INTRODUCTION

Water has been used in our life for drinking water, household products, cosmetic products, and pharmaceutical products. Clean water is also important for plants and animals for their survival. The production of high-quality products in human daily life via various methods which use many substances and chemicals has led to an issue of water pollution. The uncontrolled disposal of these products into water bodies creates long-term adverse effects on the marine life and human health. The pollutants that come from industrial processes contain intractable organic molecules or metal ions, which are persistent in water bodies make the situation become more deteriorated. Thus, it makes the river water not suitable for drinking or recreational activities. Besides that, this water pollution will also disrupt the ecosystem of the aquatic life.

The pharmaceutical industry in one of the sectors that has developed rapidly in all countries in the world due to the needs and demands which require large freshwater consumption, generating wastewater. This has led to the spread of active pharmaceutical constituents such as endocrine-disrupting compounds and other hazardous chemicals in the marine environment and can have a harmful effect on aquatic and human life. Thus, there is a need to treat this wastewater in order to get back the clean water and save water resources for the next generations. Basically, there are a few methods that can be used in wastewater remediation such as conventional methods and advanced methods. Conventional methods use chemical, physical, and biological approaches, while advanced method that can be used are Fenton reactions (i.e., a process involving Fe^{2+} and H_2O_2), sonolysis, and heterogeneous photocatalysis [1, 2].

Current advanced methods in wastewater treatment mostly involve the use of polymer-based materials in combination with the photocatalytic approach [3, 4].

This approach has become more trending due to its longer durability, efficiency, easy separation, and high recyclability. These polymer-based photocatalytic materials are recognized because of their effectiveness in degrading hazardous pharmaceutical compounds with easy separation from the media without the need for post-treatment filtration. Based on the unique properties and efficiency of the polymer-based photocatalytic materials, it is hoped that we can rely on this advanced method for wastewater remediation in order to secure more water resources in the future.

10.2   FEATURES OF POLYMERS AS GOOD SUPPORTS

Most recent studies on metal oxide nanoparticles have focused on how well they undergo photocatalysis in a powder slurry, yet they hinder large-scale or industrial practice and do not meet the purified water process and sustainable management [5, 6]. This is because catalyst recovery requires energy-intensive separation of the suspension using photocatalytic reactions. Moreover, due to the limited adsorption capability of these metal oxide catalysts towards the target pollutants, reactive oxygen species (ROS) are scavenged by background organic material, which reduces overall performance. Thus, immobilizing photocatalytic nanoparticles in materials with high target-pollutant adsorption capacities has the added advantage of facilitating simple catalyst recovery, aside from putting priority pollutants close to the catalyst sites and therefore increasing the photoactivity [7].

Numerous substances have been investigated for immobilizing photocatalysts and include metals, zeolites, fabric, glass, ceramics, cellulosic materials, wood, and polymers. The selection of acceptable supports for catalysts is mostly influenced by properties including toughness, mechanical and thermal stability, ease of availability, chemical inertness, non-toxicity, low density, and low cost. Aside from that, sustained performance and recyclability are also required in the selection, especially for industrial implementations [8, 9]. Above all, of the various support materials that have been employed, polymer-based substrates stand out and have drawn a lot of attention in recent years due to their numerous benefits, especially since they are simple to handle and manufacture (e.g., simple mixing, extrusion, moulding, etc.) [7, 10]. Aside from that, polymers are a viable option due to the fact that they are low in cost and exhibit good properties of mechanical strength, chemical inertness, low density, and also stability [11].

Polymers are among the most commonly utilized materials worldwide owing to their advantageous characteristics, including high strength, low weight, and low cost [6, 10]. Furthermore, the functional groups dispersed along the polymer backbone are another appealing feature, as they serve as anchoring sites for other materials [12]. A polymer-supported membrane has the photocatalyst immobilized inside or on the surface of the substrate. Various synthetic and natural polymers have been utilized as the bases for membranes that incorporate a variety of photocatalysts [13]. As previously reported, various types of polymers have been used as support, such as polystyrene, polyvinyl chloride, expanded polystyrene, polyaniline, poly(methyl methacrylate), poly(tetrafluoro ethane), poly(3-hexylthiophene), poly(styrene)-copoly(4-vinylpyridine), low density polyethylene, carbon nitride polymer, polycaprolactone, and polycarbonate, to name a few [11]. This has greatly increased the stability of these compounds with regards to low surface area and high porosity, hydrophilicity, and surface roughness, as well as their functionality with other materials in the degradation of contaminants.

10.2.1 SURFACE AREA AND POROSITY

Polymers are among significant supporting media owing to the production procedure of polymers, which is simple and affordable. Numerous methods, including sol–gel, electrophoresis, spray pyrolysis, photo-etching, and hydrothermal procedures supported by polymers, can be used to immobilize nano-photocatalysts [14]. These methods may produce one or more layers of photocatalysts on the substrate's surface. To enhance the interaction between the semiconductor and light, porosity is a crucial property to have. Porous polymers that are photo-actively promoted are very desirable for customized organic synthesis, environmental remediation, and scaled-up heterogeneous photocatalysis [15]. It is possible to produce very porous materials with impressive surface areas and tuneable surface structures by changing linkers using functional groups [16, 17].

Even though the catalysts in powder form have a bigger surface area and are significantly more effective at removing pollutants than the ones supported by polymers, these types of photocatalysts face some significant challenges when oxidizing the pollutants [13]. Higher loading of the photocatalyst amounts on the substrate membrane may result in a decrease in surface area, owing to the agglomeration of the particles [14, 18]. Therefore, prior to the final immobilization, an adjustment of the amount of photocatalyst loading is necessary. Composites made of porous granular materials with qualities like outstanding hydraulic characteristics are another example of a technology that suggests a potential remedy to the technological restrictions of direct usage in large-scale water treatment and purification [19].

10.2.2 HYDROPHILICITY AND SURFACE ROUGHNESS

Surface roughness is an essential factor to take into account when trying to improve the photocatalytic performance of a film photocatalyst. An increasing surface roughness would enhance the wettability of the films and consequently improve the film photocatalyst's engagement with oxygen, water, and the organic contaminants for the reaction to take place [20, 21]. Moreover, hydrophilicity is induced by the hydroxyl, carboxyl, and ether groups that are held in the matrix. Hence, superior wettability with water as the reaction medium might efficiently speed up the interactions between the catalyst and reactants, leading to better performance. The excellent water dispersibility of composites ultimately promotes higher free ·OH radical generation [22]. Additionally, it was claimed that the hydrophilic properties of the photocatalyst's immobilized layers enhance the adsorption of hazardous organic molecules from the aqueous solution. Mainly owing to these inherent qualities of thin/film photocatalysts, photocatalytic technologies have become commercially viable [14].

Pristine polymers generally exhibit hydrophobic properties. Given that no single polymer can fulfil every requirement for a specific application, combining materials with functional features can produce a composite material that can successfully attract both polar and nonpolar organic contaminants and deliver them to the photo-catalysts for total degradation [7]. The growth behaviour of the grains and the various types of nanoparticles incorporated in the polymer structure may also have an impact on the thickness of the films and their surface roughness; moreover, smaller grains are anticipated to have lower surface roughness and to be less dense [3].

10.2.3   FUNCTIONALITY WITH OTHER MATERIALS

Metal oxides are widely employed in the field of environmental remediation due to their inherent adsorption/desorption, photo-removal, redox, acid-base, and magnetic features. When combined with polymeric supports, these characteristics increase the mechanical strength, stability, and hydraulic capabilities of the latter in filtration systems. Different kinds of metal oxides composites exist, including those that utilize magnetically active polymeric particles and polymer-supported nanoparticles distributed in a polymeric matrix, both of which have the potential to be recycled for the removal of pollutants from water. Additionally, after being used for different remediation purposes like oil/water separation and the removal of harmful inorganic and organic compounds, some metal oxides such as Fe_3O_4 possess magnetic properties that are extensively exploited for the nanocomposite's recovery depending on different organic, inorganic, and polymeric matrices that are being used [19, 23].

Another method to increase photocatalytic activity is to functionalize magnetic photocatalysts using conducting polymers (CPs) [24]. When CPs are used as sensitizing agents in photocatalytic systems, the photocatalytic efficiency is boosted. This is because p-conjugated systems have a high number of electron-rich species that can be transferred in the conduction band of the semiconductor [25]. The utilization of CPs helps to tune the band gap of semiconductors, slowing down the rate of photogenerated electron–hole pair recombination [25, 26]. According to calculations using density functional theory (DFT), CPs are active photocatalysts when exposed to visible light, as they have low-band-gap materials with broad absorption spectra. To avoid photogenerated electron–hole recombination, CPs can work as p-type materials and create a p–n junction with metal oxides [26]. Accordingly, the photo-removal of pollutants using CPs and CP-based nanocomposites displays first-order reaction kinetics. Excellent photocatalytic performance has been demonstrated by the nanocomposites of CPs with metal oxides and other supporting materials under various light irradiations. The typical characteristics of hybrid materials are high photo-response and delayed charge carrier recombination. A greater surface area and different exciton–plasmon interactions are present in the hybrid system of CPs and metal oxides, which allows for significant absorption. Thus, in order to increase the effectiveness of photocatalysis, the current research has focused on the development of hybrid materials with several functionalities. The most researched photocatalysts for wastewater remediation include those based on polyaniline, poly-pyrrole, polythiophene, polycarbazole, poly(1-naphthylamine), and poly(ethylene dioxy thiophene)-based nanocomposites [26].

Another critical consideration is the selection of membranes with the highest stability because, in the case of polymeric membranes, the possibility of membrane

TABLE 10.1

Immobilization of Photocatalysts Using Various Types of Polymer-Based Supporting Materials

Photocatalyst	Supporting Material	Contaminants	References
TiO_2	Polyethersulphone membranes Polyvinylidene fluoride membranes	Diclofenac Ibuprofen	[28]
	Polyvinylidene fluoride dual-layer hollow-fibre membranes	Trimethoprim Metoprolol Carbamazepine	[29]
NiS	Magnetic Fe_3O_4@polypyrrole core/shell	Cephalexin	[23]
$ZnO/Ag_2CO_3/Ag_2O$	Polyvinylidene fluoride	Ibuprofen	[30]
Ag@ZnO/TiO_2	Polyvinylpyrrolidone membrane	Tetracycline hydrochloride	[31]
$LaFeO_3/CoFe_2O_4$	Polyaniline	Clozapine	[32]
ZnO	Polyethersulphone sheet	Acetaminophen	[5]

damage induced by continuous exposure to UV light cannot be avoided [14, 27]. It has been noted that with the presence of substances containing sulphur such as polyethersulphone and polysulphone, membranes exhibit lower UV resistance than polytetrafluoroethylene and poly(vinylidene fluoride) membranes [14, 18]. Nonetheless, the UV resistance of membranes should not be generalized from a particular polymer material but rather should be based on the findings of the actual polymer formulations (e.g., various additives), how well the photocatalyst is immobilized onto the polymer surface, and other conditions that might influence the overall physicochemical characteristics of the resulting membranes [14]. Table 10.1 displays the immobilization of several photocatalysts using various types of polymers as supporting materials.

10.3   TYPES OF POLYMER-BASED PHOTOCATALYTIC MATERIALS

A polymer is a massive macromolecule that is primarily made up of repeated subunits and has been considered an insulating compound [33]. Interestingly, a broad subclass of substances known as polymers can be classified depending on their sources, applications, structures, thermal properties, polymerization techniques, and manufacturing processes [34]. Accordingly, there are natural and synthetic polymers, depending on whether the polymer is manufactured in a laboratory or occurs naturally. This is because the classification of polymeric substances can be based on the origin of the material or the repeating units in this manner. Additionally, among the synthetic polymers, organic and inorganic polymers can be distinguished based on the existence or lack of carbon atoms in the polymer chain backbone. In regard to aspects linked to the photocatalytic characteristics, we primarily focus on synthetic and natural polymers in this chapter that have high potential in photocatalytic water treatment applications.

10.3.1 Natural Polymer-Based Photocatalytic Materials

Natural polymers produced from waste materials or renewably sourced materials can also act as attractive organic supports for inorganic semiconductor metal oxides. In order to create photocatalytic hybrid materials, polysaccharides such as lignin, cellulose, hemicellulose, chitin, chitosan, starch, and xylan can be employed as an abundant industrial raw feedstock. These substances also have great sustainability and biodegradability [34]. Natural rubber is another example of natural polymers other than cellulose, lignin, and starch. Natural polymers started to be chemically altered in the 1800s to create a variety of materials, including gun cotton, vulcanized rubber, and celluloid [35]. One of the most widely recognized biopolymers in nature, which is chitosan, it is typically produced by deacetylating chitin. It has special features like biodegradability, crystallinity, non-toxicity, biocompatibility, film-forming ability, hydrophilicity, an adsorbing nature, solubility in a variety of organic solvents, and chelation of metal ions by hydroxyl and amino functional groups [36].

10.3.2 Synthetic Polymer-Based Photocatalytic Materials

Due to the ubiquitous nature of organic monomers from the petrochemical sector, synthetic organic polymers are among more explored and used polymeric materials than inorganic polymers, and this chapter focuses mostly on these classes of polymers [35]. In contrast, inorganic polymers have been recognized for a long time, and lately, there is an increase in the number of studies in this area. Common inorganic polymers have backbones made of silicon, oxygen, phosphorus, or nitrogen. Some of the organic polymers' drawbacks include their tendency for degradation at relatively low temperatures. Thus, with the presence of oxygen or radiation, the drawbacks can be avoided via the utilization of inorganic polymers.

10.3.3 Coordination Polymers

Previously, coordination polymers were made of metal clusters and organic ligands, offering a novel type of hybrid materials that can be employed as photocatalysts. These crystalline substances have a favourable pore size and structure, a significant dispersion of active sites, and tuneable adsorption characteristics [34]. The coordination polymers offer some benefits over the conventional photocatalysts (such as ZnO and TiO_2). In addition, they also plays a significant role in photocatalytic performance, which first affects the porosity of the materials and can enhance the diffusion centre of organic molecules and create an open channel. Second, the proper selection of the organic linker in coordination polymer fabrication can allow for the significant tuneability needed for the ideal photocatalysts [37]. Accordingly, metal–organic frameworks (MOFs) and coordination polymers appear as interesting candidates for photocatalytic water treatment applications [38]. Recent research has shown that these substances are excellent photocatalysts for the degradation of organic pollutants when exposed to visible light and UV, reaching up to 90% degradation percentages for both lights. A study carried out by Yan et al. reported the results of MIL-125-NH$_2$'s optical characteristics, showing an exceptional photocatalytic performance [37].

Cui et al. studied the synthesis and characterization of a new coordination polymer, [Cd(dcbpyno)(bix)$_{1.5}$]·2H$_2$O (CP1) (dcbpyno =2,2'-bipyridine-3,3'-dicarboxylate-1,1'-dioxide, bix = 1,4'-bis(imidazol-1-ylmethyl)benzene), with in situ chemical oxidation polymerization process for a polyaniline-loaded coordination polymer composite material (named PANI/CP1). It was concluded from the study that PANI/CP1 outperforms CP1 in terms of photocatalytic performance and quantum yield when exposed to visible light [39]. General information on the structure and morphology of the particle's surface can be observed through scanning electron microscopy (SEM) (Figure 10.1). The obtained SEM image displays the differences between PANI/CP1 and CP1. While CP1 has a smooth surface with an acute fringe, PANI/CP1 has a rough, irregular surface and round fringe (Figure 10.1a, b and c). meanwhile, the PANI nanofibres have a 120 nm diameter (Figure 10.1d) [39].

In another study by Mohsen et al., they successfully fabricated a cadmium–imidazole coordination polymer (Cd–Im CP) via facile synthesis at low temperatures in ethanol/aqueous solution. It can be observed from the obtained SEM that the average size of the particles is about 2.5 μm, and they have an uneven, ribbed surface with

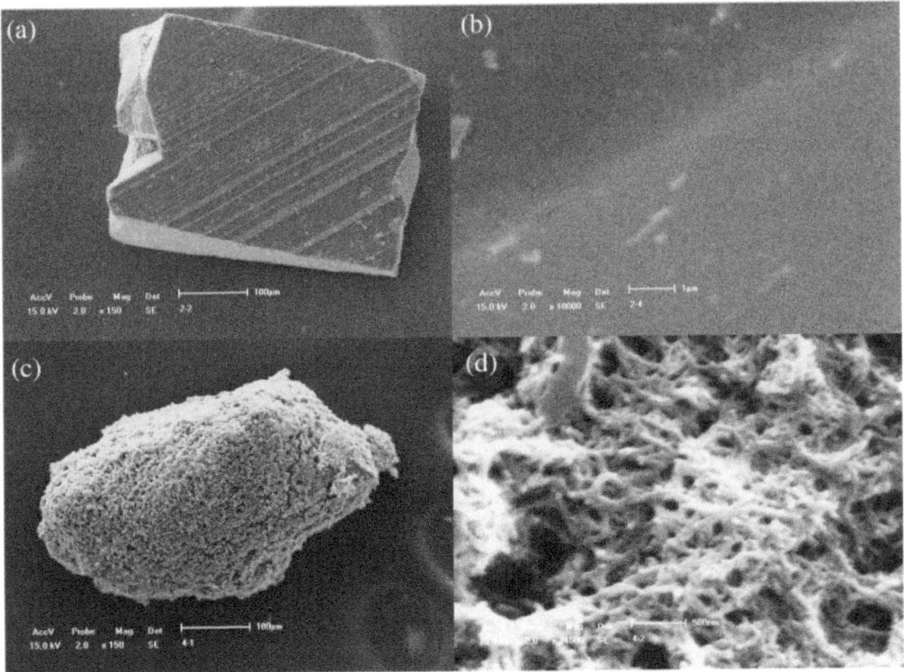

FIGURE 10.1   (a) SEM pictures of CP1; (b) CP1's surface; (c) PANI/CP1 composite; and (d) PANI nanofibers on PANI/CP1.

Source: Reproduced with permission from Ref. [39].

edges. Nevertheless, the Brunauer–Emmett–Teller (BET) analysis showed that Cd–Im CP has a relatively limited surface area with only 5.474 m^2g^{-1}, which is presumably caused by the dense packing of zig-zag linear chains. As a result, Cd–Im CP is suggested to be non-porous, which would slow down the performance of photocatalytic degradation [38]. For the photoreaction to occur, the semiconducting material's electrons need to jump from the valence band to the conducting band with sufficient photon energy, also known as band gap energy. The electrons are promoted more efficiently as the band gap value reduces [40].

In comparison, Cui et al. also reported that the band gap of CP1 is 3.33 eV. The intramolecular transition between the quinoid and benzoid segments in PANI is represented by the absorbance shift for PANI/CP1 from 700–800 to 620 nm. This suggests that PANI can effectively enhance the photocatalytic activity of CP1 in the visible light range [39]. In the study by Mohsen et al., the calculated band gap energy of Cd–Im CP was 4.55 eV. This band gap value corresponds to a wavelength of 275 nm, which is why UV lamps emitting at 254 nm were chosen for evaluations on photocatalytic degradation. Even a non-porous material like Cd–Im CP has a higher possibility of degrading due to these circumstances [38].

Fourier transform infrared spectroscopy (FTIR) analysis was employed to identify the interactions between CP1 and PANI. The significant PANI peaks for PANI/CP1 can be explained by the C–N stretching of the benzenoid and the C–N bending of the quinoid ring at 1312 and 1163 cm^{-1}, respectively. The characteristic PANI peaks at 1296 and 1143 cm^{-1} are altered with PANI/CP1 peaks at greater wavenumbers. These shifts demonstrate how well CP1 and PANI cooperate [39]. Moreover, Cui and co-workers also evaluated the photoluminescence spectra of CP1 and PANI/CP1 to study their quantum yields. The drastic reduction in PANI/CP1 emission intensity as compared to CP1 shows that the composite material demonstrates lower photogenerated electron-hole recombination rate and excellent quantum yields after being combined with PANI.

10.3.4 Conducting Polymers

In the last few years, a variety of innovative techniques have been used to create very effective polymer-based photocatalysts. Conducting polymers (CPs) were created as a result, and the finest methods and could provide good photocatalytic activity to produce energy. CPs are also electrically conducting organic polymeric materials that have functional groups with pseudocapacitance characteristics [41]. Contrary to their counterpart plastic-based polymers, CPs are typically organic macromolecules that are not thermoplastic. Despite showing a better level of conductivity, these polymers are lacking in crucial qualities such as flexibility and mechanical strength [33]. Katal et al. reported that CPs have high electrical conductivities between 2 and 100 $S\ cm^{-1}$. In an aqueous medium, the CPs exhibit significant environmental and chemical stability [42]. CPs are mechanically similar to metals and metal oxides and have metallic conductivity. The excellent optical, electronic, and electrical characteristics provided by the conjugated one-dimensional formation of CPs make them a favoured choice for photocatalytic applications [26].

Since CPs feature distinctive p-conjugated electron systems that are responsible for producing both electrons and holes, CPs-based photocatalysts have been widely studied [26, 33]. This action contributes to the introduction of electrons into the conduction band (CB) for the majority of inorganic semiconductor oxides [26]. Numerous researchers have discovered CPs doped with a variety of band gap semi-conducting metal oxides or sulphides, which have been found to be excellent pho-tocatalysts used for pollutant decomposition in wastewater [33]. Several reviews have described the basic stages implicated in the polymer semiconductor-based light-driven redox reactions, and the overall mechanism is widely known. The semiconductor absorbs energy after being exposed to light and produces charge carriers (e^- and h^+ pairs). Redox reactions are triggered by the separation of photo-generated charge and its migration to the semiconductor surface. The recombina-tion of e^- and h^+ pairs at the bulk or surface of the polymeric semiconductor is a competitive reaction. Roughly, CPs are required to have least some of the follow-ing characteristics in order to have excellent photocatalytic activity: (i) broad and powerful light absorption (visible region), (ii) productive separation of charge and charge transportation to the photocatalyst surface, and (iii) appropriate orientation of energy levels [43]. To enhance CPs' photocatalytic efficacy, many photoactive materials have been coupled with CPs, primarily polyaniline, polypyrrole, poly-acetylene, polythiophene, poly(1-naphthylamine), poly(o-phenylenediamine), and polycarbazole [26].

Bu and Chen fabricated a PANI/Ag/Silver phosphate (Ag_3PO_4) heterostructure formed via in situ deposition of Ag_3PO_4 nanoparticles on PANI chains. It was found that the nanocomposite had great photocatalytic performance and stabil-ity owing to the separation and migration efficiency of the electron–hole pairs resulting from the improvement of the interfacial electric field. Figure 10.2 illus-trates the high-resolution transmission electron microscopy (HRTEM) images of the 20 wt% PANI/Ag/Ag_3PO_4 composite at low, medium, and high magnifica-tions, where it can be seen that PANI has an uneven branch-like structure and is covered in a lot by Ag_3PO_4 particles with a diameter of less than 50 nm [44]. Researchers reported that Ag^0 growth on the surface of PANI was a result of Ag^+ being reduced to Ag^0 with PANI's aid, which was crucial for the electron and hole separation [26, 44].

10.3.4.1 Polyaniline

Among the numerous CPs, polyaniline (PANI) is the most frequently utilized and researched due to its relatively low cost, simple synthesis processes, outstanding porous structure, non-solubility in an aqueous phase, and non-toxic characteristics [33]. Pure PANI has a band gap value ranging from 2.8 to 3.2 eV [41]. PANI is also an outstanding h^+ transporter and acts as a p-type semiconductor. As PANI has a great electron carrying capacity, its photocatalytic capabilities are improved when combined with transition metal oxides, leading to the degradation of organic pol-lutants. Particularly, PANI changes the absorption maxima and reduces the metal oxides' band gap when coupled with transition metal oxides. Moreover, PANI can enhance the charge carriers' separation, resulting in an increase in the visible-light photoactivity [43].

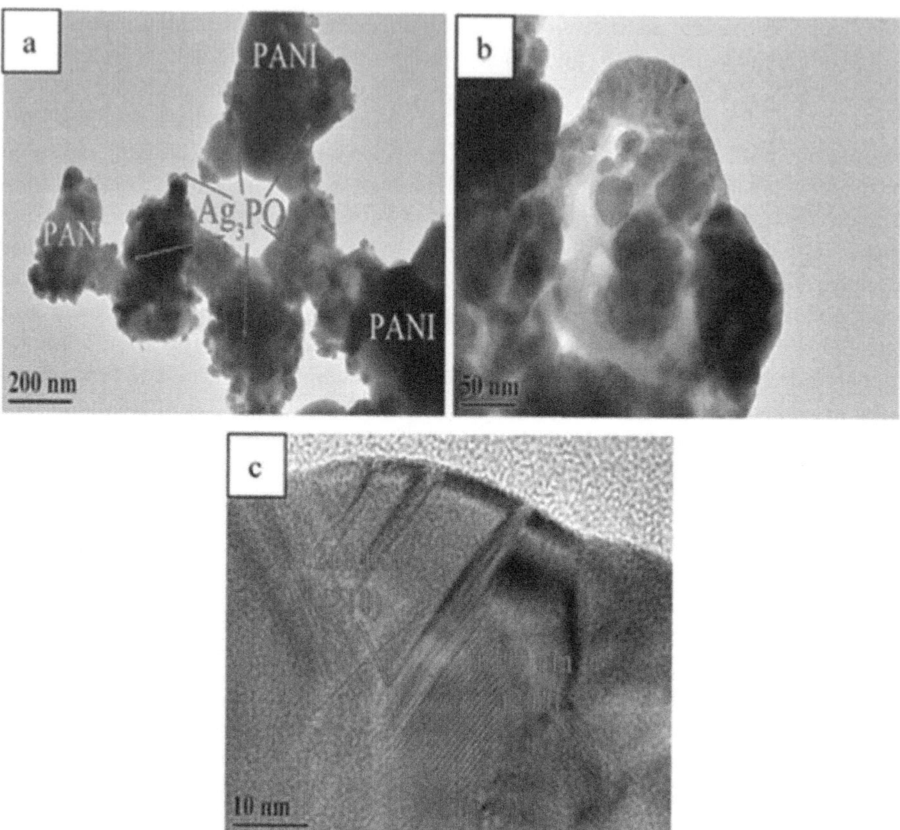

FIGURE 10.2   HRTEM images of the 20 wt % PANI/Ag/Ag$_3$PO$_4$ composite under (a) low, (b) medium, and (c) high magnifications.

Source: Reproduced with permission from Ref. [44].

The charge transfer process is caused by an interaction between the lowest unoccupied molecular orbital (LUMO) of PANI with the empty conduction band of semiconducting materials [33]. Additionally, the highest occupied molecule orbital (HOMO) and LUMO energy level positions influence the PANI chains' chemical composition in the band gap. PANI consists of C_6H_4-N repeating units that can be described by the bases emeraldine (EB), leucoemerald (LB), or pernigraniline (PB). The percentage of nitrogen atom oxidation and the development of imine structures determine how these structural units are formed. Higher conductive PANI can be produced by the approaches of oxidative doping of LB and protonation of EB with protonic acids. Based on the EB, LB, and PB chain length units in PANI, the band gap value is expected to vary from 1.53 to 2.52 eV. As for when the undoped PANI transforms to PANI-ES, the band gap value decreases (from 2.52 to 1.5 eV) [43].

In previous study, Ma et al. synthesized a hybrid PANI–TiO$_2$/rGO composite via the in situ polymerization of PANI on the surface of TiO$_2$/rGO. Although PANI increased the TiO$_2$'s visible-light absorption, it did not change the material's intrinsic excellent crystallinity [45]. As shown in Figure 10.3, the pristine PANI's morphology was revealed to be wire-like, which is advantageous for charge carrier transfer (Figure 10.3a). The PANI–TiO$_2$/rGO composites were discovered to have a similar wire-like shape (Figure 10.3b). According the findings, the introduction of PANI to this hybrid material forms a physical barrier and a conduction path, which are both greatly convenient for the effective separation and transportation of photogenerated electrons and holes [45].

In another study by Wang and co-workers, they successfully modified the surface of the CdS nanorods with different types of CPs, PANI, polypyrrole (PPy), and poly(3,4-ethylenedioxythiophene) (PEDOT) to form the core–shell composites through the proton-doped in situ polymerization method. The PANI-modified polycrystalline CdS nanospheres were found to exhibit higher photocatalytic activity and stability than the pure CdS nanospheres [46]. The SEM characterizations of CdS nanorods have a unique width of ~50 nm and length of ~10 μm (Figure 10.4). The morphologies of PANI@CdS, PPy@CdS, and PEDOT@CdS core–shell nanorods show no great difference from CdS nanorods. The sole difference is that the rapid polymerization of pyrrole during the modification stage shortens the length of the PPy@CdS core–shell nanorods. No polymeric sign can be acquired on the CdS nanorod surface, as shown in the inset of Figure 10.4a. In contrast, as illustrated in the insets of Figure 10.4b, c and d, the CPs shells for the core–shell nanorods can be vividly seen. This indicates that the CPs and CdS nanorods interact strongly, which is beneficial for preventing photocorrosion of the CdS nanorods during the photocatalytic degradation (Wang *et al.*, 2018).

The band gap energies obtained from UV-DRS of TiO$_2$, TiO$_2$/rGO, and PANI–TiO$_2$/rGO were calculated as 3.11, 2.95, and 2.56 eV, respectively. Thus, it can be concluded that PANI and rGO in the composite enhanced the light absorption of TiO$_2$ [45]. In a study by Wang et al., the band gap energies of the CdS nanorods, PANI@CdS, PPy@CdS, and PEDOT@CdS core–shell nanorods were ~2.38, ~2.39, ~2.40, and ~2.39 eV, respectively. This means that the CPs@CdS core–shell nanorods can still use the visible-light photons for photocatalytic H$_2$ production, and the CP shells

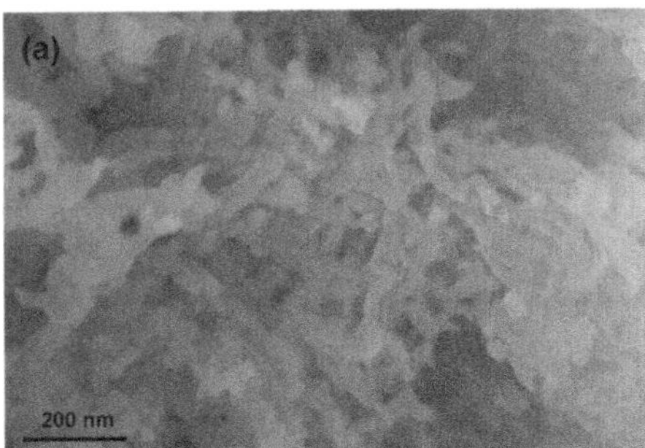

FIGURE 10.3   SEM image of (a) PANI and (b) PANI–TiO$_2$/rGO.

Source: Reproduced with permission from Ref. [45].

FIGURE 10.4 (a) CdS nanorods, (b) PANI@CdS core-shell nanorods, (c) PPY@CdS coreshell nanorods, and (d) PEDOT@CdS core-shell nanorods. The insets are the mixed SEM images of secondary electron (orange color) and backscattered electron (green color) signals6.

Source: Reproduced with permission from Ref. [46].

cannot significantly influence the intrinsic band gap of the CdS nanorods (Wang *et al.*, 2018). To sum up the study by Ma and co-workers, the PANI–TiO$_2$/rGO composite material's improved photocatalytic efficiency can be due to the quick photoinduced charge separation and effectively suppressed recombination of charge driven by the interaction of PANI, TiO$_2$, and rGO. In addition, the interfacial transfer of charge in PANI and rGO with a p-conjugated group employed for photocatalytic performance is extremely important [45]. Wang et al. also reported, from the PL spectra, that the separation capability of the photogenerated electrons and holes in the composites is as follows: PANI@CdS > PEDOT@CdS > PPy@CdS > CdS, suggesting PANI@CdS core–shell nanorods has an excellent photocatalytic performance (Wang *et al.*, 2018).

10.3.4.2 Polypyrrole

Due to its simplicity of synthesis, electrical characteristics, outstanding electrochemical features, and abundance of powerful binding or anchoring sites for future alterations, PPy is among the most explored CPs in the fields of energy, environment, and health. [43]. It also possesses the HOMO–LUMO gap or band gap value of 2.25 eV [41]. It has considerable photocatalytic performance and serves as a photosensitizer for visible light [26]. To increase photocatalytic performance, PPy could be merged with different photoactive elements to produce a photoactive nanocomposite [41]. Escobar-Villanueva et al. successfully synthesized and n–p-type heterojunction with a ZnO/PPy composite by polymerizing a pyrrole monomer via chemical oxidation in the presence of sodium dodecyl sulphate (SDS) as a surfactant onto ZnO nanoparticles [40]. The rice-like morphology was obtained via a simple precipitation reaction route with high reproducibility (Figure 10.5a). The size of these nanoparticles ranged from 70 to 227 nm in width and from 180 to 600 nm in length; 400 particles were counted, and the average width and length were 96 ± 5 and 206 ± 7 nm, correspondingly. A composite consists of ZnO nanoparticles embedded in a matrix of semiconducting PPy with a spherical structure, as can be seen in Figure 10.5b. From a previous study, a ZnO/PPy composite photocatalyst was successfully fabricated through a polymerization technique [47]. As obtained from the SEM images, the PPy displays a typically polymeric layer-like and shapeless structure in clusters. However, being doped with ZnO, the euhedral crystals with the flat, elongated shape of the ZnO is disrupted once ZnO crystals are engulfed with the polymer. In addition, to have an excellent photocatalysis performance, the surface area of the photocatalyst is crucial, since a larger surface area frequently results in higher photocatalytic activity. Silvestri et al. reported that ZnO and PPy–ZnO (25:1) specific surface areas were determined by BET analyses, and they were 4.9 and 48.0 m^2 g^{-1}, respectively. The reported value for chemically synthesized PPy is 24.4 m^2 g^{-1}, which is relatively higher compared to ZnO. This

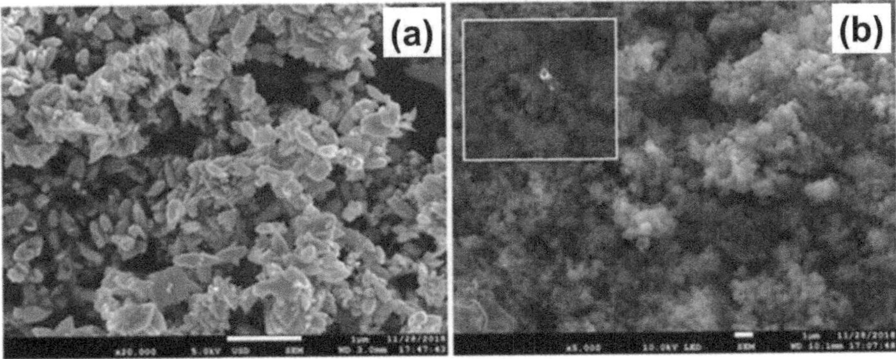

FIGURE 10.5 Micrography of (a) synthesized ZnO nanoparticles and (b) composite of ZnO/PPY. The inset corresponds to a section of the same SEM image with contrast.

Source: Reproduced with permission from Ref. [40].

suggests that the PPy–ZnO (25:1) composite's surface area is noticeably increased as an outcome from the polymerization method [47].

Through the UV-DRS analysis, the computed band gap energy of ZnO/PPy was 3.05 eV, which is slightly lower than the value previously reported for ZnO particles produced at ambient temperature. Owing to a synergistic interaction between ZnO and semiconducting PPy photocatalyst nanoparticles with a comparatively low band gap energy, the reported photodegradation efficiency and rapid reaction were achieved [40]. Nevertheless, the band gap value for the PPy–ZnO composite is approximately 1.81 eV, demonstrating a narrow gap in comparison to ZnO (3.24 eV). It is well known that strong interfacial bonds between polymers and semiconductors encourage charge transportation, and that the fundamental mechanism of the production of electron–hole pairs during irradiation is the heterojunction [47].

10.4 SYNTHESIS TECHNIQUES AND THEIR ADVANTAGES

10.4.1 CHEMICAL MIXING

The most used methods for synthesizing conducting polymer photocatalysts are facile mixing along with chemical oxidative polymerization. Facile mixing is basically mixing two or more chemical reagents. This process can yield homogeneous conducting PPy-based films with no limitations on mechanical properties or thickness, contingent only on the matrix employed [48]. It has been shown that particle size, and in particular particle agglomeration, significantly affects percolation and, consequently, the final film's conductivity. In actuality, compared to particles produced without an additive, PPy particles produced via a steric stabilizer have a higher likelihood of forming clusters. Compared to the additive-free particles, the percolation threshold would be higher for the additive-issued PPy particles because they would accumulate once in the blend [48]. Therefore, in order to create composite films that have high conductivity, the way the conducting particles are dispersed within the matrix becomes crucial. The best results were obtained with vigorous stirring (using ultrasonic or stirrer) [49]. Compared to electrochemically synthesized composites, the mixing technique yields lower-conductivity films (up to 10–1 S cm^{-1}), but it has a number of benefits. The method described allows for the production of PPy conductive films in larger quantities under more sustainable conditions than with previous techniques, from an industrial perspective [48–50]. Furthermore, the procedure is straightforward and simple to implement. This method is also cost-friendly, which explains why it is one of the most used methods for synthesizing polymer-based photocatalysts. Qing et al. synthesized C-PANI and C-PANI/BiOBr photocatalysts for 17β-oestradiol photodegradation using facile mixing [51]. For the synthesis of C-PANI, aniline and ammonium persulphate were added to a hydrochloric acid solution and stirred continuously at room temperature. The resulting product was filtered and rinsed with ethanol and deionized water, respectively. After freeze drying, the C-PANI powder (dark green powder) was obtained. For the C-PANI/BiOBr synthesis, C-PANI and bismuth(III) nitrate pentahydrate ($Bi(NO_3)_3\cdot5H_2O$) were dispersed in water and acetic acid. Then, potassium bromide (KBr) and sodium acetate (CH_2COONa) added to the solution and stirred continuously before being left to cool for 12 hours at room temperature, which allowed positively charged C-PANI (+13.9 eV) to combine with negatively charged BiOBr (−6.2 eV) via electrostatic interaction. The microscopic views of C-PANI, BiOBr, and C-PANI/BiOBr are shown in Figure 10.6. Figure 10.6a, b depicts nanosheet-like BiOBr and C-PANI, respectively, with widths of 100–200 nm and ~100 nm in length. In C-PANI/BiOBr (Figure 10.6c), the two BiOBr and C-PANI morphologies are visible, and they function well when there is good interface contact. Additionally, fringe spacings of 0.17, 0.23, and 0.27 nm were noted; these values correspond to the BiOBr crystal planes (211), (112), and (110), respectively, as seen in Figure 10.6d. The JCPDS Card 09–0393 standard card was used to fit the selected area electron diffraction (SAED) diffraction data (Figure 10.6e). The SEM image of 20% C-PANI/BiOBr (Figure 10.6f) revealed C-PANI to resemble nanofibres and BiOBr to resemble nanosheets [51].

On the other hand, PPy and AgFeO$_2$ nanohybrids was synthesized for trichlorophenol degradation using a chemical mixing method [52]. The synthesized silver ferrite was dispersed in ethanol in a round-bottom flask. With continuous stirring at 30°C, pyrrole monomer was added to the solution, followed by iron(III) chloride hexahydrate. After eight hours of sonication, the nanohybrid was passed through filters and thoroughly rinsed with acetone and distilled water to remove any unreacted pyrrole and excess ferric chloride. Aminuddin et al. also used chemical mixing method for synthesizing an Fe^{3+}–PANI–polyvinyl alcohol–glutaraldehyde (Fe-PPG)

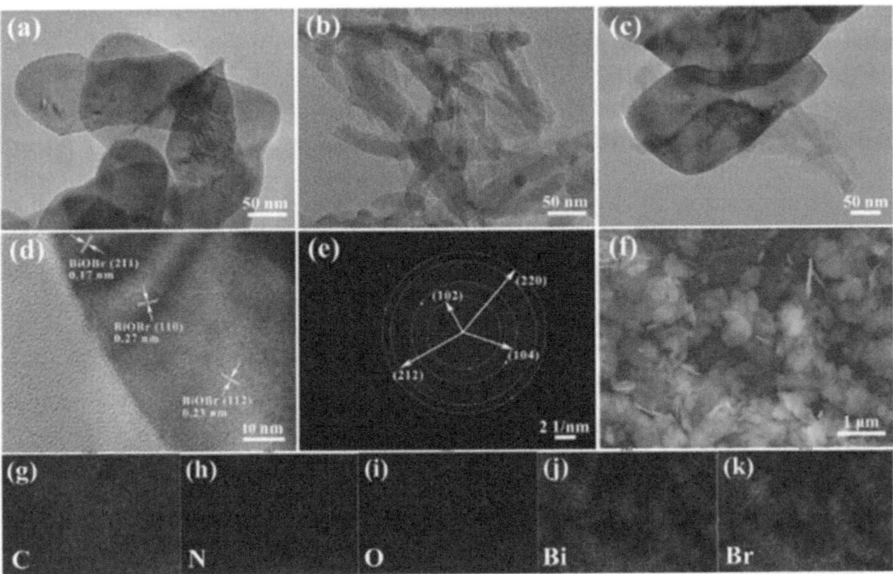

FIGURE 10.6 TEM images of (a) BiOBr, (b) C-PANI, (c) 20% C-PANI/BiOBr, (d) HRTEM images, (e) SAED images of 20% C-PANI/BiOBr, and (f)–(k) SEM image and corresponding elemental mapping images of 20% C-PANI/BiOBr.

Source: Reproduced with permission from Ref. [51].

FIGURE 10.7 SEM images of (a) PANI powder and (b) PPG at 5,000 × magnifications.

Source: Reproduced with permission from Ref. [53].

photocatalyst [53]. The elemental analysis and surface morphology of PANI powder and PPG are displayed in Figure 10.7a, b, which was obtained through a scanning electron microscope (SEM) at 5,000× magnification. The synthetic PANI powder, which is composed of particles with sharp edges and irregular granular shapes, is shown in Figure 10.7a. The synthesized PANI solution agglomerated and became larger after PG was added (Figure 10.7b). The appearance of white particles indicated a PG network all over the surface. The BET surface area of the synthesized PANI powder was 9.16 m² g⁻¹. When PVA–GLA (PG) adhesive was added to the PANI formulation, the BET surface area was drastically decreased to 3.65 m² g⁻¹. Low structure porosity was indicated by a decrease in both the total pore volume and pore size, while an increase in adsorbent density was indicated by a reduction in the total volume. The material that was immobilized by PANI immobilization on a densely packed glass plates' flat surface has resulted in a decrease in surface area, pore volume, and pore diameter. The PPG solution's surface area increased to 13.80 m² g⁻¹ which was more than the original PANI powder when iron (Fe) and silicate (Si) ions were added. This is explained by the presence of the large-surface-area material sodium metasilicate and the integration of the Fe dopant on the polyaniline polaron-lattice, both of them which increases the Fe-PPG composite's overall surface area. Because more dye molecules will adsorb onto the surface during the removal process, the composite material's greater surface area ought to have a positive impact on the rate of pollutant removal.

10.4.2 CHEMICAL OXIDATIVE POLYMERIZATION AND CROSSLINKING

Conducting polymer chemical polymerization such as PPy polymerization is a "nuts and bolts" process [49]. The chemical oxidative polymerization of PPy is a simple and quick technique that requires only a few simple instruments which can produce large quantities of fine PPy powders [54, 55]. In general, PPy polymerization takes place during the oxidation of monomer PPy in aqueous and non-aqueous media by chemical oxidants. The common procedure consists of the preparation of two individual solutions (the monomer and the oxidant) that are then mixed to begin polymerization [56]. This is the fancied method for applications in industries due to the ability to produce a large amount of PPy and varied PPy structures. The simple and quick fabrication method produces large quantities of dispersion, powder, and coatings (i.e., nanoparticles, nanofibres, etc.) (Sasso et al., 2011; Ramanavičius et al., 2006). The oxidant-to-monomer ratio (O/M) in this method influences the conductivity of PPy thin films [56]. Furthermore, the conductivity of PPy thin films can be altered by adding dopants. PPy thin film-doped camphor sulphonic acid (CSA) increases its conductivity from 4.29×10^{-9} to 1.56×10^{-8} S cm^{-1} as dopant content increases from 10 to 50% [57]. Meanwhile, chemically preparing PPy particles by quickly mixing Py monomers and oxidants FeCl$_3$ and ammonium persulphate (($NH_4)_2S_2O8$) at temperatures ranging from 25 to 170°C increases the electrical conductivity of both oxidant samples [58]. Nonetheless, because of its lower resistivity, iron(III) chloride performs better than ammonium persulphate [56].

Through chemical oxidative polymerization in an acidic aqueous medium, Uda et al. synthesized PAANIs, namely poly(2-methylaniline) (PMANI), poly(2-ethylaniline) (PEANI), and poly(2-propylaniline) (PPrANI). A 90 g sample of acidic aqueous solution was mixed with 1.0 g of AANI monomer (MANI, EANI, or PrANI), and the pH was adjusted to 2 using an HCl aqueous solution. The polymerization process was started by adding ammonium persulphate oxidant solution (monomer/1.25 M ratio, dispersed in pH 2 HCl solution) to the monomer aqueous solution after it was confirmed that the monomer had completely disintegrated. PAANIs were unable to dissolve in water and form grains. Supernatants were filtered five times after 24 hours of polymerization stirring at 250 rpm by substituting supernatants with de-ionized water quickly after decantation to remove unwanted soluble by-products (H$_2$SO$_4$, HCl, and water-soluble PAANI oligomers) [59]. SEM analysis demonstrated that PAANI grains were made up of submicrometric-sized uncommon main particles that grew in size as the alkyl group length increased (Figure 10.8). The presence of chloride and sulphur elements was confirmed by elemental analysis, indicating that PANI and PAANI had been doped with chloride ion and sulphuric acid coming from HCl and ammonium persulphate oxidant. PANIs prepared with ammonium persulphate in the presence of different acids are always partially protonated with sulphuric acid.

Sambaza et al. prepared a PANI/WO$_3$–TiO$_2$ photocatalyst for the degradation of ibuprofen by chemically oxidizing and crosslinking aniline monomers onto WO$_3$–TiO$_2$ [60]. As demonstrated in Figure 10.9, TiO$_2$ (pristine) showed varying sizes of rods (30–100 nm), whereas WO$_3$ revealed nanosphere-like microstructures. In the meantime, WO$_3$–TiO$_2$ is divided into distinct zones that are defined by rods and structures that resemble spheres that surround the rods. The WO$_3$–TiO$_2$ nanocomposite displayed nanospherical WO$_3$ heterojunction production on TiO$_2$ nanorods. The heterojunction structure of WO$_3$–TiO$_2$ surrounding a PANI layer was observed through PANI/WO$_3$–TiO$_2$. The specific surface area of WO$_3$ is 14.0974 m2/g, that of TiO$_2$ is 27.7396 m^2/g, and the combined surface area of the WO$_3$–TiO$_2$ photo-catalyst is 31.1296 m^2/g. The specific surface area of WO$_3$–TiO$_2$ was higher than

FIGURE 10.8   SEM images of (a) polyaniline (PANI), (b) poly(2-methylaniline) (PMANI), (c) poly(2-ethylaniline) (PEANI), and (d) poly(2-propylaniline) (PPrANI) grains.

Source: Reproduced with permission from Ref. [59].

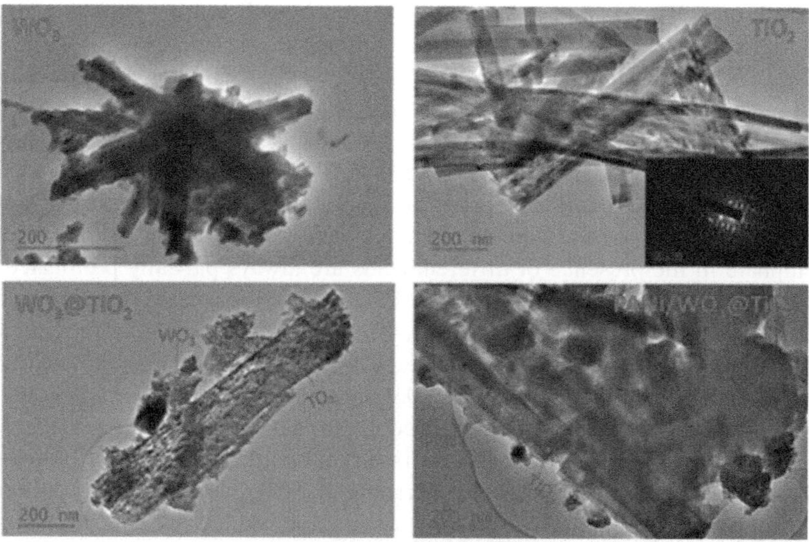

FIGURE 10.9   TEM micrographs of TiO_2, WO_3, WO_3–TiO_2, and $PANI/WO_3$–TiO_2 with EDX scan of $PANI/WO_3$–TiO_2 photocatalyst.

Source: Reproduced with permission from Ref. [60].

that of TiO_2 or WO_3. The benefit of PANI was demonstrated by its highest specific surface area of 36.1956 m^2/g for PANI/WO_3–TiO_2. Nevertheless, pore volumes slightly increased when WO_3–TiO_2 and PANI/WO_3–TiO_2 were contrasted with TiO_2 and WO_3.

Tantalum nitride–tantalum oxynitride (Ta_3N_5–TaON)-coupled organic polymers (poly (3,4-ethylene dioxythiophene)–polyaniline; PEDOT–PANI) (Ta_3N_5–TaON–PEDOT–PANI) nanohybrids were produced using the oxidative polymerization method with the addition of an acidic dopant, camphorsulphonic acid (CSA) [61]. Using a uniform magnetic stirrer, 100 mL of distilled water, freshly distilled aniline, EDOT, CSA, and a quantitative amount of Ta_3N_5–TaON nanoparticles were mixed for 30 minutes to produce an emulsion that contained PANI–CSA–PEDOT complex-loaded Ta_3N_5–TaON nanoparticles. Drop by drop, the previously cooled APS aqueous solution in distilled water was added to this, and the mixture was then left to overnight at a temperature between 0 and 5°C to finish the polymerization process. After the duration, the product underwent two or three full washes with a distilled water–methanol mixture before being vacuum dried for a full day to produce a Ta_3N_5–TaON–PEDOT–PANI nanohybrid with the colour of greenish black [61]. Figure 10.10 shows the formation of Ta_3N_5–TaON–PEDOT–PANI nanohybrids. The FESEM of Ta_3N_5–TaON (Figure 10.10a) indicates a 20 nm diameter of the particle-like structure. In addition, Figure 10.10b shows the PEDOT–PANI forming into 20 nm diameter granules. The Ta_3N_5–TaON–PEDOT–PANI sample's growth and aggregation can be directed towards the formation of regular hybrid shapes, as evidenced by the HRTEM images (Figure 10.10c and 10.10d). Furthermore, the SAED pattern of Ta_3N_5–TaON–PEDOT–PANI for the hybrid nanocomposite displayed a lower crystal character of the (023) plane (Figure 10.10e).

10.4.3 In Situ Polymerization

The preparation of a Z-scheme P-containing tungsten trioxide/polyimide (PWO/PI) photocatalyst was accomplished by in situ solid-state polymerization [62]. This method was used to increase PI's ability to be photocatalyzed and oxidized under visible light. In short, 20 mL of ethyl acetate was used to disperse 2.181 g PMDA and 2.182 g melem. The suspension was then mixed with 0.135 g PWO while being magnetically stirred to create a 3% PWO solution. Following two hours of stirring, the mixture was placed in a 30 minute ultrasonic bath at room temperature. Furthermore, after stirring for 30 minutes, the solvent gradually evaporated, and the residues were then dried for ten hours at 80°C. After that, the PWO/PI precursor was heated for a total of four hours at a rate of 10°C per minute to 300°C. After that, the resulting solid was ground into a powder known as 3% PWO/PI-300. Numerous PWO/PI composites were produced with this method. However, the same process was used to create pure PI at 300°C without the use of PWO. Figure 10.11a–d displays the TEM and SEM images of the 3% PWO/PI-300 composites. The detailed lamellae stacking morphology of PWO/PI composites is depicted in Figure 10.11d. PWO particles and PI nanoplates remain in close interaction, and PWO is wrapped in PI, as shown by the TEM images (Figure 10.11a–c). Consequently, this superior contact can help with communication and charge transfer between substrates. [62, 63]. Using an aqueous

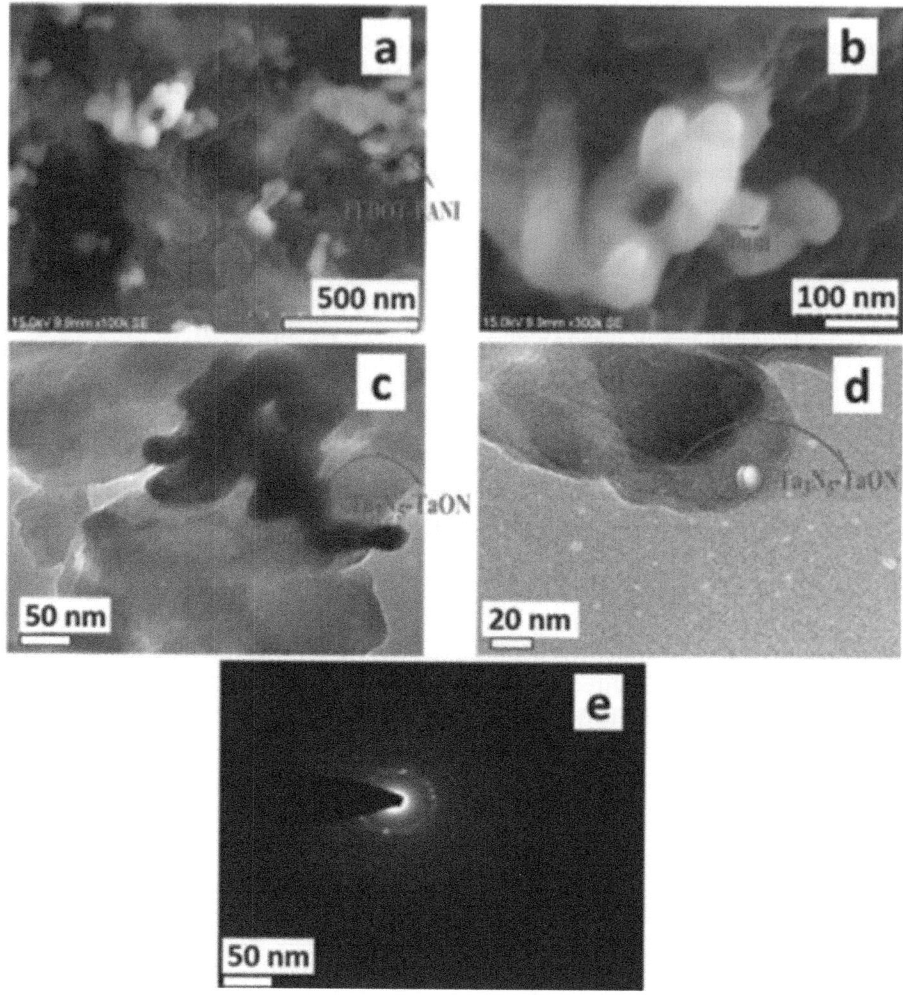

FIGURE 10.10   FESEM images (a and b), HRTEM images (c and d), and SAED pattern (e) of Ta_3N_5–TaON–PEDOT–PANI.

Source: Reproduced with permission from Ref. [61].

solution of ammonium persulphate, pyrrole was chemically oxidatively polymerized in situ to create the magnetite polypyrrole core/shell (Fe_3O_4@PPY) needed for the breakdown of cephalexin [23]. The iron oxide (Fe_3O_4) nanoparticles were ultrasonically dissolved in a solution containing ammonium persulphate. After that, the mixture was put into a flask with two necks, pyrrole was added, and it was shaken for eight hours at room temperature while being exposed to N_2. Following magnetic separation, the black product (Fe_3O_4@PPY) was dried for 24 hours at 60°C after being rinsed with deionized water and methanol.

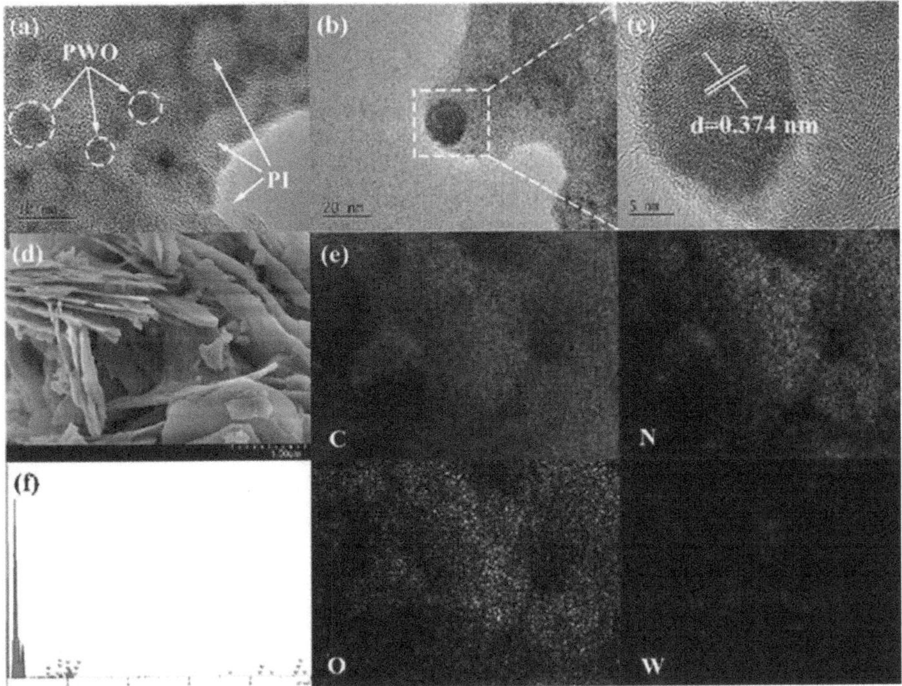

FIGURE 10.11   The images of (a–c) HR–TEM; (d) SEM–EDX mapping images of C, N, O, and W; and (f) EDS spectrum for 3%PWO/PI-300.

Source: Reproduced with permission from Ref. [62].

10.5   POLYMER-BASED MATERIALS FOR PHOTOCATALYTIC REMOVAL OF PHARMACEUTICAL RESIDUES FROM POLLUTED WATER

Water can always be polluted in various ways. Water pollution has become a major issue all over the world over time. About 70% of the surface of the earth is covered by water, but only 2.5% of that water is regarded as clean. It is a priceless resource for all species of life on earth. Minor clean water resources are used, recycled, and treated to meet the needs of living things. The use of water is not limited to domestic purposes; it also has other uses in the industrial and agricultural sectors [64]. One of the main causes of water pollution is the discharge of untreated effluents from industries like pharmaceuticals, textiles, and leather. These pharmaceutical and personal care products (PPCPs), endocrine-disrupting chemicals (EDC), and chemicals are used in agriculture and industrially [65].

Wastewater treatment always fixed on preserving the environment from the negative effects of wastewater sewage. Therefore, the befitting wastewater treatment plants (WWTP) were made to hinder these pollutants. Various treatment techniques and methods have since been implemented to improve the removal efficiency of such unknown compounds. In the long run, the production of chemicals as well as some pharmaceuticals, as well as their usage and consumption, has indeed been strongly linked to high environmental pollution and serious health consequences (Godheja et al., 2016). A large number of these chemicals are overused in personal care products, health care, hygiene, and pharmaceutical and personal care products (PPCPs).

Pharmaceutical products are designed to regulate the endocrine and immune systems, as well as cellular signal transduction, and have the potential to harm the survival of some organisms in the environment. Regardless, some illegal drugs create new categories of chemicals with unpredictable consequences and psychedelic properties in the aquatic system [66, 67]. Aquatic life may be endangered by the presence of those substances in the water system. In particular, antibiotics are a significant category of pharmaceutical drug, and their presence in pharmaceutical wastes poses a risk to both humans and aquatic life. Through the discharge of untreated wastes from pharmaceutical companies, hospitals, and other medical facilities, antibiotic compounds may enter freshwater environments. Not to mention, these pollutants create a massive proportion and have different levels of effects. PPCPs are compounds that come in a variety of types, purposes, actions, and performances and are utilized in the medical sector to improve human health. These substances in drinking water, sewage effluent, and groundwater may have genotoxic effects on unintended organisms as well as human pathogens [68].

Over the last two decades, PPCPs have been widely found in a variety of aquatic systems at concentrations from ng L^{-1} to g L^{-1}. [69–71]. Wastewater treatment facilities and agricultural runoff are two examples of the many point and non-point sources of PPCPs that are discharged into aquatic systems in combination. They have multiple impacts on ecosystems and could be very dangerous in terms of ecotoxicology [72]. This pharmaceutical waste fills up the aqueous environment via industrial and domestic wastewater [73, 74], negatively impacting aquatic organisms and potentially creating diseases by increasing antimicrobial resistance [75, 76]. Ibuprofen, a nonsteroidal anti-inflammatory used to treat toothaches, migraines, muscular pain, and inflammation, is among the most commonly prescribed analgesics worldwide [76, 77]. Antibiotics are also extensively used in human and animal medical procedures, as well as in animal feed to promote growth and to prevent and regulate the emergence of serious diseases [75].

Pharmaceutical pollutants are difficult to degrade or remove using conventional techniques due to their complex chemical structures and toxic functional groups. Because partial treatment of effluents does not completely remove particulate matter pollutants from the water through treatment plants, it has been found to be an important contributor to micropollutant contamination in aquatic matrices [78]. Additionally, during the wastewater treatment processes, PPCPs may be transformed into other potentially hazardous by-products, like disinfection by-products [79]. To degrade and remove these pollutants, intensive research has been conducted using physical, chemical, and biological methods. Though several studies have found excellent results, a significant drawback is that most of these methods can only be used up to certain pollutant concentration limits and under a particular pH, solubility, and ideal terms. In addition, common challenges involve high costs and high energy demand [80]. Among such procedures, the photocatalytic technique is thought to be the best potential candidate for organic pollutant degradation/removal. Photocatalysis can be employed for energy generation, hydrogen production, and carbon dioxide reduction in addition to wastewater treatment [81–84].

Over the past 15 years, CP nanocomposites have been popular as photocatalysts because of their comparatively narrow band gap and doping chemistry. Wastewater treatment has involved the use of polyaniline, polyethylene, polypyrrole, poly (vinyl pyrrolidone), polyfurane, polythiophene, poly (3-hexylthiophene), Nafion, poly (tetrafluoroethylene), and polyhydroxybutyrate [85–88]. Because of their ease of synthesis and adjustable band gap energy, these composites – polypyrrole, polythiophene, and polyaniline – have been employed as effective photocatalysts [89–91]. These CPs have been successfully used as photocatalysts due to their visible-light-induced photocatalytic ability and tuneable band gap energy, redox properties, harmless nature, cost effectiveness, chemical inertness, greater surface area, and doping compatibility with semiconductor nanoparticles.

Three main steps are known to be involved in photocatalysis: the photocatalyst absorbs light, the photogenerated charge carriers are separated, and interface catalytic redox reactions occur [92]. A photon with energy above the band gap excites an electron from the valence band (VB) of CP when it is exposed to any light source, moving it into the conduction band (CB) [93]. The term "electronic transition" describes this. The excited electrons move to the CP surface, where they activate the adsorbed oxygen molecules. This produces superoxide radicals and anions ($\bullet O_2^-$), which can oxidize organic pollutants in the aqueous system [94]. Conversely, organic materials may be directly oxidized by the photogenerated holes. The overall mechanism is depicted as follows:

$$CP + h\nu \rightarrow CP\ (e^-_{LUMO} \cdots h^+_{HOMO}) \tag{10.1}$$

$$e^- + O_2 \rightarrow \bullet O_2^- \tag{10.2}$$

$$\bullet O_2^- + H^+ \rightarrow \bullet OOH \tag{10.3}$$

$$2\bullet OOH \rightarrow O_2 + H_2O_2 \tag{10.4}$$

$$H_2O_2 + \bullet O_2^- \rightarrow \bullet OH + OH^- + O_2 \tag{10.5}$$

$$H_2O_2 + h^+ \rightarrow 2\bullet OH \tag{10.6}$$

Because of its lower band gap than metal oxide, CP acts as a photosensitizer in the CP–metal oxide hybrid, absorbing a wide range of visible light. These CPs absorb photons and generate electron–hole pairs, which are then transferred from the highest occupied molecular orbital (HOMO) to the lowest unoccupied molecular orbital (LUMO) via the π–π^* transition [95]. Excited electrons in the LUMO of CP chains are injected into the CB of a transition metal oxide (for example, TiO_2), where they react with water molecules that are adsorbed to form superoxide $\bullet O_2^-$ radicals, while holes react with water to form $\bullet OH$ [96–98]. The CP hybrid's mechanism is shown in Figure 10.12. Throughout the process, nanoparticles assist in reducing the rate of recombination of generated electrons and holes by gaining electrons from the polymers' CBs and transferring them to their own CBs through interfacial charge transfer. Photogenerated electrons in the conduction band will thus be unable to recombine with photogenerated holes [99].

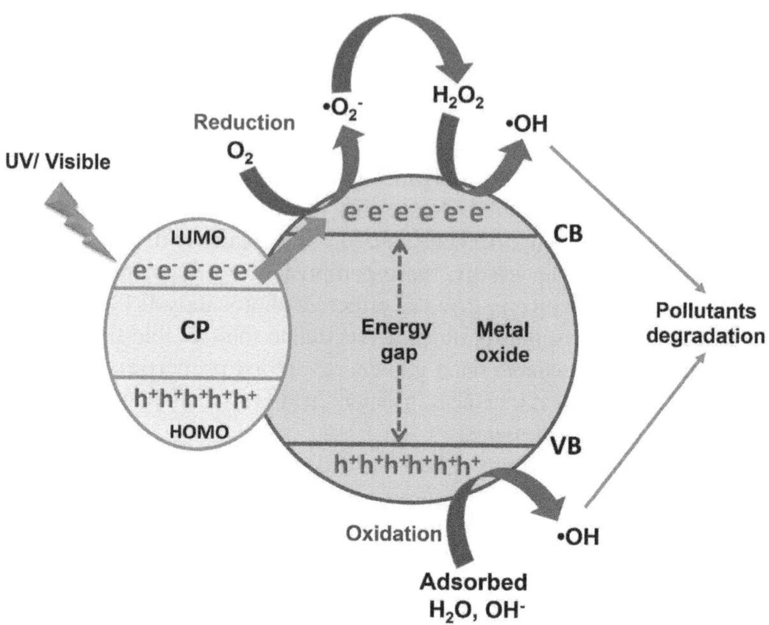

FIGURE 10.12   Illustration of photocatalysis based on a CP composite.

Source: Adapted from [95].

In a study, a reduced graphene oxide (rGO) and PANI-assisted C-doped porous ZnO (c-ZnO) (RPZ) nanocomposite photocatalyst was synthesized for the removal of amoxicillin and clavulanate potassium (ACP) [100]. The RPZ is intended as a doped combination with attachment between the PANI and c-ZnO, in which the rGO at the interaction between surface of PANI/c-ZnO acts as an electron scavenger. The interaction then increases the time of recombination, resulting in better and more efficient photodegradation of the ACP. Under 100 minutes of direct sunlight irradiation, the degradation efficiency of the nanocomposites RPZ-300, RPZ-400, and RPZ-500 were 42, 42, and 47% and the removal efficiencies were 41, 41, and 46.14% for ACP, respectively [100]. The degradation efficiency and removal efficiency were low for ACP, but this is expected when using a commercially available ACP tablet. The degradation efficiency of the RPZ-500 nanocomposite was higher than that of RPZ-300 and RPZ-400. This could be because the ZnO crystallinity phase in RPZ-500 was higher than that in RPZ-300 and RPZ-400, resulting in an enhancement in degradation performance. PANI's HOMO and LUMO bands, as well as ZnO's CB and VB, have comparable energy levels that can be combined to produce a synergistic action [101]. Given the comparable energy levels of the ZnO VB and PANI HOMO as well as the ZnO CB and PANI LUMO, interactions of chemical bonds may occur, resulting in holes in the ZnO VB to be transferred to the PANI HOMO [102]. Further, PANI has previously been reported as a good hole transporter [103, 104]. Moreover, the existence of rGO at the ZnO interface forms a Schottky barrier.

Since ZnO is an n-type semiconductor, rGO acts as electron scavenger here, and the ZnO CB's photogenerated electrons move freely to rGO [105]. The degradation results gained were fitted to pseudo-first-order reaction kinetics. The photocatalytic degradation process of all ACP nanocomposites prepared appeared to follow Langmuir Hinshelwood pseudo-first-order kinetics. The k values of the RPZ-300, RPZ-400, and RPZ-500 for ACP were 0.0052, 0.0043, and 0.0055 min^{-1}, respectively. As a result, the following findings show that the RPZ-500 nanocomposite has the highest catalytic efficiency for ACP degradation.

In the meantime, by using visible light to break down the antibiotic agent ciprofloxacin (CIP), Xu et al. (2019) examined the activity of a PANI-decorated $Bi_{12}O_{17}C_{12}$ photocatalyst [106]. Upon 120 minutes of visible light exposure, $Bi_{12}O_{17}C_{12}$ degraded by about 73.0% at a wavelength of $\lambda > 420$ nm. The composites' photoactivity rose in proportion with the PANI loading; at 120 minutes, the composite containing 0.05% PANI and $Bi_{12}O_{17}C_{12}$ showed the fastest rate of degradation of 90.4%. The apparent reaction rate constants were also indicated by a pseudo-first-order kinetic equation. Compared to pure $Bi_{12}O_{17}C_{12}$ (30%), the apparent reaction rate constant of the 0.05% $PANI/Bi_{12}O_{17}C_{12}$ composite was twice as high. The material may have somewhat agglomerated with too much PANI, as the degradation rate decreased as the PANI content was increased. Conversely, at $\lambda > 550$ nm, $PANI/Bi_{12}O_{17}C_{12}$ (64%) demonstrated substantially better photoactivity in CIP degradation than $Bi_{12}O_{17}C_{12}$ (30%) after four hours of visible-light irradiation. The 0.05% $PANI/Bi_{12}O_{17}C_{12}$ composite's reaction rate constant was 3.1 times higher than that of $Bi_{12}O_{17}C_{12}$, indicating that PANI can broaden the light response region and increase light efficiency when used as a photosensitizer. Notice that as the $PANI/Bi_{12}O_{17}C_{12}$ composite was successfully constructed, the high charge carrier separation rate and improved composite performance are the result of the good interface contacts between the PANI and $Bi_{12}O_{17}C_{12}$.

The photocatalytic efficacy of the $WO_3@TiO_2$ heterojunction supported on PANI in the degradation of ibuprofen in the presence of visible light was documented by Sambaza et al. [60]. The photocatalyst with the best performance was selected by assessing how TiO_2, WO_3, $WO_3@TiO_2$, and $PANI/WO_3@TiO_2$ degraded ibuprofen. The ibuprofen solution that was utilized had a pH of 5.8. $PANI/WO_3@TiO_2$ showed the highest degradation, with 99.8% degradation. Moreover, 90% degradation was attained by the $WO_3@TiO_2$ photocatalyst, while 50 and 46.5% degradation was attained by TiO_2 and WO_3, respectively. For $PANI/WO_3@TiO_2$, a linear correlation with an R^2 of more than 0.98 and a rate constant of 3.5×10^{-2} was discovered. The dual action of PANI and the heterojunction was credited with the better charge separation than $PANI/WO_3@TiO_2$ was able to achieve. Furthermore, compared to TiO_2, WO_3, and $WO_3@TiO_2$, the $PANI/WO_3@TiO_2$ photocatalyst surface's higher adsorption strength because of the $\pi–\pi^*$ interaction led to a longer degradation time [107, 108].

Kumar et al. reported clozapine (CZP) photodegradation using a visible-light-active polyaniline combined with lanthanum ferrite and cobalt ferrite $PANI/LaFeO_3/CoFe_2O_4$ (PLC) ternary heterojunction [32]. In total, 94.2% of CZP has been found to degrade after 120 minutes in the sun. PLC's degradation efficiency has been contrasted with that of its equivalents. PLC, $LaFeO_3/CoFe_2O_4$, $PANI/LaFeO_3$, $PANI/CoFe_2O_4$, $LaFeO_3$, $CoFe_2O_4$, and PANI have CZP degradation effectiveness values of 94.2, 74.4, 66.8, 63.4, 59.3, 49.2, and 38.5%, respectively. Furthermore, pseudo-first-order kinetics govern the photodegradation process. With a rate constant of 0.0248 min^{-1}, the PLC outperformed the other photocatalysts in comparison. PLC is superior because of its favourable heterojunction formation, visible-light-active band gap, and, most importantly, lower charge recombination rate, which promotes longer photogenerated charge carrier lifetimes. [109]. PLC absorbs radiation with the appropriate energy in accordance with its optical band gap energy when exposed to the light source.

Additionally, a tungsten disulphide–polyaniline (WS$_2$–PANI) nanocomposite was developed to aid in the nitrofurantoin (NFT) degradation process [110]. The photocatalytic activity of WS$_2$–PANI and NFT, an antibiotic, was examined. The WS$_2$–PANI nanocomposite showed outstanding performance for the 95.38% of NFT degraded in 120 min under visible-light exposure as compared to pure WS$_2$ (25.8%). This was due to the bigger pore size and enhanced charge separation efficiency of WS$_2$–PANI. Using WS$_2$ and WS$_2$–PANI photocatalysts, the rate constants for NFT were determined to be 0.00257 and 0.02135 min^{-1}, respectively. This demonstrates unequivocally that NFT degradation utilizing WS$_2$–PANI has a higher degradation efficiency (95.38%) in 120 minutes compared to pure WS$_2$ (25.8%), which is attributed to the PANI's larger pore size and superior charge separation efficiency. It was discovered that WS$_2$-PANI performed better as a photocatalyst for NFT due to its bigger pores and more efficient separation of charges than pure WS$_2$. By migrating from the CB of WS$_2$ and recombining with the holes on the VB of PANI, the photo-induced electrons follow the Z-scheme mechanism, improving the separation of the photogenerated charges (e$^-$ and h$^+$ pairs) [111]. Because of this, the combined action of WS$_2$ and PANI inhibits recombination and increases the amount of time that photogenerated charges can contribute to the reaction. WS$_2$/PANI exhibits superior photocatalytic activity (2.6 eV) due to a number of remarkable features, including a visible-light-driven band gap and high photoinduced charge separation (5.76 ns).

Other than PANI, polypyrrole (PPy) showed great results in photocatalysis. A study was conducted evaluating the photocatalytic activity of PPy–ZnO for the degradation of diclofenac (DCF) [47]. According to the photocatalytic experiments, ZnO can degrade DCF until 46%. However, the composite PPy:ZnO 25:1 degrades DCF more efficiently (81%). The presence of ZnO in PPy reduces recombination between e$^-$ and h$^+$ pairs [112–116]. The decrease in the photocatalytic activity of PPy:ZnO 5:1 can be explained by the suppression of the active sites on the PPy surface by an excess of ZnO, which can lead to defects that can serve as recombination focuses between the e$^-$ and h+ pairs [117]. The percentage of DCF degradation decreases with initial DCF concentration, suggesting that degradation promotion becomes more challenging (likely due to a reduction in the amount of photocatalytically active sites available) [115]. They ascribed the increase in efficiency to PPy's sensitizing effect. Moreover, photoexcited electrons of the PPy conduction band can move to the ZnO conduction band, according to electrical resistivity measurements. Photogenerated holes, on the other hand, shift from the ZnO valence band to the PPy valence band [118, 119]. This promotes charge separation (e$^-$ and h$^+$ pairs) and electronic transport in the PPy–ZnO composite material [116, 120]. Ultimately, favourable characteristics that lead to the degradation of DCF (81%) in 60 minutes under simulated solar irradiation include the resulting PPy–ZnO composite's large surface area, mesoporous structure, and lower band gap. A higher rate constant (k) value indicates greater photocatalytic activity; this value can be used to evaluate the samples' photocatalytic activity. Catalytic efficiency was significantly affected by the synthesis of composite PPy–ZnO since the degradation rate constant, k (min^{-1}), of PPy–ZnO is higher than that of ZnO. The synthesized PPy–ZnO composite exhibited superior photocatalytic activity and the highest constant kinetic rate [47].

Polypyrrole dispersed in a $AgFeO_2$ nanohybrid (PPy/$AgFeO_2$) for photocatalytic activity towards 2,4,6-trichlorophenol (TCP) degradation has also been investigated previously [52]. TCP degradation was carried out using 75 ppm of TCP working solution by sonophotocatalysis and was observed by UV–vis spectroscopy. The degradation rate was 47.33% in the presence of pure PPy and 45.83 wt% in the presence of $AgFeO_2$. Using 10% PPy/$AgFeO_2$, a degradation rate of about 68.14% was attained after 140 minutes of visible-light exposure. The highest degradation of TCP was observed to be 82.23% in 120 minutes with the use of 30% PPy/$AgFeO_2$, while it was found to be 99.6% in the presence of 80% PPy/$AgFeO_2$. The photocatalytic efficiency of the PPy/$AgFeO_2$ nanohybrid was observed to be higher than that of pure PPy and $AgFeO_2$. It was discovered that PPy/$AgFeO_2$ had a higher photocatalytic efficiency than both $AgFeO_2$ and pure PPy. The photocatalytic efficiency increased as the PPy loading in $AgFeO_2$ increased. The degradation kinetics data for PPy, $AgFeO_2$, and PPy/$AgFeO_2$ fit the pseudo-first-order model well. The rate constants for pure PPy and $AgFeO_2$ were calculated to be 0.005 and 0.004 min^{-1}, respectively, and the rate constant for 80%-PPy/$AgFeO_2$ was found to be the highest ($k = 0.016$ min^{-1}). Under visible-light illumination, PPy/$AgFeO_2$ was discovered to exhibit high photocatalytic activity, and TCP was discovered to degrade in just 40 minutes. Using 100 mg of catalyst, the 80% PPy/$AgFeO_2$ photocatalyst showed 84% degradation in 40 minutes for a 30 ppm TCP solution, and 150 and 200 mg showed 95% degradation in 40 minutes. Similar results were seen with 200 mg of the photocatalyst and 50 ppm and 70 ppm of TCP, which demonstrated approximate degradation of 94 and 90%, respectively. It was discovered that the degradation efficiency was high even at a low catalyst loading. Using 100, 150, and 200 mg of the photocatalyst, the rate constant values for degradation were determined to be 0.037, 0.038, and 0.072 min^{-1} for 30 ppm TCP; 0.029, 0.056, and 0.067 min^{-1} for 50 ppm TCP; and 0.013, 0.018, 0.025 min^{-1} for 70 ppm TCP, respectively. The degradation was seen in accordance with pseudo-first-order kinetics in every instance. Analysis of the degraded fragments showed that the products obtained were possibly non-toxic diols. As a result, the developed hybrids can be safely employed for the environmentally benign breakdown of dangerous organic pollutants. [52].

In order to create a PPy@Fe_3O_4 core/shell that supported a nickel sulphide (NiS) photocatalyst, iron(II,III) oxide (Fe_3O_4) and PPy were combined. [23]. This high-performance nanocomposite material was used to degrade cephalexin under UV as well as sunlight irradiation. When exposed to UV light, Fe_3O_4@PPy-NiS and NiS were present, whereas once exposed to sunlight, Fe_3O_4@PPy-NiS and NiS were present. At the photocatalysts, the respective degradation rates were 100, 69, 90, and 57%. These results showed a significant increase in efficiency for the NiS catalyst immobilized on the Fe_3O_4@PPy support. Furthermore, by employing a support to move the photocatalyst's band gap energy to a lower energy, sunlight degradation increased. On the other hand, Fe_3O_4@PPy-NiS showed 82 and 65% degradation efficiency, respectively, in 30 minutes under UV and sunshine. The rate constants of 9.3×10^{-3}, 1.7×10^{-3}, 4.0×10^{-3}, and 1.4×10^{-3} were obtained for Fe_3O_4@PPy-NiS and NiS under UV and Fe_3O_4@PPy-NiS and NiS under sunlight irradiation, respectively. The degradation percentages of the photocatalysts were 100, 69, 90, and 57%, respectively. These results demonstrated a significant improvement in the effectiveness of the NiS catalyst immobilized on the Fe_3O_4@PPy support. Moreover, the photocatalyst's band gap energy moved to a lower value with the use of the support, making sunlight degradation more noticeable. In comparison to the values provided by the other researchers, the time needed for such a degradation amount is significantly less. Amoxicillin, ampicillin, and cloxacillin are reported to degrade in aqueous solution when ZnO is used in conjunction with UV light within 180 min of reaction [121]. At the optimum conditions, Fe_3O_4@PPy-NiS degraded the pollutant to 0% under UV light and to 85% under sunlight. The findings showed that cephalexin was more effectively broken down by immobilized NiS on the magnetic support than by bulk NiS. Catalyst support was beneficial in terms of degradation rates, time, and sunlight exposure.

Anthraquinone-2-sulphonate (AQS) doped polypyrrole and functionally modified zinc indium sulphide ($ZnIn_2S_4$) were used to investigate the visible-light photocatalytic degradation of tetracycline [122]. Tetracycline was subjected to photocatalytic degradation in order to measure photocatalytic activity. The optimal photocatalytic activity was achieved by utilizing a weight ratio of 1.0% for AQS/PPy in relation to $ZnIn_2S_4$ and a molar ratio of 40 for PPy to AQS in order to create the composite catalyst. More than 98% of the tetracycline was removed after a 3.0 hour photocatalytic reaction, and the NPOC (non-purgeable organic carbon) removal rate was nearly 2.0 times higher than with pure $ZnIn_2S_4$ at the same time. Experiments on active species trapping showed that $\bullet O_2^-$, rather than $\bullet OH$ and photogenerated holes, played a significant role. The formation of $\bullet O_2^-$ was aided by the behaviour of the AQS doped in the PPy matrix, which acted as an electron acceptor–donor mediator. Tetracycline was broken down by the oxidation of active species $\bullet O_2^-$ and $\bullet OH$ via the hydroxylation reaction after being broken down by photogenerated holes through the loss of some functional groups.

The ease of synthesis, high conductivity, stability in ambient air, and biological compatibility of poly(3,4-ethylenedioxythiophene) (PEDOT) make it one of the most promising CPs with high potential for technical and commercial applications [123–125]. In the past, it was shown that nanostructured PEDOT and its composites had outstanding photocatalytic ability to break down organic pollutants [93]. In a study conducted by Kumar et al. [126], the photocatalytic removal of PPCPs using a PEDOT polymer was investigated under UVB irradiation at pH 5.8. The most effective PEDOT dosage was found to be 0.5 g/L, with metformin removal reaching >99% after 60 minutes of irradiation. A kinetic study was conducted to evaluate the photocatalytic degradation rate of PPCPs in water. Metformin removal under UVA, UVB, and UVC light irradiation was followed by pseudo-first-order reaction kinetics at pH 5.8, PEDOT dosage of 0.5 g/L, and rate constants of 0.007, 0.040, and 0.032 min^{-1}, respectively. It has been shown that for efficient photocatalytic degradation, metformin requires higher UV energy and broad spectrum [127]. Researchers examined the photocatalytic activity of a selected PPCP mixture in deionized water by PEDOT at PPCP concentrations of 1 mg/L. Using 0.5 g/L PEDOT, photodegradation of carbamazepine, benzotriazole, acetaminophen, caffeine, and metoprolol was found to be >99% under UVB light in 60 minutes. On the other hand, the removal efficiency of tris(2-chloroethyl) phosphate (TCEP) was below 30%. According to earlier studies, TCEP is comparatively fractured and degrades gradually with other PPCP mixtures [128]. It also resists photocatalysis because it contains three electron-withdrawing Cl groups, which reduces the compound's vulnerability to •OH attack [129].

On the other hand, Ledezma-Espinoza prepared a gamma (γ)-irradiated cellulose PEDOT (I-Cell-PEDOT) composite from biomass for the degradation of sulindac and carbamazepine from natural water under UVA irradiation [130]. The I-Cell sample was dissolved in iron(III) chloride ($FeCl_3$) isopropanol solution for the presence of a semiconductor. Under UVA and at pH 7, the I-Cell-PEDOT demonstrated efficient sulindac degradation in short period of time, going up to 89% within 420 minutes, which improves the remediation process and saves time. The sulindac decomposition rate from the mixture with carbamazepine was slightly similar, reaching 88%. The results demonstrated that the I-Cell-PEDOT improved the degradation of sulindac and carbamazepine from natural water systems when exposed to UVA radiation. Within seven hours of treatment, photodegradation efficiencies of 89% for sulindac and 30% for carbamazepine were achieved. In mixtures of both compounds, similar degradation results were observed. The composite's reuse as a catalyst was accepted, with similar degradation efficiencies for sulindac and carbamazepine. It has been stated that the I-Cell-PEDOT photocatalytic activity requires less energy to promote an electron to the conduction band due to the PEDOT band gap [131]. Then, at lower energy levels, electronic transitions may occur. When the photocatalyst is exposed to a light source, the semiconductor absorbs photons, which results in the generation of charge carriers [126]. When generated electron–hole pairs react with water or dissolved oxygen, reactive oxygen species (ROS) are produced, which degrade pollutants into smaller molecules [132].

Among the most widely used conjugated polymers, poly (3-hexylthiophere) (P3HT) has high mobility of charge carriers, photo-stability over time, relevant band edge positions, and thus a high capacity of visible-light harvesting [133]. Liu et al. successfully introduced a g-C_3N_4/Ag/P3HT (CN/Ag/P) heterojunction photocatalyst for the enhanced photocatalytic degradation of tetracycline (TC) under visible-light irradiation [134]. The best photodegradation efficiency is greatest when P3HT content is 0.3 wt% ($P_{0.3}$) which is approximately around 75%. CN/Ag and CN/$P_{0.3}$ exhibit greater visible-light activity than CN, while CN/Ag/$P_{0.3}$ exhibits the highest photocatalytic efficiency in TC degradation with rate constant of 0.00297 min^{-1}. The increased photocatalytic efficiency of CN/Ag and CN/P is because of the photoexcited electron transfer from g-C_3N_4 to Ag and then P3HT, as well as a reduction in photoexcited electron–hole recombination [134]. The Z-scheme system's formation and Ag's role in interface optimization are what lead to the optimum visible-light activity of CN/Ag/P3HT. Ag functions as a link within g-C_3N_4 and P3HT, facilitating the transfer of photoexcited charges from g-C_3N_4 to P3HT. Compared to the straight transfer from g-C$_3$N$_4$ to P3HT without the need of an Ag bridge, this transfer is less challenging.

10.6 CHALLENGES AND FUTURE PERSPECTIVES

In recent years, the use of polymer-based photocatalytic materials to reduce anthropogenic pollution in watercourses is gaining popularity. The photocatalytic removal of organic pollutants in aqueous media, especially from pharmaceutical industries, is a difficult road to take, especially when the photo-efficacy and stability of the fabricated catalyst materials remains a significant concern under varied circumstances. The total mineralization of those pollutants in water was not reached in most situations, and the prospective toxicity analysis of the formation by-product is still limited. If complete mineralization of pollutants is not an aim, proper toxicity assessment is critical in the photocatalytic process. Furthermore, the intent to use these materials and their technological preparedness for industrial uses and other diversified applications is still being challenged within scientific groups. Researchers are still committed to looking further into green and efficient technologies that stoke interest in freshwater resource protection, energy and cost effectiveness, environmental safety, and zero discharge of hazardous chemicals, despite the fact that these polymeric materials have recently become the focus of numerous studies.

To ensure their efficacy and safety, polymer-based photocatalytic materials must have some advantageous characteristics before they can be widely applied in industrial processes. The ability of the materials to undergo higher recyclability for continuous process, zero or less formation of toxic intermediates, and outstanding chemical and mechanical strength to withstand harsh environments is among them. Other factors include optimal light absorption capacity, suitability and selectivity, and accessible pores or channels for rapid mass transfer and adsorption. The all-inclusive study of the respective materials interacting with various organic pollutants in a complex effluent is highly desired for the development of efficient wastewater treatment plants. Therefore, the enhancement in terms of the characteristics and performance of the respective materials is necessary to achieve better understanding of the mechanism, kinetics, and thermodynamics in strengthening the environmental regulations as well as for safety measures.

10.7 CONCLUSIONS

In conclusion, this chapter highlighted an overview of the photodegradation of organic pollutants specifically originating from the pharmaceutical industry. Additionally, it is evident that extensive research has been done on this subject given the fact that pollutants from pharmaceutical wastewater are an area of growing concern due to their ubiquity in ecosystems and detrimental effects on living things. Hence, the environmental impacts of these emerging contaminants lead to the utilization of advanced materials and treatment processes such as the polymer-based heterogeneous photocatalysis approach. The photocatalytic degradation of organic pollutants specifically from pharmaceutical wastewater has been found quite efficient, with maximum efficiency and complete degradation of pollutants to harmless products. Polymer-based photocatalysts have been extensively explored in photocatalysis applications owing to their outstanding charge-transport and light-harvesting properties. A wide variety of polymer-based photocatalysts have been developed and have demonstrated their improved activities under visible light to initiate and facilitate the degradation of organic pollutants for environmental remediation applications. Based on previous literature, polymer-based photocatalysts have been combined with semiconductor photocatalysts to act as a hybrid system, resulting in a high photo-response and delayed recombination of charge carriers. Researchers have been working on the hybridization of nanomaterials with polymers which can modify the structure and physicochemical properties of polymer-based materials such as porosity, hydrophilicity, charge density, mechanical, chemical, and thermal stability. We hope that this chapter will serve as a helpful reference for those wishing to conduct further studies on polymer-based photocatalysts for pharmaceutical wastewater and may also serve as an inspiration for the creation of innovative next-generation photocatalysts.

ACKNOWLEDGEMENTS

The authors gratefully acknowledge the financial support from the Ministry of Higher Education, Malaysia, under the Fundamental Research Grant Scheme (FRGS) (Project Number: FRGS/1/2021/STG05/UITM/02/17). The authors also would like to acknowledge the technical and management support from Universiti Teknologi MARA Pahang and Universiti Teknologi MARA Selangor.

REFERENCES

[1] Vats S, Sudhakarn S, Bhardwaj A, et al. Kinetic model of photo-fenton degradation of paracetamol in an annular reactor: Main reaction intermediates and cytotoxicity studies. J Hazard Mater. 2020;124910.

[2] Pang X, Skillen N, Gunaratne N, et al. Removal of phthalates from aqueous solution by semiconductor photocatalysis: A review. J Hazard Mater. 2021;402:123461.

[3] Mohd Hir ZA, Abdullah AH. Hybrid polymer-based photocatalytic materials for the removal of selected endocrine disrupting chemicals (EDCs) from aqueous media: A review. J Mol Liq. 2022;361:119632.

[4] Mukhair HM, Abdullah AH, Zainal Z. PES-Ag3PO4/g-C3N4 mixed matrix film photocatalyst for degradation of methyl orange dye. Polymers (Basel). 2021;13:1746.

[5] Chijioke-Okere M, Halim A, Adlan Z, et al. Efficient photodegradation of paracetamol by integrated PES-ZnO photocatalyst sheets. Inorg Chem Commun. 2023;148:110377.

[6] Rafaie HA, Shohaimi NA, Ramli NIT, et al. Application of hybrid polymeric materials as photocatalyst in textile wastewater. Polym Technol Dye Wastewater. 2022;101–144.

[7] Mehmood CT, Zhong Z, Zhou H, et al. Constructing porous beads with modified polysulfone-alginate and TiO2 as a robust and recyclable photocatalyst for wastewater treatment. J Water Process Eng. 2020;38:101601.

[8] Chijioke-Okere MO, Mohd Hir ZA, Ogukwe CE, et al. TiO2/Polyethersulphone films for photocatalytic degradation of acetaminophen in aqueous solution. J Mol Liq. 2021;338:116692.

[9] Hir ZAM, Abdullah AH, Zainal Z, et al. Visible light-active hybrid film photocatalyst of polyethersulfone—Reduced TiO2: Photocatalytic response and radical trapping investigation. J Mater Sci. 2018;53:13264–13279.

[10] Howard IC, Hammond C, Buchard A. Polymer-supported metal catalysts for the heterogeneous polymerisation of lactones. Polym Chem. 2019;10:5894–5904.

[11] Srikanth B, Goutham R, Narayan RB, et al. Recent advancements in supporting materials for immobilised photocatalytic applications in waste water treatment. J Environ Manage. 2017;200:60–78.

[12] Quadrado RFN, Vitoria HFV, Ferreira DC, et al. Hybrid polymer aerogels containing porphyrins as catalysts for efficient photodegradation of pharmaceuticals in water. J Colloid Interface Sci. 2022;613:461–476.

[13] Kundu S, Karak N. Polymeric photocatalytic membrane: An emerging solution for environmental remediation. Chem Eng J. 2022;438:135575.

[14] Joseph A, Vijayanandan A. Review on support materials used for immobilization of nano-photocatalysts for water treatment applications. Inorganica Chim Acta. 2022;545:121284.

[15] Esen C, Kumru B. Photocatalyst-incorporated cross-linked porous polymer networks. Ind Eng Chem Res. 2022;61(30):10616–10630.

[16] Bai X, Chen W, Wang B, et al. Photocatalytic degradation of some typical antibiotics: Recent advances and future outlooks. Int J Mol Sci. 2022;23.

[17] Zhang J, Xiang S, Wu P, et al. Recent advances in performance improvement of metal-organic Frameworks to remove antibiotics: Mechanism and evaluation. Sci Total Environ. 2022;811:152351.

[18] Mozia S. Photocatalytic membrane reactors (PMRs) in water and wastewater treatment: A review. Sep Purif Technol. 2010;73:71–91.

[19] Rando G, Sfameni S, Galletta M, et al. Functional nanohybrids and nanocomposites development for the removal of environmental pollutants and bioremediation. Molecules. 2022;27(15):4856.

[20] Hasnan NSN, Mohamed MA, Anuar NA, et al. Emerging polymeric-based material with photocatalytic functionality for sustainable technologies. J Ind Eng Chem. 2022;113:32–71.

[21] Hir ZAM, Abdullah AH, Zainal Z, et al. Photoactive hybrid film photocatalyst of polyethersulfone-ZnO for the degradation of methyl orange dye: Kinetic study and operational parameters. Catalysts. 2017;7:313.

[22] Qiu J, Li M, Ding M, et al. Cellulose tailored semiconductors for advanced photocatalysis. Renew Sustain Energy Rev. 2022;154:111820.

[23] Torki F, Faghihian H. Sunlight-assisted decomposition of cephalexin by novel synthesized NiS-PPY-Fe3O4 nanophotocatalyst. J Photochem Photobiol A Chem. 2017;338:49–59.

[24] Javaid A, Imran M, Latif S, et al. Functionalized magnetic nanostructured composites and hybrids for photocatalytic elimination of pharmaceuticals and personal care products. Sci Total Environ. 2022;849:157683.

[25] Enesca A, Cazan C. Polymer composite-based materials with photocatalytic applications in wastewater organic pollutant removal: A mini review. Polymers. 2022;14(16):3291

[26] Zia J, Riaz U. Photocatalytic degradation of water pollutants using conducting polymer-based nanohybrids: A review on recent trends and future prospects. J Mol Liq. 2021;340.

[27] Sutisna S, Rokhmat M, Wibowo E, et al. Application of immobilized titanium dioxide as reusable photocatalyst on photocatalytic degradation of methylene blue. Adv Mater Res. 2015;1112:149–153.

[28] Fischer K, Grimm M, Meyers J, et al. Photoactive micro filtration membranes via directed synthesis of TiO2 nanoparticles on the polymer surface for removal of drugs from water. J Memb Sci. 2015;478:49–57.

[29] Paredes L, Murgolo S, Dzinun H, et al. Application of immobilized TiO2 on PVDF dual layer hollow fibre membrane to improve the photocatalytic removal of pharmaceuticals in different water matrices. Appl Catal B Environ. 2019;240:9–18.

[30] Rosman N, Norharyati Wan Salleh W, Aqilah Mohd Razali N, et al. Ibuprofen removal through photocatalytic filtration using antifouling PVDF-ZnO/Ag2CO3/Ag2O nanocomposite membrane. Mater Today Proc. 2019;42:69–74.

[31] Song J, Sun G, Yu J, et al. Construction of ternary Ag@ZnO/TiO2 fibrous membranes with hierarchical nanostructures and mechanical flexibility for water purification. Ceram Int. 2020;46:468–475.

[32] Kumar A, Chandel M, Sharma A, et al. Robust visible light active PANI/LaFeO3/CoFe2O4 ternary heterojunction for the photo-degradation and mineralization of pharmaceutical effluent: Clozapine. J Environ Chem Eng. 2021;9:106159.

[33] Shahabuddin S, Gaur R, Mukherjee N, et al. Conducting polymers-based nanocomposites: Innovative materials for waste water treatment and energy storage. Mater Today Proc. 2022;62:6950–6955.

[34] Colmenares JC, Kuna E. Photoactive hybrid catalysts based on natural and synthetic polymers: A comparative overview. Molecules. 2017;22(5):790.

[35] Rockafellar RT. Part I: Basic concepts. Convex Anal. 2015;1–40.

[36] Munyengabe A, Ndibewu PP, Sibali LL, et al. Polymeric nanocomposite materials for photocatalytic detoxification of polycyclic aromatic hydrocarbons in aquatic environments-A review. Results Eng. 2022;15.

[37] Yan JW, Wu J, Lu L, et al. Photocatalytic performances and mechanisms of two coordination polymers based on rigid tricarboxylate. J Solid State Chem. 2022;316.

[38] Mohsen M, Naeem I, Awaad M, et al. A cadmium-imidazole coordination polymer as solid state buffering material: Synthesis, characterization and its use for photocatalytic degradation of ionic dyes. J Solid State Chem. 2020;289.

[39] Cui Z, Qi J, Xu X. PANI/CPs composite material, a new type of coordination polymer based composite material: Fabrication and photocatalytic property study. Inorg Chem Commun. 2013;35:260–264.

[40] Escobar-Villanueva AG, Víctor V, Medina MO, et al. Fast photodegradation of Orange Ii Azo Dye under visible light irradiation using a semiconducting n—p heterojunction of ZnO nanoparticles/polypyrrole as catalyst. J Mater Sci: Mater Electron. 2022;31(2):1317–1327.

[41] Shanmugam M, Augustin A, Mohan S, et al. Conducting polymeric nanocomposites: A review in solar fuel applications. Fuel. 2022;325.

[42] Katal R, Farahani MHDA, Masudy-Panah S, et al. Polypyrrole and polyaniline supported TiO2 for removal of pollutants from water. J Environ Eng Sci. 2018;14(2):67–89.

[43] Saianand G, Gopalan AI, Wang L, et al. Conducting polymer based visible light photocatalytic composites for pollutant removal: Progress and prospects. Environ Technol Innov. 2022;28:102698.

[44] Bu Y, Chen Z. Role of polyaniline on the photocatalytic degradation and stability performance of the polyaniline/silver/silver phosphate composite under visible light. ACS Appl Mater Interfaces. 2014;6:17589–17598.

[45] Ma J, Dai J, Duan Y, et al. Fabrication of PANI-TiO 2/rGO hybrid composites for enhanced photocatalysis of pollutant removal and hydrogen production. 2020;156:1008–1018.

[46] Wang C, Hu Z, Zhao H, et al. Probing conducting polymers @ Cadmium Sulfide Core-shell nanorods for highly improved photocatalytic hydrogen production. J Colloids Interface Sci. 2018;521:1–10.

[47] Silvestri S, Ferreira CD, Oliveira V, et al. Synthesis of PPy-ZnO composite used as photocatalyst for the degradation of diclofenac under simulated solar irradiation. J Photochem Photobiol A Chem. 2019;375:261–269.

[48] Sasso C, Beneventi D, Zeno E, et al. Polypyrrole and polypyrrole/wood-derived materials conducting composites: A review. BioResources. 2011;6:3585–3620.

[49] Abu-Sari SM, Patah MFA, Ang BC, et al. A review of polymerization fundamentals, modification method, and challenges of using PPy-based photocatalyst on perspective application. J Environ Chem Eng. 2022;10.

[50] Goel S, Mazumdar NA, Gupta A. Synthesis and characterization of polypyrrole nanofibers with different dopants. Polym Adv Technol. 2010;21:205–210.

[51] Qing Y, Li Y, Cao L, et al. Rationally designed C-PANI/BiOBr STEP-scheme heterojunction photocatalyst for boosting photodegradation of 17β-estradiol. Sep Purif Technol. 2023;314:123545.

[52] Kashyap J, Riaz U. Facile synthesis of novel polypyrrole dispersed AgFeO2 nanohybrid with highly efficient photocatalytic activity towards 2,4,6-trichlorophenol degradation. RSC Adv. 2018;8:13218–13225.

[53] Aminuddin NF, Nawi MA, Bahrudin NN, et al. Iron ion assisted photocatalytic-adsorptive removal of acid orange 52 by immobilized TiO2/polyaniline bilayer photocatalyst. Appl Surf Sci Adv. 2021;6:100180.

[54] Gupta ND, Maity S, Chattopadhyay KK. Field emission enhancement of polypyrrole due to band bending induced tunnelling in polypyrrole-carbon nanotubes nanocomposite. J Ind Eng Chem. 2014;20:3208–3213.

[55] Ramanavičius A, Ramanavičiene A, Malinauskas A. Electrochemical sensors based on conducting polymer-polypyrrole. Electrochim Acta. 2006;51:6025–6037.

[56] Pang AL, Arsad A, Ahmadipour M. Synthesis and factor affecting on the conductivity of polypyrrole: A short review. Polym Adv Technol. 2021;32:1428–1454.

[57] Navale ST, Mane AT, Ghanwat AA, et al. Camphor sulfonic acid (CSA) doped polypyrrole (PPy) films: Measurement of microstructural and optoelectronic properties. Meas J Int Meas Confed. 2014;50:363–369.

[58] Sanches EA, Alves SF, Soares JC, et al. Nanostructured polypyrrole powder: A structural and morphological characterization. J Nanomater. 2015;2015.

[59] Uda M, Higashimoto S, Hirai T, et al. Synthesis of poly(alkylaniline)s by aqueous chemical oxidative polymerization and their use as stimuli-responsive liquid marble stabilizer. J Appl Polym Sci. 2021;138.

[60] Sambaza S, Maity A, Pillay K. Visible light degradation of ibuprofen using PANI coated WO3@TiO2 photocatalyst. J Saudi Chem Soc. 2022;26:101563.

[61] Munusamy S, Sivaranjan K, Sabhapathy P, et al. Electrochemical and photocatalytic studies of Ta3N5-TaON-PEDOT-PANI nanohybrids. Chem Phys Lett. 2021;780:138947.

[62] Meng P, Heng H, Sun Y, et al. In situ polymerization synthesis of Z-scheme tungsten trioxide/polyimide photocatalyst with enhanced visible-light photocatalytic activity. Appl Surf Sci. 2018;428:1130–1140.

[63] Li H, Tu W, Zhou Y, et al. Z-scheme photocatalytic systems for promoting photocatalytic performance: Recent progress and future challenges. Adv Sci. 2016;3.

[64] Yahya N, Aziz F, Jamaludin NA, et al. A review of integrated photocatalyst adsorbents for wastewater treatment. J Environ Chem Eng. 2018;6:7411–7425.

[65] Kodom K, Attiogbe F, Kuranchie FA. Assessment of removal efficiency of pharmaceutical products from wastewater in sewage treatment plants: A case of the sewerage systems Ghana limited, Accra. Heliyon. 2021;7:e08385.

[66] Deo RP, Halden RU. Pharmaceuticals in the built and natural water environment of the United States. Water (Switzerland). 2013;5:1346–1365.

[67] Kapelewska J, Kotowska U, Karpińska J, et al. Occurrence, removal, mass loading and environmental risk assessment of emerging organic contaminants in leachates, groundwaters and wastewaters. Microchem J. 2018;137:292–301.

[68] Durán-Álvarez JC, Avella E, Ramírez-Zamora RM, et al. Photocatalytic degradation of ciprofloxacin using mono- (Au, Ag and Cu) and bi- (Au-Ag and Au-Cu) metallic nanoparticles supported on TiO2 under UV-C and simulated sunlight. Catal Today. 2016;266:175–187.

[69] Kumar R, Sarmah AK, Padhye LP. Fate of pharmaceuticals and personal care products in a wastewater treatment plant with parallel secondary wastewater treatment train. J Environ Manage. 2019;233:649–659.

[70] Kumar R, Tscharke B, O'Brien J, et al. Assessment of drugs of abuse in a wastewater treatment plant with parallel secondary wastewater treatment train. Sci Total Environ. 2019;658:947–957.

[71] Padhye LP, Yao H, Kung'u FT, et al. Year-long evaluation on the occurrence and fate ofpharmaceuticals, personal care products, andendocrine disrupting chemicals in an urban drinking water treatment plant. Water Res. 2014;51:266–276.

[72] Lima RMAP, Alcaraz-Espinoza JJ, Da Silva FAG, et al. Multifunctional wearable electronic textiles using cotton fibers with polypyrrole and carbon nanotubes. ACS Appl Mater Interfaces. 2018;10:13783–13795.

[73] Sarafraz M, Sadeghi M, Yazdanbakhsh A, et al. Enhanced photocatalytic degradation of ciprofloxacin by black Ti3+/N-TiO2 under visible LED light irradiation: Kinetic, energy consumption, degradation pathway, and toxicity assessment. Process Saf Environ Prot. 2020;137:261–272.

[74] González-Pérez BK, Sarma SSS, Nandini S. Effects of selected pharmaceuticals (ibuprofen and amoxicillin) on the demography of Brachionus calyciflorus and Brachionus havanaensis (Rotifera). Egypt J Aquat Res. 2016;42:341–347.

[75] Ahmadi M, Ramezani Motlagh H, Jaafarzadeh N, et al. Enhanced photocatalytic degradation of tetracycline and real pharmaceutical wastewater using MWCNT/TiO2 nano-composite. J Environ Manage. 2017;186:55–63.

[76] Kumar A, Sharma G, Naushad M, et al. Visible photodegradation of ibuprofen and 2,4-D in simulated waste water using sustainable metal free-hybrids based on carbon nitride and biochar. J Environ Manage. 2019;231:1164–1175.

[77] Mohamed A, Salama A, Nasser WS, et al. Photodegradation of ibuprofen, cetirizine, and naproxen by PAN-MWCNT/TiO2–NH2 nanofiber membrane under UV light irradiation. Environ Sci Eur. 2018;30.

[78] An J, Li Y, Chen W, et al. Electrochemically-deposited PANI on iron mesh-based metal-organic framework with enhanced visible-light response towards elimination of thiamphenicol and E. coli. Environ Res. 2020;191:110067.

[79] Shamsizadeh Z, Ehrampoush MH, Dehghani Firouzabadi Z, et al. Fe3O4@SiO2 magnetic nanocomposites as adsorbents for removal of diazinon from aqueous solution: Isotherm and kinetic study. Pigment Resin Technol. 2020;49:457–464.

[80] Natarajan S, Bajaj HC, Tayade RJ. Recent advances based on the synergetic effect of adsorption for removal of dyes from waste water using photocatalytic process. J Environ Sci (China). 2018;65:201–222.

[81] Asadzadeh-Khaneghah S, Habibi-Yangjeh A. g-C3N4/carbon dot-based nanocomposites serve as efficacious photocatalysts for environmental purification and energy generation: A review. J Clean Prod. 2020;276:124319.

[82] Akhundi A, Habibi-Yangjeh A, Abitorabi M, et al. Review on photocatalytic conversion of carbon dioxide to value-added compounds and renewable fuels by graphitic carbon nitride-based photocatalysts. Catal Rev—Sci Eng. 2019;61:595–628.

[83] Akhundi A, Badiei A, Ziarani GM, et al. Graphitic carbon nitride-based photocatalysts: Toward efficient organic transformation for value-added chemicals production. Mol Catal. 2020;488:110902.

[84] Nasir MS, Yang G, Ayub I, et al. Recent development in graphitic carbon nitride based photocatalysis for hydrogen generation. Appl Catal B Environ. 2019;257:117855.

[85] Geng T, Chen H, Xu J, et al. Removal of nickels from crude oil to water by two micro-sized core-shell particles bearing poly(N-vinyl pyrrolidone). Fuel. 2019;245:181–187.

[86] Ibanez JG, Rincón ME, Gutierrez-Granados S, et al. Conducting polymers in the fields of energy, environmental remediation, and chemical-chiral sensors. Chem Rev. 2018;118:4731–4816.

[87] Chapple-McGruder T, Leider JP, Beck AJ, et al. Examining state health agency epidemiologists and their training needs. Ann Epidemiol. 2017;27:83–88.

[88] Guo J, Lee JG, Tan T, et al. Enhanced ammonia recovery from wastewater by Nafion membrane with highly porous honeycomb nanostructure and its mechanism in membrane distillation. J Memb Sci. 2019;590:117265.

[89] Steplin Paul Selvin S, Ganesh Kumar A, Sarala L, et al. Photocatalytic degradation of phodamine B using zinc oxide activated charcoal polyaniline nanocomposite and its survival assessment using aquatic animal model. ACS Sustain Chem Eng. 2018;6:258–267.

[90] Zhao H, Li G, Tian F, et al. g-C3N4 surface-decorated Bi2O2CO3 for improved photocatalytic performance: Theoretical calculation and photodegradation of antibiotics in actual water matrix. Chem Eng J. 2019;366:468–479.

[91] Moosvi SK, Majid K, Ara T. Study of thermal, electrical, and photocatalytic activity of iron complex doped polypyrrole and polythiophene nanocomposites. Ind Eng Chem Res. 2017;56:4245–4257.

[92] Liao G, Chen S, Quan X, et al. Remarkable improvement of visible light photocatalysis with PANI modified core-shell mesoporous TiO2 microspheres. Appl Catal B Environ. 2011;102:126–131.

[93] Ghosh S, Kouamé NA, Ramos L, et al. Conducting polymer nanostructures for photocatalysis under visible light. Nat Mater. 2015;14:505–511.

[94] Zhou Q, Shi G. Conducting polymer-based catalysts. J Am Chem Soc. 2016;138:2868–2876.

[95] Kumar R, Travas-Sejdic J, Padhye LP. Conducting polymers-based photocatalysis for treatment of organic contaminants in water. Chem Eng J Adv. 2020;4:100047.

[96] Muktha B, Mahanta D, Patil S, et al. Synthesis and photocatalytic activity of poly(3-hexylthiophene)/TiO2 composites. J Solid State Chem. 2007;180:2986–2989.

[97] Jana B, Bhattacharyya S, Patra A. Conjugated polymer P3HT-Au hybrid nanostructures for enhancing photocatalytic activity. Phys Chem Chem Phys. 2015;17:15392–15399.

[98] Wang D, Zhang J, Luo Q, et al. Characterization and photocatalytic activity of poly(3-hexylthiophene)-modified TiO2 for degradation of methyl orange under visible light. J Hazard Mater. 2009;169:546–550.

[99] Ahmad N, Anae J, Khan MZ, et al. Visible light-conducting polymer nanocomposites as efficient photocatalysts for the treatment of organic pollutants in wastewater. J Environ Manage. 2021;295:113362.

[100] Girija Shankar E, Aishwarya M, Khan A, et al. Efficient solar light photocatalytic degradation of commercial pharmaceutical drug and dye using rGO-PANI assisted c-ZnO heterojunction nanocomposites. Ceram Int. 2021;47:23770–23780.

[101] Wu H, Lin S, Chen C, et al. A new ZnO/rGO/polyaniline ternary nanocomposite as photocatalyst with improved photocatalytic activity. Mater Res Bull. 2016;83:434–441.

[102] Zhang H, Zong R, Zhu Y. Photocorrosion inhibition and photoactivity enhancement for zinc oxide via hybridization with monolayer polyaniline. J Phys Chem C. 2009;113:4605–4611.

[103] Kang ET, Neoh KG, Tan KL. Polyaniline: A polymer with many interesting intrinsic redox states. Prog Polym Sci. 1998;23:277–324.

[104] Shirota Y, Kageyama H. Charge carrier transporting molecular materials and their applications in devices. Chem Rev. 2007;107:953–1010.

[105] Lam SM, Sin JC, Abdullah AZ, et al. Photocatalytic TiO2/carbon nanotube nanocomposites for environmental applications: An overview and recent developments. Fuller Nanotub Car N. 2014;22:471–509.

[106] Xu Y, Ma Y, Ji X, et al. Conjugated conducting polymers PANI decorated Bi 12 O 17 Cl 2 photocatalyst with extended light response range and enhanced photoactivity. Appl Surf Sci. 2019;464:552–561.

[107] Taing J, Cheng MH, Hemminger JC. Photodeposition of Ag or Pt onto TiO2 nanoparticles decorated on step edges of HOPG. ACS Nano. 2011;5:6325–6333.

[108] Kulkarni VG, Campbell LD, Mathew WR. Thermal stability of polyaniline. Synth Met. 1989;30:321–325.

[109] Sin JC, Lam SM, Zeng H, et al. Magnetic NiFe2O4 nanoparticles decorated on N-doped BiOBr nanosheets for expeditious visible light photocatalytic phenol degradation and hexavalent chromium reduction via a Z-scheme heterojunction mechanism. Appl Surf Sci. 2021;559:149966.

[110] Fatima T, Husain S, Khanuja M. Superior photocatalytic and electrochemical activity of novel WS2/PANI nanocomposite for the degradation and detection of pollutants: Antibiotic, heavy metal ions, and dyes. Chem Eng J Adv. 2022;12.

[111] Ma X, Huo X, Hao K, et al. Visible light driven VO2/g-C3N4 Z-scheme composite photocatalysts for selective oxidation of DL-1-phenylethyl alcohol under Vis-LEDs irradiation and aerobic oxidation. ChemistrySelect. 2021;6:2101–2110.

[112] Singh R, Choudhary RB, Kandulna R. Delocalization of π electrons and trapping action of ZnO nanoparticles in PPY matrix for hybrid solar cell application. J Mol Struct. 2018;1156:633–644.

[113] Senadeera GKR, Kitamura T, Wada Y, et al. Enhanced photoresponses of polypyrrole on surface modified TiO2 with self-assembled monolayers. J Photochem Photobiol A Chem. 2006;184:234–239.

[114] Luo D, Kang Y. Synthesis and characterization of novel CaFe2O4/Bi2O3 composite photocatalysts. Mater Lett. 2018;225:17–20.

[115] Wang D, Wang Y, Li X, et al. Sunlight photocatalytic activity of polypyrrole-TiO2 nanocomposites prepared by "in situ" method. Catal Commun. 2008;9:1162–1166.

[116] Ullah H, Tahir AA, Mallick TK. Polypyrrole/TiO2composites for the application of photocatalysis. Sensors Actuators, B Chem. 2017;241:1161–1169.

[117] Jia Y, Wu C, Lee BW, et al. Magnetically separable sulfur-doped SnFe2O4/graphene nanohybrids for effective photocatalytic purification of wastewater under visible light. J Hazard Mater. 2017;338:447–457.

[118] Yan B, Wang Y, Jiang X, et al. Flexible photocatalytic composite film of ZnO-microrods/polypyrrole. ACS Appl Mater Interfaces. 2017;9:29113–29119.

[119] Techalertmanee T, Chancharoenrith S, Namkajorn M, et al. Facile synthesis of zinc-iron mixed oxide/carbon nanocomposites as nanocatalysts for the degradation of methylene blue. Mater Lett. 2015;145:224–228.

[120] Pruna A, Shao Q, Kamruzzaman M, et al. Effect of ZnO core electrodeposition conditions on electrochemical and photocatalytic properties of polypyrrole-graphene oxide shelled nanoarrays. Appl Surf Sci. 2017;392:801–809.

[121] Elmolla ES, Chaudhuri M. Degradation of amoxicillin, ampicillin and cloxacillin antibiotics in aqueous solution by the UV/ZnO photocatalytic process. J Hazard Mater. 2010;173:445–449.

[122] Gao B, Safaei Z, Babu I, et al. Modification of ZnIn2S4 by anthraquinone-2-sulfonate doped polypyrrole as acceptor-donor system for enhanced photocatalytic degradation of tetracycline. J Photochem Photobiol A Chem. 2017;348:150–160.

[123] Asplund M, Thaning E, Lundberg J, et al. Toxicity evaluation of PEDOT/biomolecular composites intended for neural communication electrodes. Biomed Mater. 2009;4.

[124] Balint R, Cassidy NJ, Cartmell SH. Conductive polymers: Towards a smart biomaterial for tissue engineering. Acta Biomater. 2014;10:2341–2353.

[125] Fabretto MV., Evans DR, Mueller M, et al. Polymeric material with metal-like conductivity for next generation organic electronic devices. Chem Mater. 2012;24:3998–4003.

[126] Kumar R, Akbarinejad A, Jasemizad T, et al. The removal of metformin and other selected PPCPs from water by poly(3,4-ethylenedioxythiophene) photocatalyst. Sci Total Environ. 2021;751:142302.

[127] Wols BA, Hofman-Caris CHM, Harmsen DJH, et al. Degradation of 40 selected pharmaceuticals by UV/H2O2. Water Res. 2013;47:5876–5888.

[128] Fernández RL, McDonald JA, Khan SJ, et al. Removal of pharmaceuticals and endocrine disrupting chemicals by a submerged membrane photocatalysis reactor (MPR). Sep Purif Technol. 2014;127:131–139.

[129] Tang T, Lu G, Wang W, et al. Photocatalytic removal of organic phosphate esters by TiO2: Effect of inorganic ions and humic acid. Chemosphere. 2018;206:26–32.

[130] Ledezma-Espinoza A, Rodríguez-Quesada L, Araya-Leitón M, et al. Modified cellulose/poly(3,4-ethylenedioxythiophene) composite as photocatalyst for the removal of sulindac and carbamazepine from water. Environ Technol Innov. 2022;27:1–15.

[131] Yuan X, Floresyona D, Aubert PH, et al. Photocatalytic degradation of organic pollutant with polypyrrole nanostructures under UV and visible light. Appl Catal B Environ. 2019;242:284–292.

[132] Xiao X, Tu S, Lu M, et al. Discussion on the reaction mechanism of the photocatalytic degradation of organic contaminants from a viewpoint of semiconductor photoinduced electrocatalysis. Appl Catal B Environ. 2016;198:124–132.

[133] Yan H, Huang Y. Polymer composites of carbon nitride and poly(3-hexylthiophene) to achieve enhanced hydrogen production from water under visible light. Chem Commun. 2011;47:4168–4170.

[134] Liu F, Nguyen TP, Wang Q, et al. Construction of Z-scheme g-C3N4/Ag/P3HT heterojunction for enhanced visible-light photocatalytic degradation of tetracycline (TC) and methyl orange (MO). Appl Surf Sci. 2019;496:143653.

11 Carbon Electrodes for Pharmaceutical Wastewater Treatment

Abdoulaye Thiam and Christian Onfray

11.1 INTRODUCTION TO CARBON ELECTRODES FOR WASTEWATER TREATMENT

11.1.1 BACKGROUND AND SIGNIFICANCE OF PHARMACEUTICAL WASTEWATER TREATMENT

Pharmaceutical wastewater is a complex mixture containing active pharmaceutical ingredients, drug intermediates, and formulation residues. Discharging such wastewater into the environment without proper treatment can have detrimental effects on aquatic ecosystems [1, 2]. The persistence of pharmaceuticals in water bodies can disrupt the ecological balance, harm aquatic organisms, and contribute to the development of drug-resistant strains. Moreover, human exposure to pharmaceuticals through drinking water or contaminated food can pose risks to health, including potential toxicity and the emergence of antibiotic resistance [3].

Recognizing the environmental and health concerns associated with pharmaceutical wastewater, its proper treatment is now seen as a matter of corporate social responsibility. The significance of pharmaceutical wastewater treatment lies in its ability to mitigate environmental impact, safeguard human health, ensure regulatory compliance, combat antibiotic resistance, promote sustainable water management practices, and uphold corporate social responsibility. However, conventional biological and physical treatment technologies, which are commonly used due to their cost effectiveness and environmentally friendly nature, have limitations in effectively removing recalcitrant and emerging contaminants present in pharmaceutical wastewater [4, 5]. Many pharmaceutical compounds resist biodegradation and can pass through these conventional processes without significant degradation [2]. This inefficiency poses risks to the environment and public health.

In this context, electrochemical wastewater treatment has proven to be a promising technology for efficiently removing pharmaceutical compounds and other recalcitrant pollutants [6, 7]. This approach involves the application of an electrical potential to induce electrochemical reactions at the surfaces of electrodes, leading to the degradation or transformation of contaminants. The selection of electrode materials is crucial, with carbon-based electrodes attracting significant attention owing

DOI: 10.1201/9781003340164-11

to their superior chemical, electronic, electrocatalytic, and mechanical properties, which depend heavily on their structures and forms [8].

11.1.2 ROLE OF CARBON ELECTRODES IN WASTEWATER TREATMENT

Carbon electrodes have long been recognized for their pivotal role in electrochemical processes, driven by the exceptional properties that make them highly sought after for industrial applications [9]. Primarily valued for their excellent conductivity and abundance of free electrons available for transfer, carbon electrodes serve as proficient conductors in electrochemistry. Moreover, they offer a multitude of advantageous features, including remarkable chemical stability, expansive surface area, customizable surface chemistry, electrochemical catalytic activity, versatility, availability, and environmental benefits [10]. These attributes collectively position carbon electrodes as indispensable components in the triumph of environmental electrochemical technology for pharmaceutical removal from water. Their significant contributions have propelled successes and advancements in environmental electrochemical technology, fostering sustainable solutions and driving progress in pharmaceutical removal.

Carbon electrodes exhibit exceptional electrical performance by their superior conductivity and abundance of free electrons, facilitating efficient electron transfer during electrochemical reactions. Their impressive chemical stability ensures enduring performance and resilience, even in harsh and corrosive environments, contributing to the longevity and reliability of electrochemical systems. Furthermore, carbon electrodes possess a substantial surface area, allowing for heightened adsorption capacity, which enhances the efficiency of the electrochemical removal of pharmaceuticals.

One notable advantage of carbon electrodes is their capability for tailorable surface chemistry. The surface of carbon electrodes can be easily modified through various techniques such as functionalization or chemical/physical processes, enabling precise control over their electrochemical properties. This tunability empowers the creation of electrode surfaces specifically tailored to exhibit enhanced reactivity, selectivity, or affinity towards specific reactions. However, it is essential to note that carbon materials are generally considered poor catalysts for both H_2 and O_2 evolution reactions [11]. As a result, in environmental electrochemical conditions, carbon electrodes are expected to function as inert electrodes. Nevertheless, this characteristic makes carbon electrodes a conspicuous choice for experimentation and implementation as both anodes and cathodes in environmental electrochemistry for pharmaceutical removal. The inert behavior of carbon electrodes allows them to serve as versatile platforms for exploring and advancing various electrochemical processes. Carbon electrode modifications will be addressed extensively in subsection 11.2.3.

Furthermore, carbon electrodes exhibit exceptional versatility and availability, derived from abundant carbon sources and renewable materials. The advancements in carbon materials have led to the development of a diverse array of structures, including graphite, various types of carbon (e.g., amorphous, nanotubes), heat-treated carbons (e.g., vitreous or glassy carbon), carbon/polymer composites, expanded carbons, and materials related to doped diamond. These carbon electrode materials can be fabricated into a wide range of forms, such as powders, plates, disks, coatings, granules, cloths, fibers, foams, or films, catering to the specific requirements of various electrochemical applications. This remarkable versatility, combined with their cost effectiveness, renders carbon electrodes highly appealing and well-suited for applications in electrochemistry for pharmaceutical removal. In addition to their technical advantages, carbon electrodes offer notable environmental benefits. They enable the electrochemical treatment of pollutants and wastewater, facilitating the removal or degradation of emerging contaminants without relying on harsh chemicals.

Figure 11.1 illustrates several applications of carbon, particularly in electrochemical technologies, demonstrating their extensive usage. It is important to note that

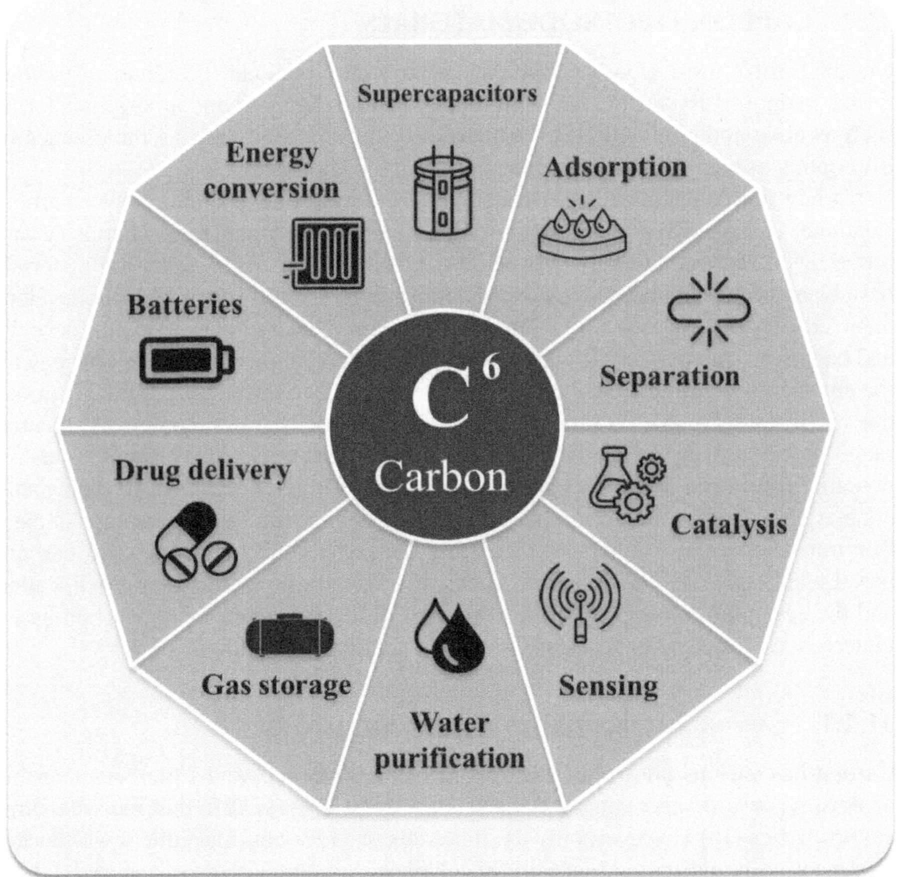

FIGURE 11.1 Applications of carbon materials in various technologies.

while the figure provides a representative overview, listing every single application is not feasible within its scope.

11.1.3 Objectives of the Chapter

This chapter focuses on discussing various electrochemical technologies that utilize carbon electrodes in different forms and structures for the treatment of pharmaceutical wastewater. It explores modifications of carbon electrodes, their performance characteristics, and practical applications in this context. Through a comprehensive analysis, we aim to gain insights into the specific roles and efficiencies of different carbon electrode materials in pharmaceutical removal. By examining the advancements in electrochemical treatment techniques, we aim to highlight the potential of carbon electrodes as practical tools for addressing the challenges associated with pharmaceutical wastewater treatment.

11.2 CARBON ELECTRODE MATERIALS

One of Earth's most plentiful elements is carbon, which can be found in various forms, including crystalline forms like graphite and diamond and amorphous forms such as coke and charcoal [12]. Furthermore, carbon, renowned as the essence of life, stands as the most utilized element in organic chemistry.

Carbon materials are diverse, encompassing forms like graphite, glassy carbon, fullerene, carbon nanotubes, carbon dots, diamond, etc., and their chemistry and properties differ dramatically between the different forms. Hence, carbon materials have been used as materials for diverse applications, such as supercapacitor development, adsorption/separation of voluminous molecules, energy conversion in fuel cells and batteries, drug delivery, gas storage, catalysis, and photocatalysis [13]. Among all the applications, carbon-based ones hold a crucial role in developing solid electrodes due to their slow oxidation kinetics. This characteristic bestows a practical potential range, particularly in the anodic field. Carbon-based electrodes also present expansive potential limits and lower background oxidation currents, in contrast to their metal counterparts. Carbon-based electrodes generally demonstrate a diverse range of electron transfer rates, contingent on the specific material employed. Moreover, carbon-based electrodes have rich surface chemistry, which allows for their modification and the design of multiple structures [14]. For all these characteristics, carbon-based materials are excellent candidates for electrochemical applications.

11.2.1 Carbon Electrode Types and Characteristics

Carbon has various allotropes, such as graphite, diamond, and fullerenes, capable of existing in a diverse range of materials, which possess different electrochemical properties [15] However, only the materials used for pharmaceutical wastewater treatment will be discussed later in this chapter. Among all carbon materials, there are two opposing groups: the ones structured mainly by sp^2-hybridized carbons, known as graphitic structure, and those composed mainly by sp^3-hybridized carbons, known as diamond structure.

11.2.1.1 Graphite and sp^2-Hybridized Carbon Materials

The structure of graphite consists of ideally infinite sheets of "graphene" arranged in stacked and parallel layers. In the structure of graphite, all the carbon atoms are sp^2 hybridized, with a C–C interplanar bond length of 1.42 Å and an interplanar spacing of 3.354 Å. Natural crystalline graphite and highly oriented pyrolytic graphite (HOPG) are the most ordered three-dimensional graphite materials. Carbon fibers also possess an sp^2-hybridized structure with diameters of 5–50 μm and are made from polymers and/or small hydrocarbons. An important variant of the graphite structure is glassy carbon (GC), produced by the thermal processing of various polymers at temperatures ranging between 1000–3000 °C in an inert atmosphere. The interplanar distance is larger than HOPG's (~ 3.6 Å). Different industrial processes fabricate carbon black; therefore, the material can have different orders and compositions. Among the different carbon materials, carbon and graphite electrodes are applicable only at limited values of potential and current, which is attributed to high corrosion rates at potentials higher than 1.7–1.9 V vs saturated calomel electrode (SCE) [16]. Corrosion leads to material loss, an increase in electrode resistance, and a decrease in stability.

11.2.1.2 Diamond and Related Materials

Diamonds have sp^3 hybridization and tetrahedral bonding, which results in high hardness and low electrical conductivity. However, the intentional introduction of impurities into a diamond can increase its conductivity enough to prepare electrodes with different electrochemical performances. The most common impurities inserted in the diamond structure are boron and nitrogen, where the former is used for the fabrication of one of the most common diamond-based electrodes, boron-doped diamond (BDD). In the production of electrodes, nitrogen gas or a source of boron is used, which leads to the formation of doped nanocrystallites (n-doped or p-doped, respectively). This seeps into the electrode with significantly higher electrical conductivity compared to undoped bulk diamonds.

The enhanced chemical stability of BDD electrodes, in comparison to graphitic and fullerene electrodes, provides a significant advantage over the more commonly used sp^2-hybridized carbon-based electrodes. BDD electrodes exhibit reduced adsorption of most solution species, making them more resilient to fouling. Moreover, BDD-type electrodes have the highest overpotential for the oxygen evolution reaction (OER) among the existing electrodes, which generates high electrocatalytic efficiency towards the oxidation of different organic compounds [17]. In this context, the properties of BDD electrodes depend on the sp^3/sp^2 ratio on their surface. A higher content of sp^3 will favor the production of radicals on the surface, and a higher content of sp^2 will favor the evolution of O_2, promoting the direct oxidation of organic compounds on its surface [18].

11.2.2 Carbon-Based Materials for Pharmaceutical Wastewater Treatment

Among their different applications, carbon-based materials have been used for the treatment of pharmaceutical wastewater. In this field, their main applications have

TABLE 11.1

Main Advantages and Disadvantages of Graphite and Diamond-Based Electrodes

Type of Electrode	Advantages	Disadvantages
Graphite-based electrodes	• Low cost • Intensive use at laboratory scale for new process investigations	• High electrode corrosion rates • Low mineralization efficiency • Low overpotential towards OER
Diamond-based electrodes	• High overpotential towards OER • High mineralization efficiency • High conductivity • High electrochemical and chemical stability • High corrosion resistance	• Expensive • Low mechanical resistance

been focused on their use as adsorbents [19] and as catalysts for electrochemical treatments of different compounds of the pharmaceutical industry in water. Specifically, carbon-based materials have been used for electrochemical treatment applications in electrochemical oxidation (EO), electro-Fenton (EF), and photoelectro-Fenton (PEF) processes. On the one hand, BDD and related materials have been widely used in the EO process, where strong oxidants, such as $^{\bullet}OH$, are produced on the anode surface. For example, Rabaaoui et al. [20] reported the use of BDD as an anode and stainless steel as a cathode for the degradation of 100 mg L^{-1} of salicylic acid (SA) by the EO process. They observed a significant concentration of hydroxyl radicals generated at the BDD surface, achieving the complete degradation of the contaminant within 2 h of treatment under optimal conditions. On the other hand, graphite felt, carbon felt, graphite, carbon-polytetrafluoroethylene (PTFE), graphite–PTFE, activated carbon fibers (ACFs), reticulated vitreous carbon (RVC), and carbon sponge were used for Fenton-based electrochemical processes (FEPs) such as EF and PEF for the in situ electrochemical generation of H_2O_2. Yahya et al. [21] studied the degradation of the antimicrobial agent ciprofloxacin (CIP) by the EF process using carbon felt as a cathode and Pt as an anode. They obtained a remarkably high degree of ciprofloxacin (CIP) degradation (100%) and mineralization (>94%) after 10 min and 6 h of treatment, respectively, under optimal conditions. Authors attributed the system's high performance to the multiple attacks of $^{\bullet}OH$ during the EF process.

Since the properties of carbon-based materials are very different from each other, it is important to study and select the accurate material for the desired application. Some of the main advantages and disadvantages of graphitic and diamond-based electrodes for electrochemical treatments are listed in Table 11.1.

11.2.3 Surface Modifications and Functionalization Techniques

Surface preparation is important for any heterogeneous process like electrochemistry. In this context, carbon materials possess a wide variety of functional groups, such as carbonyl, carboxyl, and ketone groups, which allow the interaction with different molecules (analytes) and the construction of different architectures.

Among all the carbon-based materials designed for electrochemical treatments, different strategies have been used to construct a wide variety of electrodes, such as chemisorption/adsorption, covalent bonding, assemblies of metals, metal oxides, composites, and depositions [22]. Some of them have been fabricated using carbon-based supports modified with metallic features, and some of them have been designed using metallic substrates modified with different materials, such as graphene, carbon nanotubes, etc.

11.2.3.1 Modified Carbon-Based Supports

Guo et al. [23] prepared a graphite felt (GF) by anodization and HNO_3 ultrasonic-integrated treatment for its application as a cathode for norfloxacin (NOR) degradation by the EF process. This dual-modification technique significantly enhanced the cathode's performance by creating a greater number of defective active sites (DASs), like sp^3-C, on the GF surface, as opposed to traditional single modification methods. This enhancement notably improved the two-electron ORR activity. The electrogenerated H_2O_2 was further synergistically activated with Fe^{2+} and DASs, yielding $^\bullet OH$ and $^\bullet O_2^-$ radicals for effective NOR degradation. The cathode, optimized under these conditions, demonstrated a broad operational window, both electrically (−0.4 to −0.8 V) and pH-wise (3–9), coupled with a high degradation performance, achieving 93% NOR elimination and almost 66% mineralization after 120 min of treatment. The cathode also showed good reusability. Similarly, Cui et al. [24] developed an EF–EO flow-through system using FeOCl-deposited graphite felt (FeOCl/GF) as a cathode and a Ti_4O_7 membrane as an anode, specifically targeting the degradation of oxytetracycline (OTC). The FeOCl-deposited graphite felt was prepared by immersing the GF in a 0.05 g/mL $FeCl_3H_2O$ ethanolic solution for three hours to facilitate impregnation. This was followed by a drying process at 45 °C for 12 hours to create a $FeCl_3$/GF precursor composite. The precursor then underwent calcination at 220 °C for one hour under atmospheric conditions. Results indicated that the prepared system increased the OTC degradation 2.66 times compared with the homogeneous EF–EO system. Furthermore, the flow-through configuration amplified the OTC degradation rate constant by 6.65 times relative to batch mode systems. Cruz del Álamo et al. [25] fabricated a bifunctional electrode composed of perovskite, carbon black (CB), and polytetrafluoroethylene (PTFE) for the removal of the antipyretic and analgesic drug antipyrine (ANT) by the EF process. The electrode was prepared by mixing CB, a PTFE solution, and water, followed by drying and hand pressing to enhance its mechanical strength. Different synthesis parameters, such as solvent and proportions of the reagents, were studied. The electrode showing the most promise had a composition of 20 wt% PTFE and water, exhibiting low impedance and superior H_2O_2 electrogeneration capabilities. The method of incorporation of perovskite material ($LaCu_{0.5}Mn_{0.5}O_3$) into the optimized electrode was also studied. Nevertheless, direct inclusion in the CB/PTFE formulation paste exhibited higher EF performance. A complete ANT and 92% TOC mineralization were achieved in 240 min by applying 120 mA/cm^2.

11.2.3.2 Metallic Supports Modified with Carbon-Based Materials

Recently, Nadali Pishnamaz et al. [26] designed and compared different nickel foam (Ni-F) cathodes modified with single-walled carbon nanotubes (SWCNTs), multi-walled carbon nanotubes (MWCNTs), graphitic carbon nitride (g-C_3N_4), expanded graphite (EG), carbon black (CB), and a mixture of SWCNTs and CB for the removal of organic contaminants by the EF process. They coated Ni-F with carbonaceous powder using dip coating and sonication before drying in an oven at 80 °C for 10 min. The process was repeated consecutively until coverage of 5 mg cm^{-2} Ni-F was achieved, which was subsequently covered by polyvinylidene fluoride (PVDF) to prevent damage and loss of the carbonaceous powder. Results showed that the cathode optimized by CB covered with SWCNT produced 3.3, 7.3, and 11.5 times more H_2O_2 than that with SWCNT, MWCNT, and CB alone, respectively. The system was able to remove different organic contaminants efficiently under optimal conditions. The removal efficiencies achieved were 94.1% for Methylene blue, 63.1% for tetracycline, 70.2% for phenol, 76.4% for bisphenol A, 84.0% for 4-chlorophenol, and 97.3% for amoxicillin.

As could be seen, different strategies have been performed by modifying the electrode in order to achieve surfaces with higher catalytic activities for diverse processes. Furthermore, the properties of the surface of the electrode and its efficiency towards the production of H_2O_2 and $\cdot$OH and the degradation of pharmaceutical pollutants in water can be modulated by changing the materials on the surface. Hence, a wide variety of carbon-based materials have been studied to increase the efficiency of the electrode.

11.3 MODIFICATION OF CATHODES FOR PHARMACEUTICAL REMEDIATION

The modification of the cathode is a crucial aspect of the electrochemical treatment of pharmaceutical compounds. The cathode's role in oxidation power and H_2O_2 electrogeneration efficiency within FEPs is of significant importance. In this context, carbonaceous materials have become promising candidates as materials for cathode construction since they are nontoxic, stable, and have relatively good chemical resistance, high conductivity, and high overpotential for H_2 evolution. Oloman and Watkinson did the first research in this field using graphite particles as cathodes to reduce O_2 to H_2O_2 during the decade of the 1970s [27, 28]. Various studies have been conducted using two-dimensional (2D) planar electrodes like gas-diffusion electrodes and graphite and three-dimensional (3D) electrodes such as CF, ACF, RVC, O_2-fed carbon–PTFE, carbon sponge (CS), and BDD. [29]. Among all of them, electrodes with high specific surface areas, such as 3D and gas diffusion electrodes (GDEs), have stood out because they have porous structures and a high number of active sites. These structural features facilitate the interaction of the supplied current densities with the low concentration of O_2 that could be achieved due to its poor solubility in water, which usually ranges from 40–80 mg L^{-1} at standard conditions. This interaction leads to the rapid reduction of O_2 and promotes the generation of a large amount of H_2O_2.

11.3.1 DIFFERENT MATERIALS USED FOR MODIFICATION

The nanoscale properties of the material used determine the electrode's electronic, photonic, electrocatalytic, chemical, and mechanical properties. Among the carbon-based materials, carbon-based nanomaterials have been under the spotlight over the last decades due to their optimum properties and have been used for the modification of different surfaces to modulate their properties. These materials are divided into two different groups: nanosized and nanostructured carbons. The first group includes the graphene family, such as graphene, graphene oxide (GO), and reduced graphene oxide (rGO); the carbon nanotube family (CNTs, SWCNTs, and MWCNTs); nanofibers; fullerenes; nanodiamonds; nanocoils; and nanoribbons. The second group involves new carbon materials, such as carbon fibers, ordered mesoporous carbons, and metal–organic frameworks (MOFs).

The production of H_2O_2 and $^{\bullet}OH$ is predominantly influenced by carbon atoms' electronic structure and the carbon-based electrodes' physicochemical properties. Therefore, their use as heterogeneous catalysts and cathodes for FEPs has been widely studied. Researchers have reported that carbon materials may be doped with heteroatoms to modulate their activity by enhancing their surface reactivity, electrical conductivity, stability, and number of active sites. Among different dopants, nitrogen doping is one of the most studied due to the enhancement of properties that these materials could lead to, such as an improvement in the current of oxygen reduction or reduction in the onset of overpotential. This improvement could be due to the increase in active sites and the enhancement of the hydrophobicity of the surface by improving the O_2 chemisorption. Including N atoms into the carbon structure produces changes in the charge of the material, inducing a relatively positive charge density near the carbon atoms while exhibiting electron-accepting properties [30].

11.3.1.1 Graphene and Derivatives

Graphene and its derivatives are among the most studied 2D carbon-based materials due to their unique physicochemical properties. Graphene, GO, and rGO have different morphologies (Figure 11.2), leading to different physicochemical properties. The structure of graphene is a planar 2D carbon monoatomic sheet with an sp^2 hybridization state, which is densely organized in a structure similar to a honeycomb. This structure was first reported by Novoselov and Geim in 2004 [31]. They synthesized graphene sheets from oriented pyrolytic graphite by micro-mechanical splitting. Since then, many studies have evidenced its excellent properties, such as very high specific area ($\sim 2{,}630$ m^2 g^{-1}), high mechanical strength (130 GPa tensile strength and 1,000 GPa Young's modulus), high thermal conductivity (4,850–5,300 W m^{-1} K^{-1}) and electrical conductivity (6,000 S cm^{-1}). GO is an oxygen-functionalized graphene prepared by the exfoliation of graphite oxide [32]. Many oxygen-based groups are distributed over the entire surface of GO. Hydroxyl (-OH) and epoxy (1,2-ether) groups are sited on the hexagonal array of the carbon plane, while carbonyl (C=O) and carboxyl (-COOH) groups are positioned at the edges of the graphene sheets. Oxygen-based functional groups in the carbon cathodes play

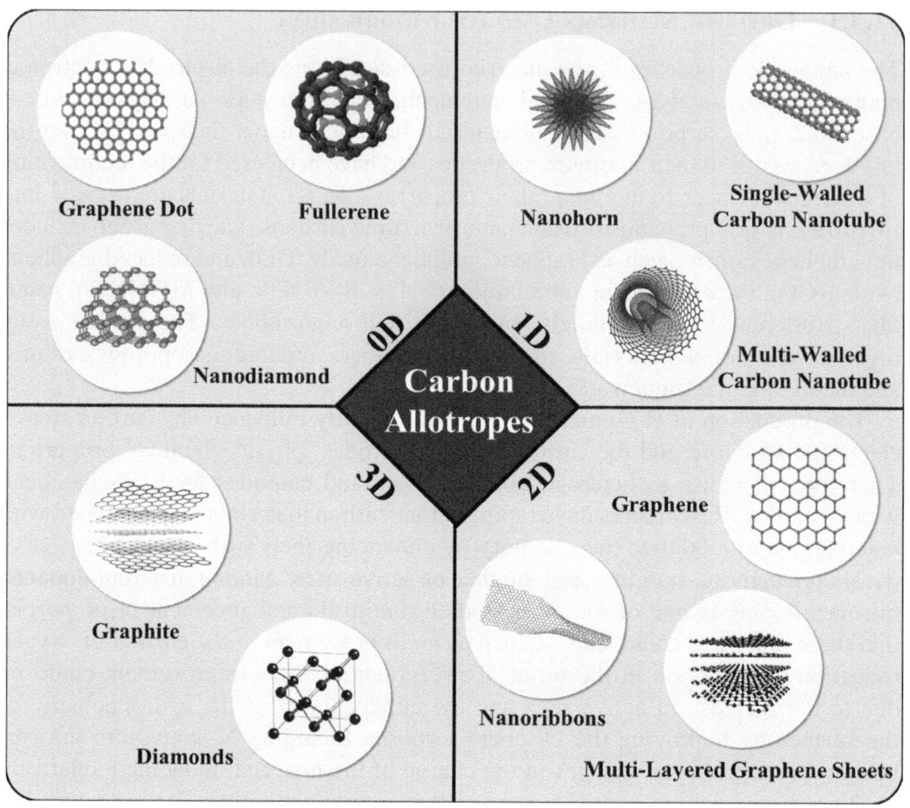

FIGURE 11.2   Allotropes carbon materials.

the role of active sites to absorb dissolved O_2 molecules, which are converted to H_2O_2 through a two-electron reduction. Although GO possesses beneficial oxygen functional groups, these groups result in high charge-transfer resistance. However, this resistance can be reduced by decreasing the excess of oxygen. [33]. Likewise, reduced graphene oxide, rGO, can be prepared by chemical, thermal, and electrochemical reduction of GO to reduce its oxygen content. Even though the rGO structure has more defects and lower conductivity compared to pristine graphene, it still exhibits sufficient conductivity for its application as an electrode material in many applications. Over the last few years, rGO has been highlighted due to its highly specific surface area, functional groups, and hydrophilicity. Therefore, graphene and its derivatives, such as GO and rGO, are suitable candidates for the potential construction of different cathodes for FEPs. Graphene-based electrodes can be obtained with different structural complexities by varying materials such as dots, fibers, films, and composites [34]. Divyapriya et al. [35] developed a cathode functionalized with ferrocene and an electrochemically reduced graphene oxide (Fc-ErGO) cathode for the degradation of ciprofloxacin and carbamazepine at a constant potential. Results indicated that ferrocene acted as an efficient catalyst for converting H_2O_2 into $\cdot OH$, facilitating the degradation of the contaminants. A removal of >99.8% was achieved for both contaminants after 60 min of the treatment. Wang et al. [36] prepared a microporous graphene aerogel (GA) with a high specific surface area for the degradation of the antibiotic ciprofloxacin by the EF process. Nearly all contaminants were removed within 90 minutes, while 91% of TOC was removed within 120 minutes. Authors attribute the activity to the abundance of macro-pores of the graphene aerogel that acted as reaction traps to accelerate the electrogeneration of H_2O_2 and decomposition into $\cdot OH$. The cathode also showed good stability and reusability. Some noteworthy studies using graphene and its derivatives has been listed in Table 11.2.

11.3.2 CARBON NANOTUBES

CNTs are one of the main carbon nanomaterials possessing unique electronic, photoelectronic, and electrochemical properties. CNTs are constructed by ordered carbon atoms with sp^2 hybridization as a one-dimensional (1D) system. CNTs are divided into two classes: the ones that are made of only one graphite layer rolled into a nanoscale tube with a diameter of approximately 1 nm (called single-walled carbon nanotubes, SWCNTs) and the ones that are produced by the rolling of two or more concentric carbon nanotubes (called multi-walled carbon nanotubes, MWCNTs). Both CNTs have shown outstanding properties, which are important for many applications, such as hydrogen storage systems, supercapacitors, fuel cells, sensors, organic photovoltaic cells, solar cells, and batteries. The conductivity of CNTs can reach up to 5,000 S cm^{-1}, which depends on the chirality of the graphitic array and diameter. The experimental specific surface area ranges between 380 and 1.587 and 180.9 and 507 m^2 g^{-1} for SWCNTs and MWCNTs, respectively. The micropore volume of SWCNTs ranges between 0.15 and 0.3 cm^3 g^{-1}, while the mesopore volume of MWCNTs ranges between 0.5 and 2 cm^3 g^{-1}. In addition to all the properties, CNTs and their derivatives favor the O_2 reduction via two electrons, making them suitable materials for FEP applications. Sadeghi et al. [37] prepared a magnetic single-walled carbon nanotube (MSWCNT) catalyst by using CNT and iron oxide nanoparticles for the degradation of diclofenac. A removal efficiency of 97.8% of the compound and 71.12% of COD (chemical oxygen demand) for 120 min of treatment was observed. The enhancement in the treatment was due to the MSWCNTs, which acted as a particle electrode favoring the H_2O_2 electrochemical production and also as a heterogeneous catalyst for the catalytic decomposition of H_2O_2 into $\cdot OH$. Moreover, the prepared material showed a high stability due to its activity being maintained for up to five cycles. Li et al. [38] fabricated a combined porous carbon nanotube filter using CNT functionalized with FeOCl for the filtration and degradation of tetracycline. The electrochemical filter achieved a 0.04 mM filtration of the contaminant at −0.8 V vs Ag/AgCl. An oxidative flux of 5.32 ± 0.41 mmol h^{-1} m^{-2} was observed at a flow rate of 1.5 mL min^{-1}. Some noteworthy research using carbon nanotubes and derivatives has been listed in Table 11.2.

TABLE 11.2

Summary of Recent Research Using Modified Cathodes with Different Carbon-Based Materials for Pharmaceutical Compounds in Water Treatment

Material	Contaminant	Process	Efficiency	REF
EEGr-CF[a]	Imatinib 35.5 mg L^{-1}	EF	100% TOC elimination, 8 h	[43]
Fc-ErGO[b]	Ciprofloxacin (CIP), 10 mg L^{-1}	EF	100% CIP removal, 15 min	[44]
γ-FeOOH GCPA[c]	Sulfamethoxazole (SMX) 0.1 mM	EF	90% TOC elimination, 5 h	[45]
CNT-GDE	Acetylsalicylic acid (ASA)	EF	100% ASA, 10 min, 62% TOC removal, 1 h	[46]
P-CNT	Metronidazole (MNZ), 200 mg L^{-1}	EF	99.5% of MNZ elimination, 30 min	[47]
FeOCl/O-CNTs/NF	Sulfamethoxazole (SMX)	EF	100% SMX degradation, 30 min	[48]
Co/Fe@NPC-500[d]	Ceftazidime (CAZ), 5 mg L^{-1}	EF	100% CAZ elimination, 60 min, 52.1% TOC removal, 90 min	[49]
Fe/Cu-NH$_2$-BDC	ceftazidime (CAZ), 5 mg L^{-1}	EF	99.5% CAZ elimination, 120 min	[50]
O-CNT@MIL-101(Fe)	Tetracycline (TC), 0.04 mM	EF	93.2% ± TC elimination, 2 h 69.0% TOC removed, 2 h	[51]
Ti/TiO$_2$NT-Au@ZIF-8 (1), Ti/TiO$_2$NT-UiO-66(Zr)NH$_2$ (2) Ti/TiO$_2$NT-Ru$_3$(BTC) (3)	Sodium diclofenac (DCF), sulfamethazine (SMT), and carbamazepine (CBZ), 200 ppb each	PEC	Elimination total of 50, 70, and 80% using (1), (2) and (3), respectively.	[52]

[a] CF modified with this electrochemically exfoliated graphene

[b] Ferrocene-functionalized graphene-coated graphite felt electrode

[c] γ-FeOOH graphene polyacrylamide carbonized aerogel

[d] MOF-derived carbon cathode containing CoFe

11.3.3 Mesoporous Carbons

Mesoporous carbons (MCs) have gained significant attention as electrode materials over the past two years. Mesoporous carbons are defined as 3D materials with pores in the range of 2–50 nm [39]. They offer significant advantages over 1D and 2D materials, such as having a high specific surface area. This feature allows for a large number of active sites and facilitates mass transport. Moreover, MCs have shown good electrical conductivity, which allows for easy electron transport, high mechanical and chemical stability and durability, and low density. The porosity of

MC materials depends on the precursors and the method of synthesis, which could be optimized in order to have hierarchical structures with a determined organized porosity. The structural features of the MC materials facilitate the reduction of O_2 and its transport to and from the active sites, achieving good yields of H_2O_2 production. These characteristics have put them under the spotlight over the last years for the construction of cathodes for FEPs. Liu et al. [40] prepared a cathode by using ACF-supported ferric citrate (Cit-Fe/ACFs) for the degradation of the anti-inflammatory drug ibuprofen (IBP). This cathode was used for in situ H_2O_2 electrogeneration from O_2 reduction and to produce ·OH from the generated H_2O_2. A 97% maximum degradation of IBP was achieved within 120 minutes of EF treatment using Cit-Fe/ACFs. The cathode exhibited good reusability, achieving nearly 85% IBP removal after six treatment cycles.

11.3.4 MOFs and MOF-Derived Materials

Over the last few years, metal–organic frameworks (MOFs) have gained attention due to their unique characteristics among all the MC materials. MOF structures possess large specific surface areas, adjustable pore size, high porosity, and structural diversity, which grant them excellent electrocatalytic properties [41]. Certain groups of MOF materials have demonstrated promising activity in degrading organic pollutants present in water. These groups include zeolite imidazole framework (ZIF) such as ZIF-8 and ZIF-67, Materials of Institute Lavoisier frameworks (MIL) such as MIL-100 and MIL-100(Fe), and Universitetet i Oslo MOFs (UIO) such as UIO-66. Wu et al. [42] prepared a high-open-site-density MOF-derived cathode by growing 2D-morphology $Mn_{0.67}Fe_{0.33}$-MOF-74 on carbon felt for removing the antibiotic sulfamethoxazole (SMX) by the EF process. Results showed that adding a second metal and changing the morphology improved the electrode's electrochemical activity, enhancing pollutant degradation. The EF reaction was ensured by the regeneration of the divalent metal ions, promoted by the electron transfer between $\equiv Fe^{II/III}$ and $\equiv Mn^{II/III}$. The optimized system achieved a high degradation efficiency of SMX, reaching 96% degradation after 90 min of treatment. Some noteworthy research using MOFs and derivatives has been listed in Table 11.2.

The studies and development of new hybrid materials for water treatment are still in process. Hence, many challenges still need to be addressed to optimize these materials and processes for large-scale industrial applications.

11.4 PERFORMANCE OF CARBON ELECTRODES IN PHARMACEUTICAL WASTEWATER TREATMENT

The effectiveness, selectivity, performance, and applicability of electrochemical processes for pharmaceutical removal depend significantly on the nature and structure of electrode materials. Carbon-based electrodes have garnered considerable attention due to the aforementioned exceptional properties and high removal efficiency of pharmaceutical compounds in the electrochemical processes [3, 53]. Their outstanding properties position carbon electrodes as a priority choice among other electrode materials for pharmaceutical wastewater treatment [54]. Carbon electrodes have been successfully utilized in diverse electrochemical processes, including EO, photoelectrocatalysis, FEPs, peroxicoagulation, and more. However, the stability of carbon electrodes is influenced by the specific electrolysis conditions. For instance, when used as an anode, graphite and cheaper forms of carbon may exhibit oxidation and surface crumbling. Moreover, the diverse performance and effectiveness observed among different carbon-based electrodes for pharmaceutical removal highlight the degradation mechanisms' complexity. For this, it is important to examine the performance and stability of carbon electrodes in diverse electrochemical processes for their easy scale-up and application at the industrial level.

11.4.1 Anodic Processes

EO is one of the most promising and simple anodic processes used in various advanced oxidation processes, especially after the development and performance optimization of carbon-based electrodes such as BDD [55]. The performance of EO strongly relies on the behavior of the anode, which can behave as active or non-active. Depending on the structure of the carbon material, carbon-based anodes can be classified as ideal non-active or active electrodes (Figure 11.3). Previous research has classified carbon-based anodes with a three-dimensional structure composed of sp^3-hybridized carbon atoms forming a tetrahedral lattice, like BDD, as non-active anodes [8]. These non-active anodes are highly effective in the mineralization of pharmaceutical compounds, leading to their efficient degradation and mineralization

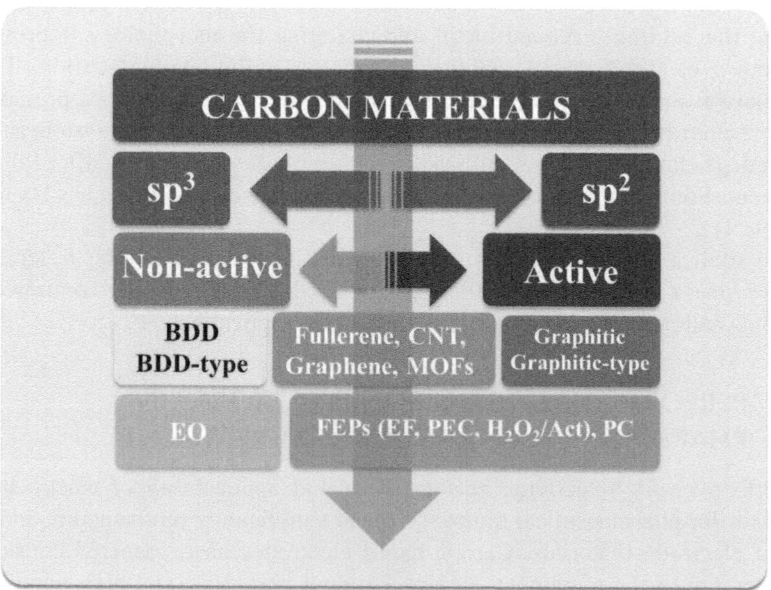

FIGURE 11.3 Classification of carbon-based electrodes and their electrochemical applications.

[56]. On the other hand, anodes based on sp^2-hybridized carbon atoms arranged in a hexagonal array, such as glassy carbon and graphite-based anodes, are classified as active anodes. These active anodes are particularly efficient in facilitating the conversion of pharmaceutical pollutants [57]. Understanding the structural differences and corresponding classifications of carbon-based anodes can give insights into their roles in electrochemical oxidation processes. Among studied carbon structures, sp^3 hybridization diamond-based electrodes have demonstrated high efficacy compared to graphite-based ones (sp^2). The proportion of diamond and graphite (sp^3/sp^2) has been evaluated for its effect on the degradation of pollutants, considering the formation of diamond with graphite impurities depending on manufacturing conditions. The results of the electrochemical degradation and mineralization of pharmaceutical compounds using BDD electrodes are strongly influenced by the sp^3/sp^2 ratios [58]. This difference is due to the higher concentration of ·OH produced on the non-active diamond carbon structure (sp^3) compared to the active anode graphite (sp^2). Higher sp^3/sp^2 ratios result in increased hydroxyl radical production, favoring pharmaceutical compounds' rapid degradation and mineralization. The boron-doped concentration and surface morphology of the substrate also influences the removal efficiency of pharmaceutical compounds. When the boron content is increased, the efficiency of removal decreases. Utilizing a boron-doped diamond layer grown on a structured silicon substrate has demonstrated a notable improvement in removal efficiency [56]. The effectiveness of EO with the promising BDD electrodes also relies on several parameters, including initial pH, flow rate of the effluent, pollutant concentration, current density, and nature of the support electrolyte, among other factors. Utilizing various support electrolytes with BDD electrodes leads to the generation of different oxidants, such as active chlorine, persulfate, percarbonate radicals, and perphosphate, which react with pharmaceutical compounds [59]. In summary, the use of sp^3-hybridized diamond-based electrodes has proven effective in efficiently removing hard-to-degrade pharmaceuticals, highlighting the promising potential of carbon-based electrodes for pharmaceutical elimination.

On the other hand, anodes based on sp^2-hybridized carbon atoms, such as graphite and carbon felt, can deviate significantly from the performance, degradation mechanism, and stability of electrodes during the electrochemical removal of pharmaceuticals. Among them, CF, activated carbon, and graphite-based electrodes are the most promising for the removal of pharmaceuticals by EO. This is attributed to their excellent electrical conductivity, porosity, high surface area, ability to offer abundant redox reaction sites, and mechanical stability at a relatively low cost [60, 61]. Several approaches have been reported to increase their electrolytic efficiency and stability, such as plasma, thermal, and chemical treatment with the deposition of metal or metal oxide and the addition of a functional group [61]. Some studies have demonstrated the potential of EO using anodes based on sp^2-hybridized carbon atoms for the effective removal of pharmaceuticals such as diclofenac, acetylsalicylic acid, hydroxychloroquine, acetaminophen, and ofloxacin. Pharmaceutical removal rates of 99.9 and 94.2% by EO have been reported by Jakóbczyk et al. [61] and Ashrafi et al. [60] by using carbon felt treated by oxygen plasma and modified by β-PbO$_2$-ZrO$_2$-MoO$_x$ respectively. Complete diclofenac removal has been reported using carbon film-supported Cu-rGO. Wan et al. [62] and Acuña-Bedoya et al. [63]

reported the complete removal of pharmaceuticals by using 3D electrochemical oxidation integrated with an adsorption process using activated carbon modified by Ti-Cu-Ni-Zn-Sb-Mn and 3D-printed activated carbon monoliths, respectively. Moreover, 3D electrochemical reactors enhance mass transfer and increase the current efficiency for pharmaceutical oxidation by using carbon-based particles, such as granular activated carbon [64]. Prepared carbon-based electrodes showed high stability and efficiency.

The efficiency, applicability, safety, and environmental cleanliness of the electrochemical oxidation process for pharmaceutical removal have been demonstrated by using upcycled secondary waste materials as electrodes. Szopinska et al. showed the conversion of harmful waste into electroactive carbon-based electrodes with an ordered architecture, highlighting their effectiveness in oxidizing acetaminophen [65]. A novel carbon mesostructured electrode was successfully fabricated using a straightforward phase inversion technique and a chemical vapor deposition reactor.

11.4.2   CATHODIC PROCESSES

The cathode plays a crucial role in water treatment through direct pollutant reduction or mediated decontamination via cathodic H_2O_2 electrogeneration. Among mediated technologies based on the production of H_2O_2, EF and PEF are considered the most efficient and promising. In these processes, the effective electrogeneration of H_2O_2 is required for more efficient degradation of pollutants. The cathode material is an essential component of two-electron oxygen reduction reactions in acid for the efficient production of H_2O_2 from the reaction (Equation 11.1). At the cathode surface, the electrochemical reduction of O_2 to H_2O_2 competes with the reduction of H^+ to H_2 and the four-electron reduction of O_2 to water (Equation 11.2). Carbonaceous cathodes have excellent electrocatalytic activity for the two-electron reduction of O_2, making them the best choice for producing H_2O_2 efficiently.

$$O_2 + 2H^+ + 2e^- \rightarrow H_2O_2 \tag{11.1}$$

$$O_2 + 4H^+ + 4e^- \rightarrow H_2O \tag{11.2}$$

Electrocatalytic H_2O_2 generation has emerged as a highly significant and topical area of research, requiring the adjustment of the oxygen reduction reaction (ORR) pathway for selective H_2O_2 formation. By harnessing electrochemistry and novel catalysts, it offers a promising avenue for environmentally friendly and energy-efficient H_2O_2 production. Extensive efforts have been dedicated to the development of electrocatalysts, encompassing both precious and non-precious metals, which have exhibited promising performance in various electrochemical applications [66–68]. However, these catalysts' high cost and toxicity hinder their practical implementation for large-scale water treatment and environmental applications. As a result, there is a growing focus on exploring inexpensive carbonaceous materials as alternative solutions. Utilizing inexpensive materials for H_2O_2 production enables easy scale-up and the development of decentralized units for pharmaceutical wastewater treatment.

In the field of pharmaceutical wastewater treatment using FEPs, the electrogenerated H_2O_2 oxidation power is significantly enhanced by introducing a small catalytic amount of Fe^{2+}. This leads to the generation of ·OH through the well-established Fenton reaction (Equation 11.3). The choice of cathode material is critical in FEPs due to its impact on the efficiency of H_2O_2 production and the continuous regeneration of Fe^{2+} from Fe^{3+} reduction at the cathode (Equation 10.4), essential for ·OH production [69]. To achieve efficient H_2O_2 production, carbonaceous materials such as graphite, BDD, CNTs, ACF, carbon sponge, CF, graphite felt, RVC, and GDEs made of carbon–PTFE are preferred [11]. Planar or 2D electrodes like graphite and BDD films exhibit limited H_2O_2 production due to the low solubility of O_2 in water when saturated with pure O_2 or air, resulting in reduced pharmaceutical removal [70]. Three-dimensional carbonaceous electrodes, including CF, ACF, RVC, carbon sponge, CNTs, and beds of graphite particles, have been employed to achieve higher reaction rates. These 3D electrodes offer advantages over 2D electrodes, such as a high surface/volume ratio and porosity, enabling improved mass transfer of dissolved O_2 and overcoming limitations like low space-time yield [11]. Using an air chamber in the electrochemical reactor enables continuous air feeding through a hydrophobic carbon–PTFE GDE, leading to significantly higher H_2O_2 concentrations. This strategy utilizes the triple gas–solid–liquid interface to aid in electron and O_2 mass transfer to the cathode surface. Alternative options like fluidized bed, packed bed, rolling tube, and porous materials have been employed for water treatment, further enhancing specific areas and improving mass transfer characteristics [46, 71].

$$Fe^{2+} + H_2O_2 + H^+ \rightarrow Fe^{3+} \ ·OH + H_2O \tag{11.3}$$

$$Fe^{3+} + e^- \rightarrow Fe^{2+} \tag{11.4}$$

Over the past two decades, carbon-based GDEs have gained significant prominence in pharmaceutical wastewater treatment using FEPs. Advancements in technology and the reduced cost of carbon materials have contributed to the increased adoption of carbon-based GDEs [72]. Among these electrodes, carbon–PTFE GDEs are widely used and extensively studied for their efficiency in H_2O_2 production and pharmaceutical removal [4]. The high removal efficiency was achieved due to the high efficiency of carbon–PTFE GDE to produce H_2O_2 and remove pharmaceuticals [73, 74]. To improve the selectivity and the chemical kinetics of ORR through the two-electron mechanism for pharmaceutical treatment, various methodologies have been developed to modify carbon-based GDE electrodes using inorganic and organic modifiers [75]. For illustration purposes, Shi et al. fabricated microtubular gas-diffusion electrodes (MGDEs) with a base of CNT hollow fibers using wet-spinning technology [75], achieving 95% elimination of IBP. After running the electrocatalytic system multiple times, its stability was tested. Similarly, Yang et al. examined the electrochemical efficiency of CNT-GDE made by a rolling method for the removal of pharmaceuticals, achieving complete degradation of acetylsalicylic acid within 10 minutes and a 62% TOC removal efficiency in one hour [46].

Efficiency upgrades in 3D systems involve the utilization of 3D electrodes and particles, which serve as both adsorbents and electrodes for continuous adsorption and regeneration processes. Carbon-based materials such as graphite, carbon fiber, and GDEs have been employed as anodes and cathodes in 3D systems, while the inter-electrode space is filled with particles such as pallet activated carbon (PAC) and manganese slag [1, 76, 77]. These particles act as particle electrodes, catalysts, and adsorbents, thereby enhancing the oxidation activity within the 3D electrochemical system. When carbon-based particles are incorporated into the 3D system, the applied electrostatic field polarizes the packed particles, forming numerous microelectrodes. This phenomenon significantly increases the surface area of the electrodes, generating a larger number of active sites and promoting strong oxidation radicals. For example, a 3D electrochemical system utilizing graphite felt with internal aeration was developed to improve the generation of H_2O_2 and facilitate the regeneration of the Fe^{3+}/Fe^{2+} cycle for the effective degradation of the pharmaceutical sulfamethoxazole [78].

To further increase the performance of carbon-based electrodes in pharmaceutical treatment, the utilization of modified and unmodified carbon-based electrodes has been explored. In recent years, bifunctional carbon-based electrodes have been developed to address the challenges associated with using powdered catalysts in homogeneous FEPs for H_2O_2 activation, such as separation difficulties and the loss of catalyst. Bifunctional carbon-based electrodes consist of two main parts: (1) the in situ electrocatalytic synthesis of H_2O_2 and (2) the activation of H_2O_2 by an immobilized catalyst to generate reactive oxygen species. Various carbonaceous materials, including CNTs [79, 80], graphene [7, 81], activated carbon [82.], carbon cloth [83.], and carbon felt [84., 85.], have been widely employed as carbon-based electrodes and supports for immobilizing Fenton catalysts such as iron and nanoparticles. For instance, in a study by Amali et al. [22], $Fe@Fe_2O_3$ core–shell nanowires were incorporated into the 3D graphene, resulting in an $Fe@Fe_2O_3/3D$-GO electrode that exhibited superior catalytic performance for the degradation of sulfasalazine pharmaceutical in aqueous solutions.

11.5　FUTURE DIRECTIONS AND CHALLENGES FOR CARBON ELECTRODE-BASED TREATMENT OF PHARMACEUTICAL WASTEWATER

Many novel and innovative strategies for carbon-based materials applied to the degradation of pharmaceuticals in water have appeared in the literature. Carbon-based materials, such as BDD as anodes and carbon felt and PTFE as cathodes, have demonstrated excellent performance in degrading pharmaceuticals and other organic compounds. They are also effective in the complete mineralization of pollutants in the solution, converting them into CO_2.

The progress of the modified carbon-based electrodes in this field opens the possibility for new technologies and processes with higher efficiency. For instance, electrodes modified with MOF-derived and porous structures allow for the application of not only electrochemical treatment but also photoelectrochemical processes, which enables the higher performance and efficiency of hybrid processes.

Likewise, the use of 3D structures, compared with 2D, permits the possibility of the enhancement of the efficiency of the process by generating a higher concentration of H_2O_2 and $^{\bullet}OH$ radicals using the same material. Furthermore, 3D structures also allow for the regeneration of the catalyst, such as Fe^{2+}/Fe^{3+}, which makes the process catalytic and permits the suitable reuse of the material.

Clearly, despite the many efforts to improve the materials and process, there are still some disadvantages related to the stability, costs (in some cases), and issues with scalability that need to be considered for their suitable application on a large scale. However, it seems that the use of carbon-based materials and their modifications are going in the correct direction for versatile, efficient, and, in some cases, environmentally friendly processes for the decontamination of pharmaceutical wastewater. Therefore, further research and development are needed to explore innovative electrode modifications that enhance the electrocatalytic and selectivity performance of carbon-based materials. This could involve the incorporation of nanomaterials, functionalization techniques, and surface engineering to improve overall performance. The successful translation of carbon electrode-based technologies from the laboratory to real-world applications requires scalable and cost-effective manufacturing processes. Developing robust and scalable electrode fabrication techniques is crucial to enable their widespread implementation in large-scale pharmaceutical wastewater treatment facilities.

11.6 CONCLUSION AND SUMMARY OF KEY POINTS

In summary, carbon-based electrodes offer tremendous opportunities for advancing sustainable technologies in water treatment and pharmaceutical degradation. Carbon-based electrodes, such as graphite, diamond, carbon nanotubes, activated carbon, carbon felt, and carbon-based gas-diffusion electrodes, have demonstrated their versatility, efficiency, and cost effectiveness, making them attractive options for achieving sustainable and accessible technologies for wastewater treatment. The importance of understanding the specific nature, structure, and form of carbon materials in order to optimize their effectiveness and performance in electrochemical processes has been demonstrated. Exploring modifications and enhancements to carbon-based electrodes can improve their performance characteristics and expand their practical applications in water treatment and pharmaceutical wastewater degradation. Looking to the future, continued research and development of carbon-based electrodes hold significant promise for addressing the challenges in pharmaceutical degradation. By optimizing the electrochemical processes and leveraging the unique properties of carbon materials, we can pave the way for more efficient, sustainable, and cost-effective solutions.

ACKNOWLEDGMENTS

The authors thank ANID (Chile) for financial support under FONDECYT Regular Project N°1210343 and Postdoctoral Project N°3220510.

REFERENCES

[1] Phan Quang HH, Nguyen TP, Duc Nguyen DD, et al. Advanced electro-Fenton degradation of a mixture of pharmaceutical and steel industrial wastewater by pallet-activated-carbon using three-dimensional electrode reactor. Chemosphere. 2022;297:134074.

[2] Olvera-Vargas H, Oturan N, Oturan MA, et al. Electro-Fenton and solar photoelectro-Fenton treatments of the pharmaceutical ranitidine in pre-pilot flow plant scale. Sep Purif Technol. 2015;146:127–135.

[3] Srinivasan R, Nambi IM. An electro-peroxone-based multi-pronged strategy for the treatment of ibuprofen and an emerging pharmaceutical wastewater using a novel graphene-coated nickel foam electrode. Chem Eng J. 2022;450:137618.

[4] Droguett C, Salazar R, Brillas E, et al. Treatment of antibiotic cephalexin by heterogeneous electrochemical Fenton-based processes using chalcopyrite as sustainable catalyst. Sci Total Environ. 2020;740:140154.

[5] Yang W, Zhou M, Oturan N, et al. Highly efficient and stable FeIIFeIII LDH carbon felt cathode for removal of pharmaceutical ofloxacin at neutral pH. J Hazard Mater. 2020;393:122513.

[6] Thiam A, Salazar R, Brillas E, et al. In-situ dosage of Fe2+ catalyst using natural pyrite for thiamphenicol mineralization by photoelectro-Fenton process. J Environ Manage. 2020;270.

[7] Amali S, Zarei M, Ebratkhahan M, et al. Preparation of Fe@Fe2O3/3D graphene composite cathode for electrochemical removal of sulfasalazine. Chemosphere. 2021;273:128581.

[8] Carolina Espinoza L, Candia-Onfray C, Vidal J, et al. Influence of the chemical nature of Boron-Doped diamond anodes on wastewater treatments. Curr Opin Solid State Mater Sci. 2021;25:100963.

[9] Besenhard JO, Fritz HP. The electrochemistry of black carbons. Angew Chem Int Ed Engl. 1983;22:950–975.

[10] Duan X, Zhang N, Wang S, et al. Carbon materials in electrocatalytic oxidation systems for the treatment of organic pollutants in wastewater: A review. In: Carbon Resources Conversion. KeAi Communications Co. Ltd.; 2023.

[11] Brillas E. A review on the photoelectro-Fenton process as efficient electrochemical advanced oxidation for wastewater remediation: Treatment with UV light, sunlight, and coupling with conventional and other photo-assisted advanced technologies. Chemosphere. 2020;250:126198.

[12] Dhawane SH, Kumar T, Halder G. Recent advancement and prospective of heterogeneous carbonaceous catalysts in chemical and enzymatic transformation of biodiesel. Energy Convers Manag. 2018;167:176–202.

[13] Ajayan S, Ramadass K, Singh G, et al. As featured in: Recent advances in functionalized micro and mesoporous carbon materials: Synthesis and applications. Chem Soc Rev. 2018;47:2680–2721.

[14] Taylor P, Uslu B, Ozkan SA. Electroanalytical application of carbon based electrodes to the pharmaceuticals. Anal Let. 2007;40(5):817–853.

[15] Mccreery RL. Advanced carbon electrode materials for molecular electrochemistry. Chem Rev. 2008;108(7):2646–2687.

[16] Kirk DW, Gattrell M. The electrochemical oxidation of aqueous phenol carbon electrode. Can J Chem Eng. 1990;68(6):997–1003.

[17] Shestakova M, Sillanpää M. Electrode materials used for electrochemical oxidation of organic compounds in wastewater. Rev Environ Sci Bio/Technol. 2017;16:223–238.

[18] Espinoza LC, Henríquez A, Contreras D, et al. Evidence for the production of hydroxyl radicals at boron-doped diamond electrodes with different sp3/sp2ratios and its relationship with the anodic oxidation of aniline. Electrochem Commun. 2018;90:30–33.

[19] Hwan C, Lim C, Kim S, et al. Surface modification of carbon materials and its application as adsorbents. J Ind Eng Chem. 2022;116:21–31.

[20] Rabaaoui N, Salah M. Anodic oxidation of salicylic acid on BDD electrode: Variable effects and mechanisms of degradation. J Hazard Mater. 2012;243:187–192.

[21] Sh M, Oturan N, El K, et al. Oxidative degradation study on antimicrobial agent ciprofloxacin by electro-Fenton process: Kinetics and oxidation products. Chemosphere. 2014;117:447–454.

[22] Svítková J, Ignat T, Švorc Ľ, et al. Chemical modification of boron-doped diamond electrodes for applications to biosensors and biosensing. Crit Rev Anal Chem. 2016;46:248–256.

[23] Guo H, Zhao C, Xu H, et al. Enhanced H2O2 formation and norfloxacin removal by electro-Fenton process using a surface-reconstructed graphite felt cathode: New insight into synergistic mechanism of defective active sites. Environ Res. 2023;220.

[24] Cui L, Sun M, Zhang Z. Flow-through integration of FeOCl/graphite felt-based heterogeneous electro-Fenton and Ti 4 O 7-based anodic oxidation for efficient contaminant degradation. Chem Eng J. 2022;450:138263.

[25] Alamo AC, Puga A, Pariente MI, et al. Activity and stability of bifunctional perovskite/carbon-based electrodes for the removal of antipyrine by electro-Fenton process. Chemosphere. 2023;334.

[26] Molla H, Pishnamaz N, Farimaniraad H, et al. Application of nickel foam cathode modified by single-wall carbon nanotube in electro-Fenton process coupled with anodic oxidation: Enhancing organic pollutants removal. J Electroanal Chem. 2023;929:117130.

[27] Watkinson AP, Oloman C. The electroreduction of oxygen to hydrogen peroxide on fluidized cathodes. Can J Chem Eng. 1975;53:268–273.

[28] Oloman C, Watkinson AP. The electroreduction of oxygen to hydrogen peroxide on fixed bed cathodes. Can J Chem Eng. 1976;54:312–318.

[29] Khataee A, Hasanzadeh A. Modified cathodes with carbon-based nanomaterials for electro-Fenton process. In: Zhou, M., Oturan, M., Sirés, I., editors. Electro-Fenton Process. The Handbook of Environmental Chemistry. Springer; 2017.

[30] Han L, Sun Y, Li S, et al. In-plane carbon lattice-defect regulating electrochemical oxygen reduction to hydrogen peroxide production over nitrogen-doped graphene. ACS Catal. 2019;9(2):1283–1288.

[31] Novoselov KS, et al. Electric field effect in atomically thin carbon films. Science. 2004;306:666–669.

[32] Shao G, Lu Y, Wu F. Graphene oxide: The mechanisms of oxidation and exfoliation. J Mater Sci. 2012;47:4400–4409.

[33] Divyapriya G, Nidheesh PV. Importance of graphene in the electro-Fenton process. ACS Omega. 2020;5(10):4725–4732.

[34] Ratajczak P, Suss ME, Kaasik F, et al. Carbon electrodes for capacitive technologies. Energy Storage Mater. 2019;16:126–145.

[35] Divyapriya G, Srinivasan R, Nambi IM. Electrochimica Acta Highly active and stable ferrocene functionalized graphene encapsulated carbon felt array—A novel rotating disc electrode for electro-Fenton oxidation of pharmaceutical compounds. Electrochim Acta. 2018;283:858–870.

[36] Wang Y, Chen J, Gao J, et al. Selective electrochemical H_2O_2 generation on the graphene aerogel for efficient electro-Fenton degradation of ciprofloxacin. Sep Purif Technol. 2021;272:118884.

[37] Sadeghi M, Hadi M, Mengelizadeh N, et al. Degradation of diclofenac by heterogeneous electro-Fenton process using magnetic single-walled carbon nanotubes as a catalyst. J Water Process Eng. 2019;31.

[38] Li Z, Shen C, Liu Y, et al. Carbon nanotube filter functionalized with iron oxychloride for flow-through electro-Fenton. Appl Catal B. 2020;260:118204.

[39] Ryoo R, Joo SH. Nanostructured carbon materials synthesized from mesoporous silica crystals by replication. Stud Surf Sci Catal. 2004;148:241–260.

[40] Liu D, Zhang H, Wei Y, et al. Enhanced degradation of ibuprofen by heterogeneous electro-Fenton at circumneutral pH. Chemosphere. 2018;209:998–1006.

[41] Li Y, Yao B, Chen Y, et al. Metal-organic frameworks (MOFs) as efficient catalysts for electro-Fenton (EF) reactions: Current progress and prospects. Chem Eng J. 2023;463:142287.

[42] Wu D, Hua T, Han S, et al. Two-dimensional manganese-iron bimetallic MOF-74 for electro-Fenton degradation of sulfamethoxazole. Chemosphere. 2023;327:138514.

[43] Yang W, Zhou M, Oturan N, et al. Electrocatalytic destruction of pharmaceutical imatinib by electro-Fenton process with graphene-based cathode. Electrochim Acta. 2019;305:285–294.

[44] Divyapriya G, Nambi I, Senthilnathan J. Ferrocene functionalized graphene based electrode for the electro-Fenton oxidation of ciprofloxacin. Chemosphere. 2018;209:113–123.

[45] Wang Y, Zhang H, Li B, et al. γ-FeOOH graphene polyacrylamide carbonized aerogel as air-cathode in electro-Fenton process for enhanced degradation of sulfamethoxazole. Chem Eng J. 2019;359:914–923.

[46] Yang H, Zhou M, Yang W, et al. Rolling-made gas diffusion electrode with carbon nanotube for electro-Fenton degradation of acetylsalicylic acid. Chemosphere. 2018;206:439–446.

[47] Xia Y, Shang H, Zhang Q, et al. Electrogeneration of hydrogen peroxide using phosphorus-doped carbon nanotubes gas diffusion electrodes and its application in electro-Fenton. J Electroanal Chem. 2019;840:400–408.

[48] Zhong S, Zhu Z, Zhou P, et al. FeOCl nanoparticles loaded onto oxygen-enriched carbon nanotubes and fickel-foam-based cathodes for the electro-Fenton degradation of pollutants. ACS Appl Nano Mater. 2022;5(9):12095–12106.

[49] Wang H, Tang C, Wang L, et al. Environmental MOF-derived Co/Fe @ NPC-500 with large amounts of low-valent metals as an electro-Fenton cathode for efficient degradation of ceftazidime. Appl Catal B. 2023;333:122755.

[50] Wang L, Tang C, Huang P, et al. Efficient degradation of ceftazidime in heterogeneous electro-Fenton process with Fe/Cu bimetal MOF-derived nitrogen-doped cathode. J Alloys Compd. 2023;945:169263.

[51] Dai Y, Yao Y, Li M, et al. Carbon nanotube filter functionalized with MIL-101 (Fe) for enhanced. Environ Res. 2022;204:112117.

[52] Candia-Onfray C, Irikura K, Calzadilla W, et al. Degradation of contaminants of emerging concern in a secondary effluent using synthesized MOF-derived photoanodes: A comparative study between photo-, electro- and photoelectrocatalysis. Chemosphere. 2023;315:137683.

[53] Isarain-Chávez E, Arias C, Cabot PL, et al. Mineralization of the drug β-blocker atenolol by electro-Fenton and photoelectro-Fenton using an air-diffusion cathode for H2O2 electrogeneration combined with a carbon-felt cathode for Fe2+ regeneration. Appl Catal B. 2010;96:361–369.

[54] Santos G de OS, Eguiluz KIB, Salazar-Banda GR, et al. Testing the role of electrode materials on the electro-Fenton and photoelectro-Fenton degradation of clopyralid. J Electroanal Chem. 2020;871:114291.

[55] Martínez-Huitle CA, Panizza M. Electrochemical oxidation of organic pollutants for wastewater treatment. Curr Opin Electrochem. 2018;11:62–71.

[56] Mordačíková E, Vojs M, Grabicová K, et al. Influence of boron doped diamond electrodes properties on the elimination of selected pharmaceuticals from wastewater. J Electroanal Chem. 2020;862.

[57] Kumar A, Omar RA, Verma N. Efficient electro-oxidation of diclofenac persistent organic pollutant in wastewater using carbon film-supported Cu-rGO electrode. Chemosphere. 2020;248:126030.

[58] Espinoza LC, Aranda M, Contreras D, et al. Effect of the sp3/sp2 ratio in boron-doped diamond electrodes on the degradation pathway of aniline by anodic oxidation. ChemElectroChem. 2019;6:4801–4810.

[59] Grilla E, Taheri ME, Miserli K, et al. Degradation of dexamethasone in water using BDD anodic oxidation and persulfate: Reaction kinetics and pathways. J Chem Technol Biotechnol. 2021;96:2451–2460.

[60] Ashrafi P, Nematollahi D, Shabanloo A, et al. A detailed electrochemical study of anti-malaria drug hydroxychloroquine: Application of a highly porous 3D multi-metal oxide carbon felt/β-PbO2-ZrO2–MoOx electrode for its electrocatalytic degradation. Electrochim Acta. 2023;458:142555.

[61] Jakóbczyk P, Skowierzak G, Kaczmarzyk I, et al. Electrocatalytic performance of oxygen-activated carbon fibre felt anodes mediating degradation mechanism of acetaminophen in aqueous environments. Chemosphere. 2022;304.

[62] Wan J., et al. Three-dimensional electrochemical degradation of p-aminophenol with efficient honeycomb block AC@Ti-Cu-Ni-Zn-Sb-Mn particle electrodes. Sep Purif Technol. 2021;267:118662.

[63] Acuña-Bedoya JD, Rangel-Sequeda JF, et al. Integration of the adsorption and electrooxidation process using 3D printed activated carbon monoliths for the degradation of pharmaceutical compounds. J Environ Chem Eng. 2022;10(4):108203.

[64] Zhan J, Li Z, Yu G, et al. Enhanced treatment of pharmaceutical wastewater by combining three-dimensional electrochemical process with ozonation to in situ regenerate granular activated carbon particle electrodes. Sep Purif Technol. 2019;208:12–18.

[65] Szopińska M, Ryl J, Pierpaoli M. Closing the loop: Upcycling secondary waste materials into nanoarchitectured carbon composites for the electrochemical degradation of pharmaceuticals. Chemosphere. 2023;313.

[66] Salmerón I, Plakas K V., Sirés I, et al. Optimization of electrocatalytic H2O2 production at pilot plant scale for solar-assisted water treatment. Appl Catal B. 2019; 242:327–336.

[67] Siahrostami S, Verdaguer-Casadevall A, Karamad M, et al. Enabling direct H2O2 production through rational electrocatalyst design. Nat Mater. 2013;12:1137–1143.

[68] Pizzutilo E, Kasian O, Choi CH, et al. Electrocatalytic synthesis of hydrogen peroxide on Au-Pd nanoparticles: From fundamentals to continuous production. Chem Phys Lett. 2017;683:436–442.

[69] Petrucci E, Da Pozzo A, Di Palma L. On the ability to electrogenerate hydrogen peroxide and to regenerate ferrous ions of three selected carbon-based cathodes for electro-Fenton processes. Chem Eng J. 2016;283:750–758.

[70] Espinoza-Montero PJ, Alulema-Pullupaxi P, Frontana-Uribe BA, et al. Electrochemical production of hydrogen peroxide on Boron-Doped diamond (BDD) electrode. Curr Opin Solid State Mater Sci. 2022;26:100988.

[71] Moratalla Á, Araújo DM, Moura GOMA, et al. Pressurized electro-Fenton for the reduction of the environmental impact of antibiotics. Sep Purif Technol. 2021;276.

[72] Salmerón I, Oller I, Plakas K V., et al. Carbon-based cathodes degradation during electro-Fenton treatment at pilot scale: Changes in H2O2 electrogeneration. Chemosphere. 2021;275.

[73] El-Ghenymy A, Rodríguez RM, Brillas E, et al. Electro-Fenton degradation of the antibiotic sulfanilamide with Pt/carbon-felt and BDD/carbon-felt cells: Kinetics, reaction intermediates, and toxicity assessment. Environ Sci Pollut Res. 2014;21:8368–8378.

[74] Garcia-Segura S, Cavalcanti EB, Brillas E. Mineralization of the antibiotic chloramphenicol by solar photoelectro-Fenton: From stirred tank reactor to solar pre-pilot plant. Appl Catal B. 2014;144:588–598.

[75] Shi K, Wang Y, Xu A, et al. Efficient degradation of ibuprofen by electro-Fenton with microtubular gas-diffusion electrodes synthesized by wet-spinning method. J Electroanal Chem. 2021;897:115615.

[76] Long Y, Feng Y, Li X, et al. Removal of diclofenac by three-dimensional electro-Fenton-persulfate (3D electro-Fenton-PS). Chemosphere. 2019;219:1024–1031.

[77] Robles I, Moreno-Rubio G, Garciá-Espinoza JD, et al. Study of polarized activated carbon filters as simultaneous adsorbent and 3D-type electrode materials for electro-Fenton reactors. J Environ Chem Eng. 2020;8.

[78] Li D, Zheng T, Liu Y, et al. A novel Electro-Fenton process characterized by aeration from inside a graphite felt electrode with enhanced electrogeneration of H2O2 and cycle of Fe3+/Fe2+. J Hazard Mater. 2020;396:122591.

[79] Mohseni M, Demeestere K, Du Laing G, et al. CNT microtubes with entrapped Fe3O4 nanoparticles remove micropollutants through a heterogeneous electro-Fenton process at neutral pH. Adv Sustain Syst. 2021;5.

[80] Su P, Song G, Wang X, et al. Novel self-sustained flow-through electrochemical advanced oxidation processes using nano-Fe 0 confined B, N co-doped carbon nanotubes/graphite felt cathode: Higher mineralization efficiency under wider pH. Sep Purif Technol. 2023;318:123944.

[81] Barranco-López A, Moral-Rodríguez AI, Fajardo-Puerto E, et al. Highly graphitic Fe-doped carbon xerogels as dual-functional electro-Fenton catalysts for the degradation of tetracycline in wastewater. Environ Res. 2023;228.

[82] Gholizadeh AM, Zarei M, Ebratkhahan M, et al. Phenazopyridine degradation by electro-Fenton process with magnetite nanoparticles-activated carbon cathode, artificial neural networks modeling. J Environ Chem Eng. 2021;9:104999.

[83] Emeji IC, Ama OM, Khoele K, et al. Electro-Fenton degradation of selected antiretroviral drugs using a low-cost iron-modified carbon-cloth electrode. Electrocatalysis. 2021;12:327–339.

[84] Muzenda C, Arotiba OA. Improved magnetite nanoparticle immobilization on a carbon felt cathode in the heterogeneous electro-Fenton degradation of aspirin in wastewater. ACS Omega. 2022;7:19261–19269.

[85] Yang W, Zhou M, Oturan N, et al. Highly efficient and stable FeIIFeIII LDH carbon felt cathode for removal of pharmaceutical ofloxacin at neutral pH. J Hazard Mater. 2020;393:122513.

12

Metallic Electrodes for Pharmaceutical Removal

Electrocoagulation and Electrooxidation

*Farshid Ghanbari, Ali Yaghoot-Nezhad,
Kun-Yi Andrew Lin, and Aydin Hassani*

12.1 PHARMACEUTICAL POLLUTANTS: ENVIRONMENTAL IMPACT AND REMEDIATION

Pharmaceutical wastewater is becoming a major source of concern as emerging contaminants are being detected in environmental samples worldwide [1]. This is due to the increasing use of drugs, which can have harmful effects on human life, animals, and plants when they are discharged into the wastewater stream [2]. Notably, these pharmaceuticals are capable of eliciting genotoxic, mutagenic, and eco-toxicological effects, making the eradication of these micro-contaminants from industrial and municipal effluents a pressing necessity, despite being an arduous challenge [3]. Industrial and community sources of wastewater generate a substantial amount of pollutants, thereby altering the physicochemical characteristics of wastewater, including chemical oxygen demand (COD), total organic carbon (TOC), pH, biochemical oxygen demand (BOD), suspended solids, and total nitrogen and phosphorus [4]. The primary contributors to water pollution caused by pharmaceuticals include drug manufacturing plants, animal waste in livestock farming, and human waste discharged from hospitals or domestic activities. The most common classes of pharmaceuticals frequently contaminating surface and groundwater have already been recognized as nonsteroidal anti-inflammatory drugs (NSAIDs), beta-blockers, antibiotics, antipsychotics, hormones, and lipid regulators [5]. The drugs that are most often detected in the aforementioned classes are diclofenac, ibuprofen, propranolol, β-lactams, macrolides, amoxicillin, tetracyclines, carbamazepine, diazepam, olanzapine, 17-α-ethinyl estradiol, 17-β-estradiol, bezafibrate, and gemfibrozil. It should be noted that in aqueous environments, the presence of said pharmaceutical agents has been recurrently measured at concentrations spanning from multiple ng/L to mg/L levels [6, 7]. The most important concern with pharmaceutical pollutants in the environment is the potential for antibiotic-resistant strains to emerge in natural bacterial populations. Antibiotic overuse is the main reason for antibiotic resistance, threatening the treatment of various infectious diseases caused

DOI: 10.1201/9781003340164-12

281

by antibiotic-resistant pathogenic microorganisms [8]. Active ingredients in pharmaceuticals that bring about the intended effects of a drug, as well as the microorganisms (e.g., bacteria and fungi) that grow on the remnants, can have severe negative impacts on the environment. Consequently, pharmaceutical wastewater can be categorized into two separate groups. The first category encompasses waste produced by pharmaceutical companies as well as the secondary waste generated from recycling and treatment facilities. The second category involves medical waste from hospitals and households that significantly pollutes sewage systems. The latter is a primary source of active pharmaceutical residues in local sewers and eventually in urban wastewater treatment plants. It should be noted that wastewater disinfection facilities are typically not adequately equipped to handle such specialized waste streams [9, 10]. Furthermore, the degradation-resistant nature of pharmaceuticals has rendered standard wastewater treatment methods ineffective. It is worth noting that although conventional biological and physical treatment methods are considered eco-friendly solutions for wastewater treatment, their potential drawbacks, namely prolonged treatment times and incomplete elimination of recalcitrant pharmaceutical substances, would necessitate the exploration of alternative approaches [11]. This calls for a thorough evaluation of the prevailing practices and technologies to identify gaps and areas that require improvement. In response, a plethora of studies were conducted to identify optimal techniques for wastewater decontamination, and electrochemical processes have been identified as some of the most interesting approaches by researchers. The following methods are considered efficient electrochemical processes, including electrochemical oxidation (EO), electrocoagulation (EC), electro-Fenton, and electro-reduction processes [12]. EC and EO, in particular, have emerged as promising methods, leveraging metallic electrodes and electric currents to effectively eliminate pollutants from wastewater.

The effectiveness of an electrochemical process is contingent on the type of the electrode utilized. A wide range of materials can be employed as electrodes in electrochemical approaches (EC and EO), including iron, aluminum, Pt, and mixed metal oxide (MMO) electrodes, also called dimensionally stable anodes (DSAs), consisting of catalytic materials, such as metal oxides, applied to the metal surface [13]. These materials serve to not only enhance the electrocatalytic properties of the electrode but also increase its effective surface area. MMOs play a pivotal role in multiple scientific domains, such as physics, chemistry, biology, and biosensing. Electrodes coated with MMOs electrodes, including titanium/iridium/nickel/ruthenium(IV) oxides, have been recognized as suitable electrodes for the EO process to eliminate organic compounds. These electrodes with considerable effective surface areas could confer numerous active sites for catalytic reaction, resulting in the generation of hydroxyl radicals ($^{\bullet}$OH) [14]. Additionally, electrochemical processes would enable the conversion of harmful organic pollutants into non-harmful substances through redox reactions or the separation of organic contaminants from aqueous solution and transfer into a solid phase, thus exemplifying sustainable technology. Electrochemical processes are characterized by their simplicity, low chemical consumption, rapid reaction times, and high selectivity in water purification [15]. Several review articles and book chapters have been published regarding the use of EC and EO processes in the removal of various pollutants and pharmaceutical compounds [12, 15–19]. EC applications for pharmaceutical wastewater treatment [20, 21] and recent developments in EO for the decontamination of pharmaceuticals have been studied [18, 22, 23]. However, there is no study based on metallic electrodes for the separation/degradation of pharmaceutical compounds from polluted water.

In the following chapter, electrochemical processes based on metallic electrodes, including EC and EO ($^{\bullet}$OH-mediated oxidation and electrogeneration of active chlorine (EAC)), are discussed regarding the elimination of pharmaceutical compounds, and the corresponding recent studies are scrutinized.

12.2 METALLIC ELECTRODES FOR WATER AND WASTEWATER TREATMENT

The use of metals for the removal of pharmaceutical compounds has been extensively studied in water and wastewater treatment processes. In general, metals can act as catalysts, coagulants, and reductants. In this way, transition metals (Co, Fe, Mn, and Cu) can be used as electron donors to activate various peroxides. Noble metals (Pt, Ir, and Ru) and various metal oxides (TiO_2, PbO_2, SnO_2, IrO_2, and RuO_2) can generate reactive species to abate pollutants due to their photocatalytic and electrocatalytic properties [24]. Iron and aluminum are classical coagulants in water and wastewater treatment, and various species formed from these salts have been used to remove soluble and colloidal contaminants. Zero-valent metals such as zero-valent iron (ZVI) and zero-valent aluminum can remediate chlorinated organic pollutants through a reduction mechanism [25]. In electrochemical processes, the electrode material plays a key role in the performance and mechanism of elimination of organic pollutants. Electrochemical processes can be classified into two categories: (i) electrocoagulation (EC) and (ii) electrooxidation (EO). In fact, electrodes including iron and aluminum are corroded (as the anode), and coagulant agents (Fe^{2+} and Al^{3+}) are released into the solution, while in electrooxidation, stable electrodes (metals/metal oxides) destruct organic pollutants through the generation of reactive species [12]. Therefore, in this chapter, EC and EO processes with a focus on metallic electrodes are discussed. The mechanism and theory of EC, modified EC, and EO for the removal of pharmaceutical pollutants from contaminated water are presented to stimulate researchers in conducting novel works and to create interesting ideas for future studies.

12.3 METALLIC ELECTRODES IN ELECTROCOAGULATION (EC) FOR PHARMACEUTICAL REMOVAL

Electrocoagulation (EC) is a complicated process that involves many chemical and physical phenomena occurring in an electrochemical cell, including the generation of coagulant agents, destabilization of particulate suspensions and pollutants, and formation of flocs. The EC process is an effective process that was patented by J.T. Harries in the United States of America (U.S.A.) in 1909 [26]. The EC process has been successfully applied for treating a variety of pollutants and wastewater such as heavy metals, urban wastewater, industrial wastewater, oily wastewater, nitrogen and phosphorous, and pharmaceutical compounds [27–29]. In the EC process, the coagulating ions or metallic ions (typically Fe and Al) are electrochemically generated 'in situ'. This happens in three consecutive steps. First is the generation of coagulant agents by electrolytic oxidation of the sacrificial anode (Fe, Al, and stainless steel). Second is the destabilization of the pollutants and particulate suspension. Third is the formation of flocs through the aggregation of the destabilized pollutants. There are several reactions in EC process [27]:

At the anode:

$$M \rightarrow M^{n+} + ne^-$$

(12.1)

$$2H_2O \rightarrow 4H^+ + O_2 + 4e^-$$

(12.2)

At the cathode:

$$3H_2O + 3e^- \rightarrow 1.5H_2 + 3OH^-$$

(12.3)

$$2H_2O + 2e^- \rightarrow H_2 + 2HO^-$$

(12.4)

The amount of metal dissolved in the solution by the corrosion of the anode (Fe and Al) can be theoretically calculated according to the following equation [23].

$$m = \frac{ItM}{zF}$$

(12.5)

where m is metal mass, t is electrolysis time, I is applied current, M is the atomic weight of iron or aluminum, F is Faraday's constant (96 485 C/mol), and z is the valance of metals (Fe and Al).

Metals released into water rapidly convert to their hydroxide forms with poor solubility, thereby swiftly precipitating. Similar to chemical coagulation or classical coagulation in water treatment, the EC mechanism can be explained by the separation of organic pollutants through co-precipitation/precipitation and complexation mechanisms. In fact, the aggregation of particles in colloidal systems is improved by adding coagulating agents (Fe and Al). The aggregation process is elucidated by the Derjaguin–Landau–Verwey–Overbeek (DLVO) theory. The DLVO theory describes that the interaction of van der Waals and electrostatic forces (resulting from the presence of an electrical double layer at the particles' surfaces) is the main reason for the formation of an aggregate, in which the van der Waals force is an attractive force while the electrostatic force is repulsive. Hence, the attractive forces must prevail over the repulsive forces to form the aggregate [27]. Zeta potential is a parameter to describe the stability of colloids and is indeed the potential difference between the dispersion medium and the stationary layer of water attached to the dispersed particle. For van der Waals forces to prevail over repulsion forces, or, in other words, for the destabilization of particles, zeta potential should approach 0 mV. It has been reported that zeta potential in stable colloidal systems can be more than +30 mV and less than −30 mV [19].

In the EC process, destabilization mechanisms can be classified into four mechanisms, including (i) compression of the diffuse double layer, (ii) adsorption and charge neutralization, (iii) sweep floc, and (iv) inter-particle bridging.

(i) Compression of the diffuse double layer around the charged species takes place by the interactions of ions produced by the oxidation of the sacrificial anode (cations). The presence of ions diminishes the repulsive forces between particles. (ii) Cations released from sacrificial anodes neutralize the negative charges of particles. Indeed, cations reduce the electrostatic inter-particle repulsion to the extent that the van der Waals attraction prevails, resulting in the aggregation of particles. (iii) In the sweep floc mechanism, the floc formed by the coagulation process makes a sludge that traps and binds the colloidal particles that remain in the aqueous medium. Hence, organic and inorganic pollutants adsorb onto flocs formed and then precipitate. This mechanism usually occurs under high applied current (high coagulant dosage). (iv) Polymerized metals with high molecular weight can form a link between particles through charge–charge interactions and hydrogen bonding, resulting in enlarging particles for the consequently easy precipitation. [16, 23, 30]. In spite of the main reactions in the EC process, several side reactions may take place in the EC reactor, affecting the performance of the process: (i) the electro-reduction of pollutants, especially nitrate and some heavy metals (Cr^{6+}), at the cathode, in which nitrate and Cr^{6+} may be reduced into nitrite and Cr^{3+}, respectively; (ii) the electrodeposition of metals on the surface of the electrode; (iii) the electrogeneration of chlorine if chloride salt is used as the electrolyte; and (iv) an increase in solution pH through the formation of OH^- or the consumption of H_3O^+ and H^+.

The EC process, compared to chemical coagulation, has several advantages, including [16, 19, 27]:

1 The ability to operate in a wide range of pH, as EC does not require pH control.
2 Coagulants are electrochemically generated; thus, no counter anions are added to the wastewater.
3 Very effective for eliminating various pollutants from wastewater.
4 Rapid separation (short time for sedimentation or floatation).
5 The cost of electrodes is lower than that of salts used in the coagulation process.
6 Lower sludge formation resulting from the lower amount of coagulants.
7 The potential for treating oily water.
8 EC can be used for small-scale industries and rural areas with solar panels to provide electricity.

The main operating parameters of EC are pH, current density, reaction time, electrode material, and space between electrodes. Several review articles and book chapters have focused on operating parameters. Therefore, a summary of the main factors is described in the following text.

12.3.1 Iron Electrodes for EC to Remove Pharmaceutical Pollutants

EC processes with iron electrodes have been widely used for the removal of various organic and inorganic pollutants. In recent years, EC has been applied for the purification of pharmaceutical pollutants. In EC-iron process, the iron anode is oxidized, and ferrous and ferric ions are released into the solution. However, it has been reported that the amount of direct ferric ions released is scant [31]. The conversion of Fe(II) to Fe(III) is significantly dependent on the solution pH and oxygen. In the presence of oxygen and high concentrations of H$^+$, Fe(II) is converted into Fe(III), while Fe(OH)$_2$ is formed in the presence of hydroxide ions (neutral and alkaline condition) (Equations 12.6 and 12.7). After that, Fe(OH)$_2$ is converted into Fe(OH)$_3$ in the presence of oxygen (Equation 12.8) [32].

$$4Fe^{2+} + O_2 + 2H_2O \rightarrow 4Fe^{3+} + 4OH^- \tag{12.6}$$

$$Fe^{2+} + 2OH^- \rightarrow Fe\left(OH\right)_2 \tag{12.7}$$

$$4Fe\left(OH\right)_2 + O_2 + 2H_2O \rightarrow 4Fe\left(OH\right)_3 \tag{12.8}$$

Based on the solution pH, several iron-complexes are formed in the solution. Fe(OH)$^{2+}$, Fe(OH)$_2^+$, and Fe(OH)$_3$ species are present in acidic solutions, while Fe(OH)$_6^-$ and Fe(OH)$_4^-$ ions are formed under alkaline conditions. Overall, anionic and cationic species of iron can contribute to the coagulation mechanism [16]. In this way, cefazolin (CFZ) was removed by the EC-iron process. Under highly acidic conditions (pH < 3), the existing Fe ions are soluble, which is an undesirable form for the coagulation process, while a favorable result was obtained at pH 5–9 [33]. It has been reported that the type of electrolyte affects the morphology of the particles generated from the sacrificial anode. Spherical particles were formed when chloride and sulfate ions were used as electrolytes. Moreover, higher applied current resulted in the production of spherical particles. In addition, bicarbonate electrolytes produced a coagulant with a high surface area and stable zeta potential particles. Hence, the highest removal efficiency was obtained in the presence of the carbonate and bicarbonate ions [34]. Moreover, the presence of chloride ions can lead to the generation of reactive chlorine species (RCS) near the anode. Indeed, the EC process is assisted by the oxidation of organic pollutants with electrogenerated chlorine. Zhou et al. reported that Cl$^\bullet$ and ClO$^\bullet$ were generated in the EC process for the removal/degradation of metronidazole (MNZ) [35].

12.3.2 Al Electrodes for EC to Remove Pharmaceutical Pollutants

Al electrodes have been broadly used in the EC process for the removal of various organic pollutants such as pharmaceutical compounds. An Al electrode as the anode is oxidized into Al ions (Al^{3+}) (Equation 12.9), while hydrogen gas is generated at the cathode. The overall reaction of aluminum hydroxide generation by Al electrodes is presented in Equation 12.10.

$$Al \rightarrow Al^{3+} + 3e^- \tag{12.9}$$

$$Al^{3+} + 3H_2O \rightarrow Al(OH)_3 + 3H^+ \tag{12.10}$$

Several monomers of Al species, including $Al(OH)^{2+}$, $Al(OH)_2^+$, $Al(OH)_3$, and $Al(OH)_4^-$, based on the solution pH are produced in the electrochemical cell, among which the formation of amorphous $Al(OH)_3$ finally occurs through complex precipitation mechanisms from the soluble monomeric and polymeric cations [27]. $Al(OH)_3$ is dominant at pH 4–9, while Al^{3+} and $Al(OH)_4^-$ are the primary species at pH values under 3 and over 10. However, it should be noted that the solution pH increases during the EC process because of hydroxide ion generation. $Al(OH)_3$ has a high surface area for the adsorption and entrapment of particles and organic pollutants. Moreover, the solution pH affects the size of hydrogen bubbles. Typical bubble sizes in EC are in the range of 20–70 µm. The smaller-sized bubbles are usually produced at a neutral pH [36]. Generally, the higher applied current leads to the enhanced generation of Al species and consequently improves the performance of the EC process. In fact, applied current in the EC process plays the role of coagulant dosage in the coagulation process. Moreover, it has been reported that the higher the applied current, the greater the variation of dissolved oxygen (DO), and a 0.5 mg/L DO was provided, which is considered a pretreatment for biological treatment [28]. Table 12.1 demonstrates the performance of the EC process for eliminating pharmaceutical compounds using Fe and Al electrodes.

TABLE 12.1

Pharmaceutical Removal by the EC Process Using Iron and Aluminum Electrodes

Anode/Cathode	Pollutant (concentration mg/L)	Experimental Conditions	Removal (%)	Ref.
Fe/Fe	Metronidazole (21.6)	pH = 8.3, current density = 6 mA/cm², and time = 14.8 min	98.6	[37]
Fe/Fe	Ciprofloxacin (60)	pH = 7.5, current density = 15 mA/cm², time = 20 min, and electrolyte = NaCl (70 mM)	100	[38]
Fe/Fe	Metronidazole (25)	pH = 6.5, current density = 40 A/m², time = 30 min, and electrolyte = NaCl (1600 mg/L)	99.25	[35]
Fe/Al	Cefazolin (25.7)	pH = 7.55, current density = 16 mA/cm², and time = 33 min	86.7	[33]
Fe/Fe	Flurbiprofen (5)	pH = 6.5, current density = 2.5 mA/cm², and time = 30 min	82	[39]
Steel/steel	Ampicillin (50)	pH = 6–6.5, current density = 18 A/m², and time = 36 min	4.3	[29]

(Continued)

TABLE 12.1 (*Continued*)

Pharmaceutical Removal by the EC Process Using Iron and Aluminum Electrodes

Anode/Cathode	Pollutant (concentration mg/L)	Experimental Conditions	Removal (%)	Ref.
Steel/steel	Doxycycline hyclate (50)	pH = 7.0, current density = 18 A/m^2, and time = 36 min	98.3	[29]
Fe/Fe+H$_2$O$_2$	Azithromycin (190 as COD)	pH = 3.0, H$_2$O$_2$ = 2 mM, current density = 20 mA/cm^2, and time = 60 min	95.6	[40]
Fe/Fe+S$_2$O$_8^{2-}$	Oxcarbazepine (5.04)	pH = 7.0, S$_2$O$_8^{2-}$ = 0.5 mM, current density = 5 mA/cm^2, and time = 10 min	75	[41]
Al/stainless steel	Amoxicillin (4)	pH = 7–8, current density = 1.8 mA/cm^2, and hydraulic retention time = 19 h	~80	[42]
Al/Al	Dexamethasone (0.1)	pH = 6.5–8, applied current = 1.0 A, and time = 45 min	38	[28]
Al/Al	doxycycline hyclate (100)	pH = 6.5–8, current density = 5.39 mA/cm^2, and time = 60 min	>95	[36]
Al/Al	Amoxicillin (10)	pH = 2.5, applied current = 0.7 A, and time = 60 min	74.5	[43]

12.3.3 Advanced Electrocoagulation for the Removal of Pharmaceutical Pollutants

To improve the performance of the EC process, several strategies have been suggested and developed in recent years. Ultraviolet, ultrasound, various oxidants, and photocatalysis can be combined with the EC process to enhance its performance (Figure 12.1). In this way, various oxidants (persulfate, hydrogen peroxide, and ozone) are externally added to the EC process to generate free radicals like hydroxyl radicals, sulfate radicals, and ozonide radicals. These oxidants are activated by two mechanisms in the EC reactor. First, ferrous ions generated at the anode can decompose peroxygens based on Fenton chemistry. Hence, iron electrodes should be used in this method for the generation of ferrous ions to activate H$_2$O$_2$, S$_2$O$_8^{2-}$, HSO$_5^-$, and O$_3$. Second, the reduction of oxidants at the cathode can occur through an electron transfer mechanism in which various cathode materials can be used (Equations 12.11 and 12.12).

$$\text{Electrogenerated } Fe^{2+} + \text{oxidant} \rightarrow Fe^{3+} + \text{free radicals} \qquad (12.11)$$

$$\text{Oxidant} + e^- \left(\text{at the cathode}\right) \rightarrow \text{free radicals} \left(SO_4^{\bullet-} \text{ and } {}^{\bullet}OH\right) \qquad (12.12)$$

It has been reported that H$_2$O$_2$ was added to EC (with an iron electrode as the anode) for the degradation of azithromycin (AZM) (based on COD) [40]. In this

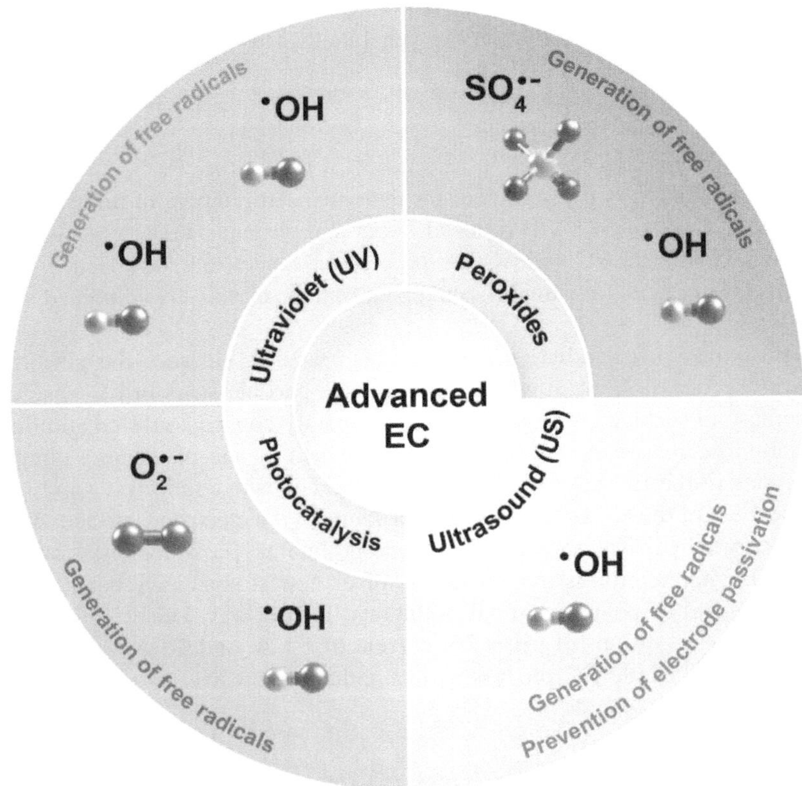

FIGURE 12.1   Advanced EC process assisted by various methods.

work, the optimum solution pH was obtained at pH = 3, and considering that the ideal solution pH is 2.8 for Fenton oxidation, free radicals had the main role in the degradation of AZM.

Moreover, an increase in applied current enhanced the COD removal efficiency significantly. In another study, EC (Al/steel) assisted with ozone was applied for the degradation of ibuprofen (IBP). This study indicated that the role of coagulation for IBP removal was insignificant (<1% contribution), while hydroxyl radical was the major oxidative agent in the EC/O$_3$ process [44]. In the EC/peroxydisulfate (S$_2$O$_8^{2-}$) process, both hydroxyl and sulfate radicals contributed to oxcarbazepine (OXC) degradation since sulfate radicals can be converted into hydroxyl radicals in the presence of H$_2$O and OH$^-$ [41].

The EC process can be integrated with photocatalysis by applying a photocatalyst in the EC reactor. In fact, degradation (photocatalysis) and separation (coagulation) are simultaneously applied for the removal/degradation of organic pollutants. A photocatalyst (Fe-doped titania nanotube arrays) was also used to improve the performance of the EC process through its photocatalytic activity. This combination (EC + photocatalysis) significantly enhanced ciprofloxacin (CIP) degradation as well as H$_2$ production [45].

The EC process can be coupled with UV irradiation for the improvement of removal efficiency. UV irradiation can generate hydroxyl radicals based on the following equation.

$$Fe(OH)^{2+} + UV \rightarrow Fe^{2+} + {}^{\bullet}OH \tag{12.13}$$

However, UV alone is not sufficient for the enhanced removal of pollutants in the EC process. In this way, COD removals from real pharmaceutical wastewater were 31.96 and 35.88% for EC and EC/UV processes, respectively [46]. Hence, UV and oxidants (H_2O_2, persulfate anions, and ozone) can be effectively combined with the EC process.

Ultrasound (US) irradiation has been also used to enhance the performance of the EC process. Accordingly, US can reduce particle sizes in EC, resulting in an increase in surface area and adsorption capacity consequently. In addition, US irradiation decreases electrode passivation, which is one of the most important limitations in the EC reactor. EC destructs the deposited solid layer and reduces the thickness of the electrical double layer found at the electrode surface. Also, US may generate hydroxyl radicals to degrade pollutants via water dissociation [32, 47]. US/EC application is increasing; however, few studies have been conducted on the removal of pharmaceutical pollutants. In this way, US/EC removed 86% of CIP under conditions of pH = 7.5, current of 1.5 A, and 30 min reaction time. Compared to the sole EC process, a 14% enhancement was observed in EC/US process [48].

12.4 ELECTROOXIDATION USING METALLIC ELECTRODES FOR THE DEGRADATION OF PHARMACEUTICAL POLLUTANTS

12.4.1 ANODIC OXIDATION

Electrooxidation (EO) is an effective process for the decontamination of wastewater with a variety of mechanisms and methods. EO can be classified into two methods, including reactive chlorine species (RCS)-based methods and reactive oxygen species (ROS)-based methods. However, both reactive species can be concurrently produced and applied in the electrochemical processes. The type of electrode and electrolyte plays an important role in the EO process. In the absence of chloride, the EO process is known as anodic oxidation (AO), which is based on two mechanisms: (i) direct electron transfer to the anode and (ii) oxidation with ${}^{\bullet}OH$ formed at the anode through water discharge (Figure 12.2).

The hydroxyl radical formed is highly dependent on the type of electrode. The term 'active' electrode is related to ${}^{\bullet}OH$ chemisorbed at the surface of anode (M), in which organic pollutants are selectively oxidized into smaller compounds such as carboxylic acids; this phenomenon is known as electrochemical conversion. Conventional electrodes for the ${}^{\bullet}OH$ chemisorbed mechanism are Pt, IrO_2, and RuO_2. Hydroxyl radicals formed strongly interact with the surface of the anode (M(${}^{\bullet}OH$)), which in turn is transformed into 'active oxygen' [17, 18].

$$M + H_2O \rightarrow M({}^{\bullet}OH) + H^+ + e^- \tag{12.14}$$

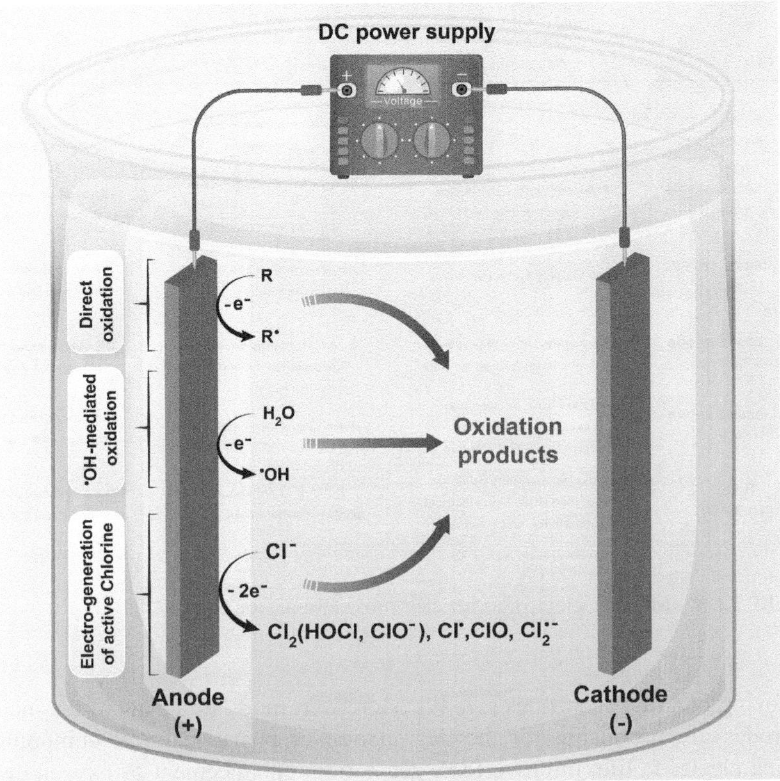

FIGURE 12.2 The mechanism of electrooxidation for the generation of reactive species.

$$M\left(\,^{\bullet}OH\right) \to MO + H^{+} + e^{-} \tag{12.15}$$

$$MO + R \to M + RO \tag{12.16}$$

The 'non-active' electrode is related to the $^{\bullet}OH$ physisorbed at the surface of the anode degrading organic pollutants, which is called electrochemical combustion. Boron-doped diamond (BDD), PbO_2, Ti_4O_7, and SnO_2 are effective electrodes for this mechanism, in which there is a weak interaction between the anode surface and hydroxyl radicals directly attacking organic compounds in electrochemical cells [17]. Moreover, other ROS can be generated in the AO process, namely O_3 and H_2O_2, which are produced through water discharge at the anode and the recombination of $^{\bullet}OH$ mechanisms, respectively [15].

$$3H_2O \to O_3 + 6H^{+} + 6e^{-} \tag{12.17}$$

$$2M\left(\,^{\bullet}OH\right) \to 2M + H_2O_2 \tag{12.18}$$

Several metallic electrodes have been employed in the AO process, including PbO_2, Ti_4O_7, SnO_2, IrO_2, RuO_2, and Pt. Figure 12.3 shows the pros and cons of

	Metallic electrodes for electrooxidation		
Electrode	**Advantages**	**Disadvantages**	**Overpotential**
Pt (active)	High stability and conductivity	Low performance and oxygen evolution	The overpotential for O_2 evolution (1.6 V vs SHE)
RuO_2-based anode (active)	Low cost, high stability, and electrocatalytic	Low performance, selective oxidation of organic pollutant and oxygen evolution	The overpotential for O_2 evolution (1.47 V vs SHE)
IrO_2-based anode (active)	High stability and conductivity	Low performance and oxygen evolution	The overpotential for O_2 evolution (1.52 V vs SHE)
PbO_2-based anode (non-active)	Inexpensive, easy to prepare, good conductivity, high stability	Low current efficiency and high toxicity (leached Pb)	The overpotential for O_2 evolution (1.9 V vs SHE)
SnO_2-based anode (non-active)	Doped SnO_2 anode show excellent conductivity high oxygen and chlorine	Service time is short, not stable	The overpotential for O_2 evolution (1.9 V vs SHE)
Ti_4O_7 (non-active)	High oxygen evolution potential, Cheaper than BDD, high electrical conductivity, and stability	Difficult to prepare and needs to further investigation	The overpotential for O_2 evolution (1.8-2.3 V vs SHE)

FIGURE 12.3 Metallic electrodes for electrooxidation processes.

employing metallic electrodes [17, 49]. Moreover, modified forms of the mentioned electrodes are promising for the degradation of pharmaceutical compounds. To prolong electrode life, improve ROS generation, enhancement of oxygen evolution potential (OEP), increase catalytic capacity, and elevate stability, various metals (Cu, Al, Co, Sb, La, etc.) have been doped on metallic anodes [50–54]. Several configurations of electrodes and their modification have been synthesized and applied for the oxidation of pharmaceutical pollutants. Recently, Ti_4O_7 has been synthesized as a promising electrode that can be considered as an alternative non-active electrode for water discharge [55]. In non-active electrodes, besides $\cdot OH$, H_2O_2, and O_3, other peroxides like peroxydisulfate ($S_2O_8^{2-}$), peroxydicarbonate ($C_2O_6^{2-}$), and peroxydiphosphate ($P_2O_8^{2-}$) may be generated in the solution in which sulfate, bicarbonate, and phosphate are used as electrolytes [56].

12.4.1.1 Operating Parameters for AO (Anode Materials, pH, and Current Density)

Pharmaceutical oxidation by metallic anodes is highly dependent on the type of anode materials, determining reactive species for the degradation of organic compounds. As mentioned earlier, active electrodes (Pt, RuO_2, and IrO_2) have been used for the remediation of contaminated water with pharmaceuticals. Although the performance of these electrodes for the degradation of target pollutants is desirable, the mineralization degree of organic pollutants for active electrodes is insignificant. Wu et al. used a Ti/RuO_2-IrO_2 anode for the degradation of tetracycline (TC). In total, 98% of TC was degraded within a 20 min period, while only 11% of COD was removed at the same time [57]. Conversely, metallic non-active electrodes (PbO_2, SnO_2, and

Ti_4O_7) can mineralize pharmaceutical compounds markedly. The non-active electrodes can effectively incinerate organic compounds and convert them into CO_2 and water. Wang et al. studied CIP degradation using a non-active electrode (SnO_2–Sb/Ti). A high degree of mineralization (86% COD and 70% TOC) was achieved for 50 mg/L CIP after 120 min electrolysis time. Hence, the non-active electrodes are effective anodes for the detoxification and mineralization of pharmaceutical compounds [58]. In another study, by applying only 120 mA, 69, 36, and 41% of amoxicillin (AMX) were mineralized by Ti_4O_7, DSA, and Pt anodes during 8 h of electrolysis time, respectively, indicating that the non-active electrode (Ti_4O_7) has a good potential for the mineralization of organic pollutants [59]. A 3D-RuO_2 electrode exhibited an excellent performance compared to flat RuO_2 for the degradation of pyrazole. This was associated with strong electro-adsorption capacity (270.3 µg/cm^2), resulting in the enhanced mass transfer of pyrazole to the electrode surface [60].

In the AO process, the current density is a key parameter for the generation of ROS and the degradation of organic pollutants. In AO, electrons are transferred from the anode's surface to the pharmaceutical compounds adsorbed on the anode's surface, and pharmaceutical pollutants are thereby oxidized. Hence, higher current density enhances the removal rate efficiency through higher ROS generation [50, 51]. The TC degradation rate was dramatically increased when the current density was increased from 0.5 to 3 mA/cm^2 [55]. Moreover, excess applied current may result in a reduction in the degradation rate due to the oxidation of hydroxyl radicals to oxygen at the anode [61]. It should be noted that determining the optimal current density can save excess electrical power and reduce process costs. Also, an increase in current density may decline the current efficiency of the AO process, especially for active electrodes. In this way, current efficiency was sharply dropped by increasing the current density in the Pt anode, while this phenomenon was insignificant when a non-active electrode was used for AMX degradation [62]. Not only does the solution pH affect the pollutant species, the power of the hydroxyl radicals also depends on the solution pH based on the Nernst equation. It has been frequently reported that an acidic condition is optimal for the degradation of organic pollutants, in which the redox potential of hydroxyl radical is 2.8 V (pH = 1). On the other hand, the oxidation rate of pharmaceutical pollutants also corresponds to their pK_a values. For example, the pK_a of acetaminophen (APAP) is 9.5; therefore, more hydroxyl radicals are required for APAP degradation under acidic conditions. However, under alkaline conditions, the hydroxyl group in the molecular structure of pharmaceutical compounds is ionized and can be rapidly cleaved in the presence of lower hydroxyl radicals [63]. Indeed, the degradation of pharmaceutical compounds at various pH values can be explained by the protonation and deprotonation of pharmaceuticals.

The solution pH can also affect the OEP of anodes. It has been reported that the OEP value of a Pt/Ti anode is 1.2 V (vs. Ag/AgCl) for acidic conditions (pH = 3 and 5), while its value is 0.9 V (vs. Ag/AgCl) at pH 7 and 9. Indeed, the formation of oxygen by water molecules is barricaded under acidic conditions, which is conducive to the AO process. On the other hand, a lower OEP value at pH levels of 7–9 decreases the lifetime of hydroxyl radicals by hindering the oxygen transfer from the •OH to the target molecule to oxidize pharmaceuticals [18, 64]. Table 12.2 shows the performance of the AO process for the removal of pharmaceutical pollutants.

TABLE 12.2

Electrooxidation of Pharmaceutical Compounds (AO and EAC)

Anode/Cathode	Pollutant Concentration (mg/L)	Electrolyte	Experimental Conditions	Removal (%)	Ref.
Anodic oxidation (AO)					
Pt	Diclofenac (175)	50 mM KH_2PO_4 + 50 mM Na_2SO_4 + NaOH	pH = 3.5, current density = 46.1 mA/cm^2, and time = 360 min	82.2	[80]
$Cu\text{-}PbO_2$/Ti	Amoxicillin (177.3 mg/L)	Na_2SO_4	pH = 3.5, current density = 46.1 mA/cm^2, and time = 30 min	96.3	[50]
$Co\text{-}PbO_2$/Ti	Metronidazole (100 mg/L)	Na_2SO_4 (0.01 M)	pH = 6.4, current density = 10 mA/cm^2, and time = 120 min	78	[52]
$Al\text{-}PbO_2$/Ti	Aspirin (500)	Na_2SO_4 (0.025 M)	Applied current = 0.7 A and time = 150 min	83.2	[53]
$Ti\text{-}TiO_2\text{-}PbO_2$/steel	Ampicillin (105)	Na_2SO_4 (0.1 M)	pH = 4, current density = 50 mA/cm^2, and time = 45 min	100	[81]
$Ti\text{-}SnO_2\text{-}Sb\text{-}Er\text{-}PbO_2$/steel	Sulfamethoxazole (10)	Na_2SO_4 (0.02 M)	pH = unadjusted, current density =12 mA/cm^2, and time = 30 min	99.6	[51]
$Ti/SnO_2\text{–}Sb$ tube/steel	Abacavir (5)	Na_2SO_4	pH = 7, current density = 0.2 mA/cm^2, and time = 10 min	97	[61]
$Ti/TiO_2/SnO_2\text{-}Sb\text{-}Cu$/Pt wire	Ceftazidime (5)	Na_2SO_4 (1 g/L)	pH = 6, current density = 1 mA/cm^2, and time = 60 min	83.4	[54]
$Ti/SnO_2\text{-}Sb/La\text{-}PbO_2$	Isoniazid (10)	Na_2SO_4 (0.025 M)	Current density = 10 mA/cm^2, and time = 60 min	~100	[82]
IrO_2/steel	Caffeine (20)	Na_2SO_4	Applied current = 2 A, and time = 40 min	39	[83]
Ti_4O_7/steel	Tetracycline (50)	Na_2SO_4 (0.1 M)	pH = 6, current density = 3 mA/cm^2, and time = 120 min	98	[55]

Electrogeneration of active chlorine (EAC)

RuO$_2$/stainless steel	5-Fluorouracil (50)	NaCl (85 mM)	pH = 7, current density = 5 mA/cm^2, and time = 30 min	96.9	[73]
Ti/RuO$_2$–ZrO$_2$/Ti	Acetaminophen (40)	NaCl (100 mM)	Flow = 11 mL/min, applied current = 100 mA, and time = 10 min	~100	[84]
SnO$_2$/Sb$_2$O$_3$/stainless steel	Norfloxacin (100)	NaCl (1.64 g/L)	Current density = 114 mA/cm^2, and time = 240 min	~90	[85]
Ti/RuO$_2$–ZrO$_2$/stainless steel	Salicylic acid (25)	NaCl (50 mM)	pH = 4, current density = 5 mA/cm^2, and time = 50 min	93	[79]
Ti/(RuO$_2$)$_{0.5}$(IrO$_2$)$_{0.5}$/stainless steel	Ciprofloxacin (20)	NaCl (60 mM)	pH = 5, current density = 20 mA/cm^2, and time = 120 min	100	[74]
Ti-RuO$_2$/Ti-RuO$_2$	Ofloxacin (50)	NaCl (2 g/L)	pH = 2, applied current = 1 A, and time = 30 min	88.6	[86]
Ti-RuO$_2$/stainless steel	Chloroxylenol (100)	NaCl (50 mM)	pH = 6–6.4, current density = 20 mA/cm^2, and time = 180 min	100	[87]
Ti-IrO$_2$/stainless steel	Carbamazepine (10)	NaCl (10.5 mM)	Applied current = 1.17 A and time = 10 min	63	[70]

12.4.2   Electrogeneration of Active Chlorine (EAC)

The chloride ion is one of the most abundant ions in water resources; therefore, its presence is unavoidable in contaminated water. Chloride ions can be the most inexpensive electrolyte for electrochemical processes. Chlorine as a conventional oxidant can be electrochemically generated in the electrochemical cell. Indeed, chlorine is the most common chemical oxidant and is anodically generated through the oxidation of chloride ions (Equation 12.19). Chlorine is hydrolyzed in an electrochemical cell to form hypochlorous acid (HOCl) (Equation 12.20). HOCl is a weak acid that is in equilibrium with hypochlorite ion (ClO^-) (pK_a = 7.54 at 25 °C) (Equation 12.21). It should be noted that HOCl and ClO^- are typical species of chlorine-based water treatment in which most water resources are within the pH range of 6–9. According to the solution pH and temperature, different concentrations of HOCl and ClO^- are formed in an aqueous solution [65]. In addition, excess chlorine may react with chloride ions and produce trichloride (Cl_3^-) (Equation 12.22). Moreover, chlorine hemioxide (Cl_2O) and $HOCl^+$ are also produced in the solution. However, these species are predominant at pH < 4. Hence, their concentrations are very low in the chlorination of water [12, 65].

$$2Cl^- \rightarrow Cl_2 + 2e^- \tag{12.19}$$

$$Cl_2 + H_2O \rightleftharpoons HOCl + Cl^- + H^+ \qquad k = 1.3\text{-}5.1 \times 10^{-4} M^2 \tag{12.20}$$

$$HOCl \rightleftharpoons ClO^- + H^+ \qquad pKa = 7.54 \tag{12.21}$$

$$Cl_2 + Cl^- \rightleftharpoons Cl_3^- \qquad k = 0.191 \tag{12.22}$$

The redox potentials of Cl_2, HOCl, and ClO^- are 1.36, 1.49, and 0.89 V/SHE, respectively. Since Cl_2, HOCl, and ClO^- are respectively dominant at pH = 3, a pH range of 3–8, and pH > 8, it can be stated that acidic conditions can be favorable for electrooxidation by chlorine species based on their redox potentials. There are several side reactions (anodic and cathodic) involved in the loss of ClO^-/HOCl. ClO^- ions are anodically oxidized into ClO_2^-, ClO_3^-, and ClO_4^- consecutively when a non-active anode is used (Equations 12.23–12.25) [12, 66].

$$ClO^- + H_2O \rightarrow ClO_2^- + 2e^- + 2H^+ \tag{12.23}$$

$$ClO_2^- + H_2O \rightarrow ClO_3^- + 2e^- + 2H^+ \tag{12.24}$$

$$ClO_3^- + H_2O \rightarrow ClO_4^- + 2e^- + 2H^+ \tag{12.25}$$

In addition, ClO^- can be reduced to chloride ions at the cathode. Moreover, some parasite reactions can reduce the performance of electrochemical oxidation, in which the recombination of ClO^- occurs [12].

$$ClO^- + H_2O + 2e^- \rightarrow Cl^- + 2OH^- \tag{12.26}$$

$$ClO^- + 2HOCl \rightarrow ClO_3^- + 2Cl^- + 2H^+ \qquad (12.27)$$

$$ClO^- + ClO^- \rightarrow 2Cl^- + O_2 \qquad (12.28)$$

In addition to Cl_2, HOCl, and ClO^-, it has been reported that chlorine and dichlorine radicals contribute to the degradation of organic pollutants in that chlorine radicals ($Cl^{\bullet}$) are produced through the oxidation of chloride ion at the anode, and dichlorine radicals ($Cl_2^{\bullet-}$) are formed by the reaction of chloride ions with $Cl^{\bullet}$ (Equations 12.29 and 12.30) [67, 68]. However, this mechanism in electrochemical oxidation with an active anode needs further investigation.

$$Cl^- - e^- \xrightarrow{\text{Anode}} Cl^{\bullet} \qquad (12.29)$$

$$Cl^- + Cl^{\bullet} \rightarrow Cl_2^{\bullet-} \qquad (12.30)$$

Another method for the generation of HOCl has been proposed in which HOCl is formed at the surface of the anode based on the following reaction.

$$MO_x\left(^{\bullet}OH\right) + Cl^- \rightarrow MO_x\left(HOCl\right) + e^- \qquad (12.31)$$

$$MO_x\left(HOCl\right) + Cl^- \rightarrow MO_x + Cl_2 + HO^- \qquad (12.32)$$

Luna-Trujillo et al. proposed generating of $ClO^{\bullet}_{ads}$ in the Ti/RuO_2–IrO_2 anode based on the following reactions. In fact, $^{\bullet}OH_{ads}$ and $O^{\bullet}_{ads}$ are generated at the surface of anodes, contributing to the generation of $HOCl_{ads}$ and $ClO^{\bullet}_{ads}$ in the presence of chloride ions [69].

$$MO_x\left(^{\bullet}OH\right) \rightarrow MO_x\left(O^{\bullet}\right) + H^+ + e^- \qquad (12.33)$$

$$MO_x\left(O^{\bullet}\right) + Cl^- \rightarrow MO_x\left(ClO^{\bullet}\right) + e^- \qquad (12.34)$$

$$MO_x\left(ClO^{\bullet}\right) + Cl^- \rightarrow MO_x\left(O^{\bullet}\right) + Cl_2 + e^- \qquad (12.35)$$

The HOCl generated at the surface of the anode competes with hydroxyl radicals to degrade pharmaceutical pollutants. Although HOCl has a lower redox potential than hydroxyl radicals, it is more selective than hydroxyl radicals and can react with electron-rich moieties more rapidly. Therefore, both mechanisms of hydroxyl radicals and electrogeneration of active chlorine (EAC) may simultaneously contribute to the degradation of pharmaceutical compounds. The mechanisms initiated by HOCl for the degradation of organic pollutants are one-electron oxidation, H-abstraction, and adding to unsaturated C–C bonds, while the main mechanisms of hydroxyl radicals include the latter two [70]. Although RCS are powerful oxidants, their addition to organic compounds results in the production of organochlorinated compounds, which are toxic and recalcitrant. In fact, the most important drawback of chlorine-based processes is the production of toxic and stable by-products (organochlorinated compounds), which are highly dependent on the concentration of chlorine, the composition of water, reaction time, the solution pH, and the nature of the pollutant [71].

12.4.2.1   Operating Parameters for EAC (Anode Materials, Current Density, pH, and Cl⁻ Concentration)

Pharmaceutical degradation by EAC depends on various parameters, including electrode materials, current density, pH, and Cl⁻ concentration. Several studies have claimed that active anodes (IrO_2, RuO_2, and Pt) are effective electrodes for chlorine evolution thanks to their electrocatalytic properties [66]. However, it has been reported that PbO_2 and SnO_2 also exhibited high potential for EAC. In this way, Pt showed good performance for chlorine generation, which was comparable to BDD [72]. In another study, RuO_2 was much better than an PbO_2 anode for the degradation of 5-fluorouracil. It should be mentioned that non-active electrodes benefit from ·OH-mediated electrooxidation for the enhanced degradation of organic pollutants. Therefore, both ROS and RCS can contribute to the oxidation of organic pollutants when chloride ion is used as an electrolyte. Apart from electrode materials, the morphological structure of the anode is also influential on the EAC. The flat surface structure reduces the accessibility of the electrolyte to the anode surface; thus, the electrochemical production of chlorine is decreased. Liu et al. showed that hierarchically ordered macroporous RuO_2 generated a greater chlorine concentration compared to flat RuO_2 [73]. In a study, 2D and 3D electrodes were compared for the degradation of CIP. A significant improvement was observed at the 3D electrode (92.9% CIP removal efficiency) compared to the 2D electrode (57.5% CIP removal efficiency) [74].

The solution pH is an influential factor in chlorine species produced in the EAC process. The solution pH affects the generation rate, efficiency, and stability of chlorine species. Accordingly, when pH < 1, a mixture of Cl_2 and HOCl is dominant, where Cl_2 has a greater contribution. When 1 < pH < 4, HOCl is predominant over Cl_2; at 4 < pH < 7.5, there is a mixture of HOCl and ClO⁻ where HOCl dominates; at pH > 7.5, the dominant chlorine species is ClO⁻ [75]. Liu et al. reported that as the pH value was raised from 2 to 11, the rate of 5-fluorouracil degradation was reduced noticeably [73]. A Sb_2O_3-doped Ti/RuO_2-ZrO_2 anode was used for the EAC process to degrade cefadroxil under various pH values. The cefadroxil degradation rate followed this order based on the solution pH: 2 > 5.5 >> 8 >> 11 [76]. Similar to the AO process in the absence of chloride ions, the current density is a key parameter for the EAC, in which an increase in current density boosts the chlorine generation and the degradation of pharmaceutical compounds consequently. Accordingly, CIP degradation time was reduced from 10 to 5 min by increasing current density from 5 to 10 mA/cm² [77]. However, higher current density subsequently increases energy consumption. Therefore, current density should be carefully governed to balance degradation efficiency and energy consumption for the elimination of pharmaceutical compounds [78]. Higher chloride ion contents enhance the formation of active chlorine species and boost the rapid degradation of pharmaceutical pollutants. Excessive chloride ions may inhibit the degradation of pharmaceutical pollutants since chloride ions may be adsorbed on the anode surface, hindering the mass transfer of the pharmaceuticals toward the anode, which limits degradation efficiency. Moreover, the increase in chloride ions may enhance the corrosion of the electrodes, which in turn decreases the performance. It should be noted that the presence of excess Cl⁻ results in several parasite reactions for the scavenging of ·OH and consequently predominating RCS with less reactivity. Salicylic acid degradation was insignificant when chloride ions increased from 50 to 80 mM [79]. However, a wide range of chloride concentrations (20–100 mM) have been used in EAC processes for the degradation of pharmaceuticals. Also, an increase in electrolyte concentration reduces electrical energy consumption significantly, while it can increase the cost of chemicals. Hence, the optimization of NaCl concentration (electrolyte) is a key point to reach a balance between efficiency and cost. Table 12.2 depicts the application of the EAC process for the removal of pharmaceutical pollutants.

12.4.3 Advanced ·OH-Mediated AO and EAC for the Degradation of Pharmaceutical Pollutants

The AO process is improved by UV irradiation, especially when AO produces H_2O_2, O_3, and persulfate, in which the enhanced generation of sulfate radicals and hydroxyl radicals occurs through the cleavage of the O–O bond of peroxygens. Several works have reported these mechanisms by a BDD anode, while there are only a few studies using metallic anodes for advanced AO to abate pharmaceutical pollutants. EAC can be promoted by applying other methods to enhance the degradation process and shorten the reaction time. Regarding the in situ electrochemical generation of chlorine, UV irradiation can activate chlorine to generate hydroxyl radicals and chlorine radicals based on the following reactions [88].

$$HOCl + UV \rightarrow {}^{\bullet}OH + Cl^{\bullet} \tag{12.36}$$

$$OCl^- + UV \rightarrow O^{\bullet -} + Cl^{\bullet} \tag{12.37}$$

$$O^{\bullet -} + H_2O \rightarrow {}^{\bullet}OH + HO^- \tag{12.38}$$

Recently, EAC/UV process with a RuO_2/IrO_2-Ti anode has been used for the degradation of trimethoprim, ciprofloxacin, metoprolol, and carbamazepine. The results showed that the EAC/UV process was more effective than EAC and Cl_2/UV processes for the degradation of pharmaceuticals. The generation of free radicals (·OH, Cl·, and $Cl_2^{\bullet -}$) enhanced the degradation process. Moreover, the production of chlorate as a hazardous by-product in the EAC/UV process was less than in the Cl_2/UV process. EAC/UV reduced the electrical energy per order (EE/O) significantly compared to Cl_2/UV and EAC processes. As a conclusion, the electrogeneration of chlorine by NaCl is an economical method compared to the use of chlorine solution [78]. In another study, CIP was completely degraded within a short time (5 min) in the EAC/UV process with a Ti/TiO_2-RuO_2 anode. This study confirmed that ·OH, Cl·, and $Cl_2^{\bullet -}$ contributed to CIP degradation in the EAC/UV process [77]. Zhang et al. reported that ClO· and $Cl_2^{\bullet -}$ were the major agents produced in the EAC/UV process for the degradation of pharmaceutical compounds [89]. Recently, US irradiation has been combined with electrooxidation for the degradation of a variety of pollutants, while few studies have focused on pharmaceutical removal using metallic electrodes. US can break down water molecules to produce hydroxyl radicals and hydrogen radicals. In US/AO, US irradiation may liberate the primary hydroxyl radicals from the surface of the anode to bulk solution (Equation 12.39). The US waves can expedite the mass transfer rate on the anode surface, clean the surface incessantly, and activate the electrode surfaces consequently. In fact, US allows the hydroxyl radicals to diffuse into the solution rapidly, increasing the chance of the direct reaction of pharmaceutical pollutants with $^{\bullet}$OH [90, 91]. However, considering the very short lifetime of HO$^{\bullet}$, more contribution of HO$^{\bullet}$ in the medium is very likely associated with the breaking of water molecule using US irradiation.

$$M\left(^{\bullet}OH\right)_{adsorbed} +))) \rightarrow M + {}^{\bullet}OH_{Bulk} \tag{12.39}$$

Around 91% of ofloxacin was eliminated using AO/US with a Ti/RuO$_2$ electrode, while a 35% synergistic effect was observed for this combination [90]. This synergy was around 11% when the Ti/PbO$_2$ anode was combined with the US process for carbamazepine removal [92]. US-assisted electrooxidation was applied for a real pharmaceutical wastewater using a Ti/RuO$_2$ anode to remove COD. There was a synergistic effect for a US/electrooxidation process in which chlorine evolution occurred in the solution owing to the generation of HOCl and Cl$_2$. Furthermore, 69% of COD was removed, while phytotoxicity was significantly reduced after treatment [93]. Recently, a particle electrode was added into a conventional two-dimensional (2D) electrode reactor, in which the particle electrode was filled between the anode and cathode. In the 3D electrode reactor, the particle electrode extremely enhanced the electrode area to react effectively with ROS and pollutants, thereby reducing the mass transfer distance and increasing the conductivity of the solution [94]. Various materials have been used for this purpose, among which carbon-based materials (activated carbon) are conventionally employed in the 3D electrode reactor. To increase catalytic activity, various metals are loaded on activated carbon-based materials. Mn and Co were loaded on granular activated carbon (GAC) to enhance AMX degradation. A dramatic enhancement in AMX degradation was observed in Mn-Co/GAC compared to GAC and the 2D electrode reactor [94]. The honeycomb block of AC@Ti-Cu-Ni-Zn-Sb-Mn was filled between the Ti/RuO$_2$-IrO$_2$-Ta$_2$O$_5$ anode and the stainless steel cathode to improve the degradation of P-aminophenol. The presence of various metals increased the catalytic degradation of P-aminophenol compared to AC, in which 99.87% removal efficiency was obtained after 120 min reaction time [95]. Ti-Sn-Sb/γ-Al$_2$O$_3$ exhibited high catalytic performance in the 3D electrode reactor for oxytetracycline degradation with a dose of 25 g/L [96]. Chloramphenicol [97], estriol [98], and diclofenac [99] were eliminated by the particle electrodes of Ti-Sn/γ-Al$_2$O$_3$, graphite, and GAC, respectively. Overall, using enhanced AO and EAC methods for the degradation of pharmaceuticals should have a good cost benefit.

12.5 EC AND EO APPLICATIONS WITH METALLIC ELECTRODES FOR REAL WASTEWATER TREATMENT

The application of EC and EO for the treatment of real wastewater containing pharmaceuticals has been less studied. The presence of organic and inorganic compounds in real effluent may affect the performance of EC and EO processes. In addition, the assessment of electrical energy consumption (EEC) can be an indication for the feasibility studies of EC and EO on a large scale. In general, municipal wastewater, hospital wastewater, and industrial pharmaceutical wastewater have been considered effluents containing pharmaceutical compounds. In this way, EC and photo/EC with iron electrodes were utilized for the treatment of pharmaceutical wastewater. To achieve a 4% increase in COD removal efficiency, EEC was significantly increased (~3 times, 90.8 kWh/kg COD) for the EC/photo process, which was associated with UV irradiation [46]. In another work, EC and EC/H_2O_2 were sequentially applied for real pharmaceutical wastewater treatment, and 2 h of EC and 1 h of EC/H_2O_2 could remove only 26.84% TOC with an EEC of 347.68 kWh/kg TOC [100]. However, a low EEC (1.5 kWh/m^3) for the EC process has been reported to remove 78% of COD from hospital wastewater [101]. In the case of EO, EAC with a Ti/RuO_2 anode eliminated 81% of COD from actual pharmaceutical wastewater, while high electrical energy was consumed (~2000 kWh/kg COD) [102]. EO processes for the treatment of real wastewaters usually need a longer reaction time compared to EC processes. A Ti/Ta_2O_5-SnO_2 anode was employed for the oxidation of carbamazepine from a biological effluent. Complete degradation (EEC of 109.4 kWh/m^3) was obtained in the absence of a supporting electrolyte [103]. In electrochemical processes, EEC is a key factor because most of the process cost is related to electrical energy and the electrode, since real wastewater may provide enough electrolyte for the reaction time. EC sludge is another problem that should be properly managed. EC sludge is an Fe/Al-rich material that can be reused and recovered for various utilizations. EC sludge can be applied as a raw material for manufacturing building blocks and ceramic pigments [104], a catalyst for the activation of oxidants [105], an adsorbent, and a fertilizer [106]. Iron-based EC sludge can be a win–win strategy for the catalytic activation of oxidants in sequential processes after the EC method.

12.6 FUTURE PROSPECTS AND CHALLENGES

Electrochemical processes can be employed for the treatment of pharmaceutical wastewater in the foreseeable future. In the last decade, with the increase in the use of renewable energies, providing electricity is not an unsolvable problem for electrochemical processes. Therefore, nowadays, the application of electrochemical processes is considered a practical and feasible method for the treatment of contaminated water. Another matter in EO is the electrode and its related issues. The synthesis of metallic electrodes for AO is costly in terms of the instruments and metals used. Moreover, the passivation and fouling of electrodes are the main limitations for the removal of organic pollutants. In the case of the EC process, the generation of Al- and Fe-rich sludge is an important challenge. Hence, sludge management, including recovery, dewatering, and disposal methods, are serious subjects of future investigation. The stability of the electrodes is a crucial factor for scaling up EO processes. New materials and nanotechnology can develop new electrodes with high porosity to improve electrocatalytic features and stability to reduce the leaching of toxic metals. In fact, catalytic activity can be manipulated by tailoring the structure of the electrodes to modify the surface of the electrode. Numerous studies are expected to focus on nanomaterial-assisted electrochemical processes for wastewater treatment. In the case of electrolytes, pharmaceutical wastewater probably contains high total dissolved solids (TDS), which in turn can reduce energy consumption and the need for chemicals (as supporting electrolytes). However, the selection of supporting electrolyte and electrode materials is a key point for electrochemical remediation since both factors determine the mechanism of the generation of reactive species. In EAC, degradation pathways and monitoring by-products are very important because of the generation of halogenated organic compounds as by-products of a chloride-based electrolyte. Indeed, the toxicity evaluation and measurement of halogenated compounds, chlorate, and perchlorate should be considered. In addition, the treatment of real pharmaceutical wastewater should be examined by electrochemical processes to demonstrate the performance of said processes to eliminate low concentrations of pharmaceutical pollutants due to mass transfer limitation and competition reactions with other pollutants. Combining AO, EAC, and EC processes with other methods (UV, US, 3D electrodes, oxidants, and photo-anodes) is increasing to accelerate the removal process. However, the cost of these processes and the toxicity of by-products need to be studied profoundly. No chemicals and reagents, easy automation, and remote controllability cause electrochemical processes to be considered effective processes for decentralized wastewater treatment. The application of electrochemical processes on a large scale for this purpose is conceivable.

ACKNOWLEDGMENT

This project has been supported by Research Center for Environmental Contaminants (RCEC), Abadan University of Medical Sciences (Iran), under Contract No. 1401RCEC1537.

REFERENCES

[1] Ghanbari F, Yaghoot-Nezhad A, Wacławek S, et al. Comparative investigation of acetaminophen degradation in aqueous solution by UV/Chlorine and UV/H_2O_2 processes: Kinetics and toxicity assessment, process feasibility and products identification. Chemosphere. 2021;285:131455.

[2] Margot J, Rossi L, Barry DA, et al. A review of the fate of micropollutants in wastewater treatment plants. WIREs Water. 2015;2:457–487.

[3] Carvalho JFD, Moraes JEFD. Treatment of simulated industrial pharmaceutical wastewater containing amoxicillin antibiotic via advanced oxidation processes. Environ Technol. 2021;42:4145–4157.

[4] Ikhlaq A, Kasprzyk-Hordern B. Catalytic ozonation of chlorinated VOCs on ZSM-5 zeolites and alumina: Formation of chlorides. Appl Catal B: Environ. 2017;200:274–282.

[5] Szymonik A, Lach J, Malińska K. Fate and removal of pharmaceuticals and illegal drugs present in drinking water and wastewater. Ecol Chem Eng S. 2017;24:65–85.

[6] Li J, Li W, Liu K, et al. Global review of macrolide antibiotics in the aquatic environment: Sources, occurrence, fate, ecotoxicity, and risk assessment. J Hazard Mater. 2022;439:129628.

[7] Iancu V-I, Radu G-L, Scutariu R. A new analytical method for the determination of beta-blockers and one metabolite in the influents and effluents of three urban wastewater treatment plants. Anal Methods. 2019;11:4668–4680.

[8] Hassani A, Scaria J, Ghanbari F, et al. Sulfate radicals-based advanced oxidation processes for the degradation of pharmaceuticals and personal care products: A review on relevant activation mechanisms, performance, and perspectives. Environ Res. 2023;217:114789.

[9] Sim WJ, Kim HY, Choi SD, et al. Evaluation of pharmaceuticals and personal care products with emphasis on anthelmintics in human sanitary waste, sewage, hospital wastewater, livestock wastewater and receiving water. J Hazard Mater. 2013;248–249:219–227.

[10] Eniola JO, Kumar R, Barakat MA, et al. A review on conventional and advanced hybrid technologies for pharmaceutical wastewater treatment. J Clean Prod. 2022;356:131826.

[11] Crini G, Lichtfouse E. Advantages and disadvantages of techniques used for wastewater treatment. Environ Chem Lett. 2019;17:145–155.

[12] Martínez-Huitle CA, Brillas E. Decontamination of wastewaters containing synthetic organic dyes by electrochemical methods: A general review. Appl Catal B: Environ. 2009;87:105–145.

[13] Carretero DS, Huang C-P, Tzeng J-H, et al. The recovery of sulfuric acid from spent piranha solution over a dimensionally stable anode (DSA) Ti-RuO$_2$ electrode. J Hazard Mater. 2021;406:124658.

[14] Fu R, Zhang P-S, Jiang Y-X, et al. Wastewater treatment by anodic oxidation in electrochemical advanced oxidation process: Advance in mechanism, direct and indirect oxidation detection methods. Chemosphere. 2023;311:136993.

[15] Martínez-Huitle CA, Rodrigo MA, Sirés I, et al. A critical review on latest innovations and future challenges of electrochemical technology for the abatement of organics in water. Appl Catal B: Environ. 2023;328:122430.

[16] Mollah MYA, Schennach R, Parga JR, et al. Electrocoagulation (EC)—science and applications. J Hazard Mater. 2001;84:29–41.

[17] Panizza M, Cerisola G. Direct and mediated anodic oxidation of organic pollutants. Chem Rev. 2009;109:6541–6569.

[18] Zhang J, Zhou Y, Yao B, et al. Current progress in electrochemical anodic-oxidation of pharmaceuticals: Mechanisms, influencing factors, and new technique. J Hazard Mater. 2021;418:126313.

[19] Moussa DT, El-Naas MH, Nasser M, et al. A comprehensive review of electrocoagulation for water treatment: Potentials and challenges. J Environ Manage. 2017;186:24–41.

[20] Zaied BK, Rashid M, Nasrullah M, et al. A comprehensive review on contaminants removal from pharmaceutical wastewater by electrocoagulation process. Sci Total Environ. 2020;726:138095.

[21] Ahmad A, Priyadarshini M, Das S, et al. Electrocoagulation as an efficacious technology for the treatment of wastewater containing active pharmaceutical compounds: A review. Sep Sci Technol. 2022;57:1234–1256.

[22] da Silva SW, Welter JB, Albornoz LL, et al. Advanced electrochemical oxidation processes in the treatment of pharmaceutical containing water and wastewater: A review. Curr Pollut Rep. 2021;7:146–159.

[23] Sirés I, Brillas E. Remediation of water pollution caused by pharmaceutical residues based on electrochemical separation and degradation technologies: A review. Environ Int. 2012;40:212–229.

[24] Jaafarzadeh N, Ghanbari F, Zahedi A. Coupling electrooxidation and Oxone for degradation of 2,4-Dichlorophenoxyacetic acid (2,4-D) from aqueous solutions. J Water Process Eng. 2018;22:203–209.

[25] Zhang Y-F, Zhang C-H, Xu J-H, et al. Strategies to enhance the reactivity of zero-valent iron for environmental remediation: A review. J Environ Manage. 2022;317:115381.

[26] Vik EA, Carlson DA, Eikum AS, et al. Electrocoagulation of potable water. Water Res. 1984;18:1355–1360.

[27] Garcia-Segura S, Eiband MMSG, de Melo JV, et al. Electrocoagulation and advanced electrocoagulation processes: A general review about the fundamentals, emerging applications and its association with other technologies. J Electroanal Chem. 2017; 801:267–299.

[28] Arsand DR, Kümmerer K, Martins AF. Removal of dexamethasone from aqueous solution and hospital wastewater by electrocoagulation. Sci Total Environ. 2013; 443:351–357.

[29] Baran W, Adamek E, Jajko M, et al. Removal of veterinary antibiotics from wastewater by electrocoagulation. Chemosphere. 2018;194:381–389.

[30] Nabgan W, Saeed M, Jalil AA, et al. A state of the art review on electrochemical technique for the remediation of pharmaceuticals containing wastewater. Environ Res. 2022;210:112975.

[31] Sasson MB, Calmano W, Adin A. Iron-oxidation processes in an electroflocculation (electrocoagulation) cell. J Hazard Mater. 2009;171:704–709.

[32] Nidheesh PV, Scaria J, Babu DS, et al. An overview on combined electrocoagulation-degradation processes for the effective treatment of water and wastewater. Chemosphere. 2021;263:127907.

[33] Bajpai M, Katoch SS, Kadier A, et al. Treatment of pharmaceutical wastewater containing cefazolin by electrocoagulation (EC): Optimization of various parameters using response surface methodology (RSM), kinetics and isotherms study. Chem Eng Res Des. 2021;176:254–266.

[34] Govindan K, Angelin A, Kalpana M, et al. Electrocoagulants characteristics and application of electrocoagulation for micropollutant removal and transformation mechanism. ACS Appl Mater Interfaces. 2020;12:1775–1788.

[35] Zhou R, Liu F, Du X, et al. Removal of metronidazole from wastewater by electrocoagulation with chloride ions electrolyte: The role of reactive chlorine species and process optimization. Sep Purif Technol. 2022;290:120799.

[36] Zaidi S, Chaabane T, Sivasankar V, et al. Electro-coagulation coupled electro-flotation process: Feasible choice in doxycycline removal from pharmaceutical effluents. Arab J Chem. 2019;12:2798–2809.

[37] Ahmadzadeh S, Dolatabadi M. Electrochemical treatment of pharmaceutical wastewater through electrosynthesis of iron hydroxides for practical removal of metronidazole. Chemosphere. 2018;212:533–539.

[38] Yoosefian M, Ahmadzadeh S, Aghasi M, et al. Optimization of electrocoagulation process for efficient removal of ciprofloxacin antibiotic using iron electrode; kinetic and isotherm studies of adsorption. J Mol Liq. 2017;225:544–553.

[39] Barışçı S, Ulu F, Sillanpää M, et al. Evaluation of flurbiprofen removal from aqueous solution by electrosynthesized ferrate(VI) ion and electrocoagulation process. Chem Eng J. 2015;262:1218–1225.

[40] Yazdanbakhsh AR, Massoudinegad MR, Eliasi S, et al. The influence of operational parameters on reduce of azithromyin COD from wastewater using the peroxi-electro-coagulation process. J Water Process Eng. 2015;6:51–57.

[41] Bu L, Zhou S, Shi Z, et al. Iron electrode as efficient persulfate activator for oxcarbazepine degradation: Performance, mechanism, and kinetic modeling. Sep Purif Technol. 2017;178:66–74.

[42] Ensano BMB, Borea L, Naddeo V, et al. Removal of pharmaceuticals from wastewater by intermittent electrocoagulation, water, 2017.

[43] Oulebsir A, Chaabane T, Tounsi H, et al. Treatment of artificial pharmaceutical wastewater containing amoxicillin by a sequential electrocoagulation with calcium salt followed by nanofiltration. J Environ Chem Eng. 2020;8:104597.

[44] Jin X, Xie X, Liu Y, et al. The role of synergistic effects between ozone and coagulants (SOC) in the electro-hybrid ozonation-coagulation process. Water Res. 2020; 177:115800.

[45] Muttaqin R, Pratiwi R, Ratnawati, et al. Degradation of methylene blue-ciprofloxacin and hydrogen production simultaneously using combination of electrocoagulation and photocatalytic process with Fe-TiNTAs. Int J Hydrogen Energy. 2022;47:18272–18284.

[46] Farhadi S, Aminzadeh B, Torabian A, et al. Comparison of COD removal from pharmaceutical wastewater by electrocoagulation, photoelectrocoagulation, peroxi-electrocoagulation and peroxi-photoelectrocoagulation processes. J Hazard Mater. 2012; 219–220:35–42.

[47] Al-Qodah Z, Al-Shannag M. On the performance of free radicals combined electrocoagulation treatment processes. Sep Purif Rev. 2019;48:143–158.

[48] Samarghandi MR, Shahbazi Z, Bahadori R, et al. The survey of ultrasound—electrocoagulation process in removal of Ciprofloxacin from aqueous through central composite design. J Behdasht dar Arseh. (i.e., Health in the Field). 2018;6:9–19.

[49] Jiang Y, Zhao H, Liang J, et al. Anodic oxidation for the degradation of organic pollutants: Anode materials, operating conditions and mechanisms. A mini review. Electrochem Commun. 2021;123:106912.

[50] Hu J, Bian X, Xia Y, et al. Application of response surface methodology in electrochemical degradation of amoxicillin with Cu-PbO$_2$ electrode: Optimization and mechanism. Sep Purif Technol. 2020;250:117109.

[51] Wang Y, Zhou C, Wu J, et al. Insights into the electrochemical degradation of sulfamethoxazole and its metabolite by Ti/SnO$_2$-Sb/Er-PbO$_2$ anode. Chin Chem Lett. 2020;31:2673–2677.

[52] Dai Q, Zhou J, Weng M, et al. Electrochemical oxidation metronidazole with Co modified PbO$_2$ electrode: Degradation and mechanism. Sep Purif Technol. 2016; 166:109–116.

[53] Xia Y, Dai Q, Weng M, et al. Fabrication and electrochemical treatment application of an Al-Doped PbO$_2$ electrode with high oxidation capability, oxygen evolution potential and reusability. J Electrochem Soc. 2015;162:E258.

[54] Li X, Duan P, Lei J, et al. Fabrication of Ti/TiO$_2$/SnO$_2$-Sb-Cu electrode for enhancing electrochemical degradation of ceftazidime in aqueous solution. J Electroanal Chem. 2019;847:113231.

[55] Liang S, Lin H, Yan X, et al. Electro-oxidation of tetracycline by a Magnéli phase Ti$_4$O$_7$ porous anode: Kinetics, products, and toxicity. Chem Eng J. 2018;332:628–636.

[56] Feng L, van Hullebusch ED, Rodrigo MA, et al. Removal of residual anti-inflammatory and analgesic pharmaceuticals from aqueous systems by electrochemical advanced oxidation processes. A review. Chem Eng J. 2013;228:944–964.

[57] Wu J, Zhang H, Oturan N, et al. Application of response surface methodology to the removal of the antibiotic tetracycline by electrochemical process using carbon-felt cathode and DSA (Ti/RuO$_2$—IrO$_2$) anode. Chemosphere. 2012;87:614–620.

[58] Wang Y, Shen C, Zhang M, et al. The electrochemical degradation of ciprofloxacin using a SnO$_2$-Sb/Ti anode: Influencing factors, reaction pathways and energy demand. Chem Eng J. 2016;296:79–89.

[59] Ganiyu SO, Oturan N, Raffy S, et al. Sub-stoichiometric titanium oxide (Ti$_4$O$_7$) as a suitable ceramic anode for electrooxidation of organic pollutants: A case study of kinetics, mineralization and toxicity assessment of amoxicillin. Water Res. 2016; 106:171–182.

[60] Liu S, Liu R, Zhang Y, et al. Development of a 3D ordered macroporous RuO$_2$ electrode for efficient pyrazole removal from water. Chemosphere. 2019;237:124471.

[61] Zhou C, Wang Y, Chen J, et al. High-efficiency electrochemical degradation of antiviral drug abacavir using a penetration flux porous Ti/SnO$_2$—Sb anode. Chemosphere. 2019;225:304–310.

[62] Sopaj F, Rodrigo MA, Oturan N, et al. Influence of the anode materials on the electrochemical oxidation efficiency. Application to oxidative degradation of the pharmaceutical amoxicillin. Chem Eng J. 2015;262:286–294.

[63] Karaoğlu AG, Öztürk D, Akyol A, et al. PCT degradation with electrooxidation (EO$_x$) and ultrasound (US) hybrid process using different type electrodes: BDD, Ti/PbO$_2$ and Ti/Pt. Sep Purif Technol. 2023;311:123313.

[64] Zambrano J, Park H, Min B. Enhancing electrochemical degradation of phenol at optimum pH condition with a Pt/Ti anode electrode. Environ Technol. 2020;41:3248–3259.

[65] Deborde M, von Gunten U. Reactions of chlorine with inorganic and organic compounds during water treatment—Kinetics and mechanisms: A critical review. Water Res. 2008;42:13–51.

[66] Martínez-Huitle CA, Ferro S. Electrochemical oxidation of organic pollutants for the wastewater treatment: Direct and indirect processes. Chem Soc Rev. 2006;35:1324–1340.

[67] Yang Y, Shin J, Jasper JT, et al. Multilayer heterojunction anodes for saline wastewater treatment: Design strategies and reactive species generation mechanisms. Environ Sci Technol. 2016;50:8780–8787.

[68] Cho K, Qu Y, Kwon D, et al. Effects of anodic potential and chloride Ion on overall reactivity in electrochemical reactors designed for solar-powered wastewater treatment. Environ Sci Technol. 2014;48:2377–2384.

[69] Luna-Trujillo M, Palma-Goyes R, Vazquez-Arenas J, et al. Formation of active chlorine species involving the higher oxide MO$_{x+1}$ on active Ti/RuO$_2$-IrO$_2$ anodes: A DEMS analysis. J Electroanal Chem. 2020;878:114661.

[70] García-Espinoza JD, Mijaylova-Nacheva P, Avilés-Flores M. Electrochemical carbamazepine degradation: Effect of the generated active chlorine, transformation pathways and toxicity. Chemosphere. 2018;192:142–151.

[71] Sánchez-Montes I, Santos GOS, dos Santos AJ, et al. Toxicological aspect of water treated by chlorine-based advanced oxidation processes: A review. Sci Total Environ. 2023;878:163047.

[72] Murugananthan M, Latha SS, Bhaskar Raju G, et al. Anodic oxidation of ketoprofen—an anti-inflammatory drug using boron doped diamond and platinum electrodes. J Hazard Mater. 2010;180:753–758.

[73] Liu R, Wang L, Wu R, et al. Active-chlorine-mediated oxidation of 5-fluorouracil on a hierarchically ordered macroporous RuO$_2$ electrode. Chemosphere. 2022;301:134728.

[74] Simões AJA, Dória AR, Vieira DS, et al. Electrochemical degradation of ciprofloxacin using a coupled 3D anode to a microfluidic flow-through reactor. J Water Process Eng. 2023;51:103443.

[75] Dionisio D, Rodrigo MA, Motheo AJ. Electrochemical degradation of a methyl paraben and propylene glycol mixture: Interference effect of competitive oxidation and pH stability. Chemosphere. 2022;287:132229.

[76] Moreno-Palacios AV, Palma-Goyes RE, Vazquez-Arenas J, et al. Bench-scale reactor for Cefadroxil oxidation and elimination of its antibiotic activity using electro-generated active chlorine. J Environ Chem Eng. 2019;7:103173.

[77] Yañez-Rios AE, Carrera-Crespo JE, Luna-Sanchez RM, et al. The influence of pH and current density on an UV254 photo-assisted electrochemical process generating active chlorine and radicals for efficient and rapid ciprofloxacin mineralization compared to individual techniques. J Environ Chem Eng. 2020;8:104357.

[78] Zhang Y, Wang H, Li Y, et al. Removal of micropollutants by an electrochemically driven UV/chlorine process for decentralized water treatment. Water Res. 2020;183:116115.

[79] Liu Y-H, Chen H-F, Kuo Y-S, et al. Degradation of salicylic acid using electrochemically assisted UV/chlorine process: Effect of operating conditions, reaction kinetics, and mechanisms. J Environ Chem Eng. 2022;10:108714.

[80] Brillas E, Garcia-Segura S, Skoumal M, et al. Electrochemical incineration of diclofenac in neutral aqueous medium by anodic oxidation using Pt and boron-doped diamond anodes. Chemosphere. 2010;79:605–612.

[81] Boukhchina S, Akrout H, Berling D, et al. Highly efficient modified lead oxide electrode using a spin coating/electrodeposition mode on titanium for electrochemical treatment of pharmaceutical pollutant. Chemosphere. 2019;221:356–365.

[82] Qian X, Peng K, Xu L, et al. Electrochemical decomposition of PPCPs on hydrophobic Ti/SnO_2-Sb/La-PbO_2 anodes: Relationship between surface hydrophobicity and decomposition performance. Chem Eng J. 2022;429:132309.

[83] Gholami Shirkoohi M, Tyagi RD, Vanrolleghem PA, et al. Modelling and optimization of psychoactive pharmaceutical caffeine removal by electrochemical oxidation process: A comparative study between response surface methodology (RSM) and adaptive neuro fuzzy inference system (ANFIS). Sep Purif Technol. 2022;290:120902.

[84] Flores-Terreros RR, Serna-Galvis EA, Navarro-Laboulais J, et al. An alternative approach to the kinetic modeling of pharmaceuticals degradation in high saline water by electrogenerated active chlorine species. J Environ Manage. 2022;315:115119.

[85] Carrillo-Abad J, Mora-Gómez J, García-Gabaldón M, et al. Effect of the CuO addition on a Sb-doped SnO_2 ceramic electrode applied to the removal of Norfloxacin in chloride media by electro-oxidation. Chemosphere. 2020;249:126178.

[86] Kaur R, Kushwaha JP, Singh N. Electro-oxidation of Ofloxacin antibiotic by dimensionally stable Ti/RuO_2 anode: Evaluation and mechanistic approach. Chemosphere. 2018;193:685–694.

[87] Abidi J, Clematis D, Samet Y, et al. Influence of anode material and chlorides in the new-gen solid polymer electrolyte cell for electrochemical oxidation—Optimization of Chloroxylenol degradation with response surface methodology. J Electroanal Chem. 2022;920:116584.

[88] Yaghoot-Nezhad A, Wacławek S, Madihi-Bidgoli S, et al. Heterogeneous photocatalytic activation of electrogenerated chlorine for the production of reactive oxygen and chlorine species: A new approach for Bisphenol A degradation in saline wastewater. J Hazard Mater. 2023;445:130626.

[89] Zhang Y, Zhan J, Wang B, et al. Integration of ultraviolet irradiation with electrochemical chlorine and hydrogen peroxide production for micropollutant abatement. Chem Eng J. 2022;430:132804.

[90] Patidar R, Srivastava VC. Mechanistic insight into ultrasound-induced enhancement of electrochemical oxidation of ofloxacin: Multi-response optimization and cost analysis. Chemosphere. 2020;257:127121.

[91] Yang B, Zuo J, Li P, et al. Effective ultrasound electrochemical degradation of biological toxicity and refractory cephalosporin pharmaceutical wastewater. Chem Eng J. 2016;287:30–37.

[92] Tran N, Drogui P, Brar SK. Sonoelectrochemical oxidation of carbamazepine in waters: Optimization using response surface methodology. J Chem Technol Biotechnol. 2015;90:921–929.

[93] Patidar R, Srivastava VC. Ultrasound-induced intensification of electrochemical treatment of bulk drug pharmaceutical wastewater. ACS ES&T Water. 2021;1:1941–1954.

[94] Ma J, Gao M, Liu Q, et al. High efficiency three-dimensional electrochemical treatment of amoxicillin wastewater using Mn—Co/GAC particle electrodes and optimization of operating condition. Environ Res. 2022;209:112728.

[95] Wan J, Zhao F, Meng Y, et al. Three-dimensional electrochemical degradation of p-aminophenol with efficient honeycomb block AC@Ti-Cu-Ni-Zn-Sb-Mn particle electrodes. Sep Purif Technol. 2021;267:118662.

[96] Sun W, Sun Y, Shah KJ, et al. Electrochemical degradation of oxytetracycline by Ti-Sn-Sb/γ-Al$_2$O$_3$ three-dimensional electrodes. J Environ Manage. 2019;241:22–31.

[97] Sun Y, Li P, Zheng H, et al. Electrochemical treatment of chloramphenicol using Ti-Sn/γ-Al$_2$O$_3$ particle electrodes with a three-dimensional reactor. Chem Eng J. 2017;308:1233–1242.

[98] Shen B, Wen X-H, Huang X. Enhanced removal performance of estriol by a three-dimensional electrode reactor. Chem Eng J. 2017;327:597–607.

[99] BAcuña-Bedoya JD, Alvarez-Pugliese CE, Castilla-Acevedo SF, et al. Degradation of diclofenac aqueous solutions in a 3D electrolytic reactor using carbon-based materials as pseudo third electrodes in fluidized bed, anodic and cathodic configurations. J Environ Chem Eng. 2022;10:108075.

[100] Başaran Dindaş G, Çalışkan Y, Çelebi EE, et al. Treatment of pharmaceutical wastewater by combination of electrocoagulation, electro-Fenton and photocatalytic oxidation processes. J Environ Chem Eng. 2020;8:103777.

[101] P R, K C, S A, et al. Optimization of pharmaceutical wastewater treatment by electrocoagulation and dual coagulation. Water Environ J. 2022;36:525–540.

[102] Babu BR, Venkatesan P, Kanimozhi R, et al. Removal of pharmaceuticals from wastewater by electrochemical oxidation using cylindrical flow reactor and optimization of treatment conditions. J Environ Sci Health A. 2009;44:985–994.

[103] Gurung K, Ncibi MC, Shestakova M, et al. Removal of carbamazepine from MBR effluent by electrochemical oxidation (EO) using a Ti/Ta$_2$O$_5$-SnO$_2$ electrode. Appl Catal B: Environ. 2018;221:329–338.

[104] Tezcan Un U, Onpeker SE, Ozel E. The treatment of chromium containing wastewater using electrocoagulation and the production of ceramic pigments from the resulting sludge. J Environ Manage. 2017;200:196–203.

[105] Ghanbari F, Zirrahi F, Olfati D, et al. TiO$_2$ nanoparticles removal by electrocoagulation using iron electrodes: Catalytic activity of electrochemical sludge for the degradation of emerging pollutant. J Mol Liq. 2020;310:113217.

[106] Rajaniemi K, Tuomikoski S, Lassi U. Electrocoagulation sludge valorization: A review. Resources. 2021;10:127.

13 Boron-Doped Diamond Electrodes for Pharmaceutical Removal

*Letícia G.A. Costa, Letícia Milena Gomes da Silva,
José Eudes L. Santos, Tabata Natasha Feijoó
Zambrano, Amanda D. Gondim, Lívia N.
Cavalcanti, Elisama Vieira dos Santos,
and Carlos A. Martínez-Huitle*

13.1 INTRODUCTION

Hydrophilic chemical compounds with biological activity are denominated "pharmaceuticals," and their derivatives exhibit substantial variation in structure, function, behavior, and activity [1, 2]. These compounds undergo metabolic processing in the bodies of humans and animals after their ingestion, and some of them are entirely deteriorated while others are partly decomposed and consequently released. Then, they are introduced into the environment via sewage leaks or the discharge of wastewater, where they harm aquatic life and the quality of both surface and groundwater [3].

Another way that pharmaceuticals pollute aquatic systems, is during the steps related to their production (e.g., fermentation, extraction, chemical synthesis, formulation, and packing). However, other contaminants are also present in the wastewaters from chemical synthesis and fermentation (solvents, refractory organics, pharmaceutical residue, and salts [4]). Therefore, it is important to follow the preparation and discharge of these kinds of compounds when produced. Recently, pharmaceuticals have been identified and quantified by analytical methods in a wide range of environmental matrices, including surface water, sub-surface water, drinking water, waste effluents, and sediments.

In the case of water pollution by pharmaceuticals, there are two main situations that should be considered: monitoring by analytical methods and remediation technologies. In the former, the pharmaceuticals have been qualitatively identified and quantified in a wide range of environmental matrices [5]. Meanwhile, several traditional techniques have been tested for wastewater treatment, but these have been insufficient in the latter.

Two of the main issues in wastewater treatment are the utilization of low-cost technologies and ensuring safety and health requirements in the recycled treated effluents.

The utilization of low-cost technologies and assurance of high water quality are issues in wastewater treatment processes. In this frame, the adsorption process is an

DOI: 10.1201/9781003340164-13

"

excellent alternative [6–9] for water pollution prevention and wastewater management because of its cheap cost, minimal maintenance requirements, high removal efficiency, and lack of unwanted by-products. However, materials science and engineering areas have demonstrated that advanced alternatives can be developed to remove pharmaceuticals from water matrices with comparable techno-economic features other than adsorption.

13.2 DANGERS TO THE ENVIRONMENT

Recently, it was determined that the wastewater treatment to eliminate pharmaceuticals from water is also dependent on the physicochemical properties of these compounds, which is completely related to the specific biological effects of pharmaceuticals because of their functional structural groups. Consequently, the pharmaceuticals vary in a wide range of types, like antibiotics, hormones, analgesics (anti-inflammatory drugs and opioids), anti-depressants, lipid regulators, and anti-cancer, cytostatic, anti-seizure, beta-blocker, and anti-diabetic agents, and so on.

The drugs are designed, in terms of their intrinsic biological power, to favor easy absorption in living organisms and active interaction with them to induce continuous activity in biological systems [10] at low concentrations. Continuous biological activity for a longer period of time (non-biodegradability) is an important scope in the synthesis of pharmaceuticals; however, this has the potential to pose serious biological risks to the equilibrium of the ecosystem and human health [11]. For example, different activities can significantly contribute to the release of pharmaceuticals into the environment, such as human excretion, improper disposal of expired or unused medication, and wastewaters from hospitals, pharmaceutical factories, research facilities, fisheries, poultry, and livestock farming [12]. Water and soil contamination are the results of sewage being dumped into bodies of water or being reused for irrigation purposes. As a result, the pharmaceutical transfer cycle can lead to hazards that are either synergistic or antagonistic [13] with potential short-term and long-term effects. On the other hand, the drugs are designed to influence a particular organism, but prolonged exposure to low environmental concentrations can have the same effects on different kinds of organisms [14]. From another environmental point of view, the cytotoxic effects could be chronic rather than acute due their lipophilicity, which makes most pharmaceuticals persistent in water, even at low concentrations. Therefore, the chronic side effects of pharmaceuticals are species- and context-specific. Possible molecular, histological, developmental, behavioral, reproductive, and mortal effects of pharmaceuticals on aquatic organisms are all part of the ecotoxicology of pharmaceuticals [15].

Looking at the literature and the environmental investigations, the most common pharmaceuticals found in aquatic environments across the globe are the antibiotics. Antibiotic pollution is harmful because it slows the growth of algae and cyanobacteria and it contributes to the spread of bacteria that are resistant to antibiotics. Some examples have been shown in Chapter 2 along with several negative impacts that the antibiotics and other pharmaceuticals have provoked, such as hormones and endocrine-disrupting chemicals, analgesics and anti-inflammatory drugs (diclofenac, ibuprofen, ketoprofen, paracetamol, naproxen), anti-cancer drugs (5-fluorouracil, cyclophosphamide, methotrexate, bleomycin, and vincristine), anti-epileptic drugs, anti-depressants, anti-viral drugs, blood lipid regulators, β-blockers, and so on.

The pharmaceutical transformation products that are produced during water treatment or in the environment through biodegradation, photodegradation, and oxidation are a major scientific concern because they pose a significant risk to the ecosystem. The by-products of pharmaceutical processing can be just as harmful, if not more so, than the original, unadulterated drugs. Therefore, innovative and effective water treatment technologies should be proposed and developed for the elimination of pharmaceuticals and by-products from different water matrices.

13.3 ELECTROCHEMICAL OXIDATION-BASED TECHNOLOGIES

EO-based techniques have emerged as a highly effective treatment method for various types of organic pollutants. These processes are considered versatile and energy efficient for eliminating pollutants and pathogens from water [16–19], and these offer several advantages compared to conventional sewage treatment methods. These can function at ambient temperature and pressure, require minimal or no use of chemicals, are highly adaptable, and can easily adjust to variations in the composition, flow mode, and reactors [20, 21]. Due to their versatility, EO-based technologies have been extensively proposed for water remediation. Additionally, these can be easily integrated or combined with other techniques. EO-based processes have been utilized in various applications such as water purification, sewage cleanup, groundwater treatment, soil effluent remediation, and disinfection [16, 22–24]. Additionally, other advanced oxidation methods, such as electro-Fenton (EF) processes, sulfite and sulfate radical advanced oxidation, ultrasonication, and ultraviolet irradiation (UV), can be integrated with this technology [25–28].

EO, which is the base oxidation approach for the more complex EO-based technologies, is the most attractive point-of-use application for decentralized wastewater treatment because of its compact system and adaptable structure. The effectiveness of EO in eliminating organic pollutants has been demonstrated through various experiments conducted in laboratory-scale, pilot-scale, and full-scale reactors [16, 20, 29]. Therefore, this simple and effective electrochemical technology has been translated to compose other advanced EO-based technologies.

The investigations performed with different anodic materials and organic pollutants in different synthetic and real water matrices have provided comprehensive analysis on fundamental principles and electrochemical/chemical reactions and have allowed researchers to explore the influence of different operational variables on the process performance. These investigations have also allowed researchers to determine degradation mechanisms (degradation pathways and potential by-products), thermodynamic properties (kinetics, oxidation rates, enthalpy, entropy and activation energies), and techno-economic factors (energy consumption, costs, and technology readiness level (TRL)). Consequently, the investigation of the performance of EO-based technologies has been a great challenge in the last years by treating simulated water matrices (synthetic effluents) and by using real wastewaters due to the complex composition. More precisely, the treatment of real wastewaters can favor a process of transformation or oxidation, which can lead to the creation of hazardous organic compounds or not [19, 30–32]. For example, the electrolysis of effluents containing chloride may lead to the production of harmful chlorate and perchlorate [21, 30, 33]. The key emphasis of electrochemical treatment in real wastewater scenarios has been on the elimination of dissolved organic carbon and organic matter [19, 34], with limited attention given to a reduction in specific organic components.

Recent advancements in EO-based processes have centered around sophisticated electrode and reactor designs, as well as the scaling up of these methods from pilot to full scale. Additionally, there has been a strong emphasis on utilizing renewable energy as the primary driving force behind EO-based treatments [18, 35, 36]. The primary objective of these studies was to bridge the technological disparities between the initial development and widespread implementation of this technology. Nevertheless, the applicability of EO-based oxidation technologies is limited by the high cost of several electrocatalytic materials, the production of hazardous organic by-products from the oxidation of organic pollutants in the effluent, and the electrical energy requirements. Perhaps a significant obstacle to fully commercializing of this technology is the operational expense, including the costs of electrodes and energy requirements, as well as the installation fee [37, 38].

13.4 FUNDAMENTALS OF EO

EO is the most investigated technology among the electrochemical advanced oxidation processes (EAOPs) due to its simplex degradation mechanism as well as its technological structures and prototypes. The efficacy of this electrolytic technique in completely converting a diverse array of organic compounds in different water samples, whether synthetically generated or derived from real wastewater, has been proven. Nevertheless, the effectiveness in water disinfection and decontamination is influenced by the electrocatalytic material (specifically, the anode), along with various operational factors. As a result of these disadvantages, the system has not been able to be fully implemented on an industrial level. However, in the last decade, these limitations have also created new possibilities and challenges [39].

The anode material governs the nature, amount, and oxidative capacity of the electrogenerated oxidants, but the operation conditions significantly impact the adaptability, automation, energy consumption, cost, and design of the devices. Thus, it is crucial to understand the oxidations mechanisms (organic pollutants are completely broken down or selectively converted into other molecules), operating conditions, and the fundamental principles of the electrocatalytic material role [40–44].

An important scheme compiles the route reaction information (Figure 13.1) that describes the electrochemical/chemical reactions involved in the EO process [43], allowing researchers to differentiate between active and non-active anodes [45]. The nature of the anodic materials is predicated on the oxygen evolution potential of each electrode and the interaction between the anode surface and the hydroxyl radicals produced electrochemically from water discharge, and consequently, it is possible to determine that the selectivity and efficiency of the process are influenced by the composition of the electrode material.

The heterogeneous generation of physisorbed hydroxyl radicals (M(·OH)), see reaction (*a*) in Figure 13.1, at the surface anodes (which are indicated as M) occurs

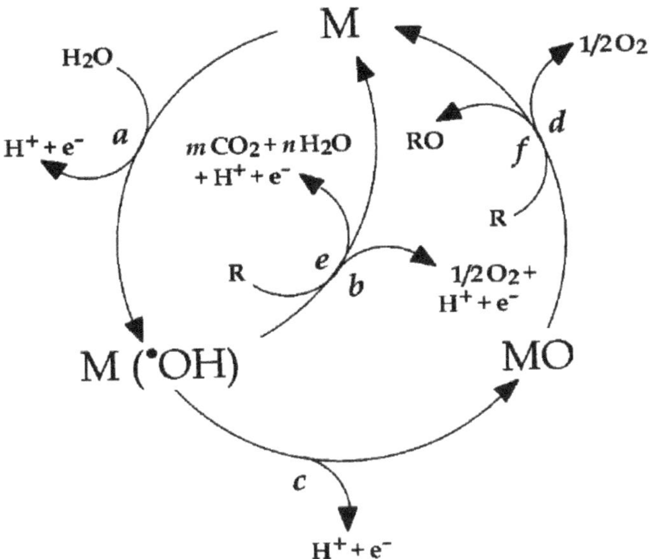

FIGURE 13.1   Scheme of the indirect oxidation of organic compounds with simultaneous oxygen evolution: Water discharge to hydroxyl radicals (path *a*); oxygen evolution by electrochemical oxidation of hydroxyl radicals (*b*); formation of the higher metal oxide (*c*); oxygen evolution by chemical decomposition of the higher metal oxide (*d*); oxidation of the organic compound, R, via hydroxyl radicals (*e*); oxidation of the organic compound via the higher metal oxide (*f*).

Source: Adapted with permission from ref [44]. Copyright Electrochemical Society Inc. 2003.

due to the electrolysis of water. Afterwards, M(•OH) can then interact either strongly or weakly with the anode surface [46], depending on the nature of this material (active or inactive). When M(•OH) is produced via reaction (*a*), these oxidants remain available to react with organic matter in solution; then, the anode is considered non-active due to a weak interaction between the •OH and the electrode surface. Meanwhile, M(•OH) forms higher oxides or superoxides via Equation 13.1, which are denoted as MO in this model, because a strong interaction of the oxidants with the anode surface is attained. These anodes are named as active electrodes and favor the electrochemical conversion of organic matter in solution.

$$M(\cdot OH) \rightarrow MO + H^+ + e^- \tag{13.1}$$

MO, an intermediate species nominated as chemisorbed "active oxygen," is formed when higher oxidation states at the M surface are accessible [46, 47]. Then, as a side reaction, it plays a mediator role through the reaction in Equation 13.2 in the selective oxidation of organic compounds, which takes place concurrently with oxygen evolution (Equation 13.3) and affects the effectiveness of the EO process.

$$MO + R \rightarrow M + RO \tag{13.2}$$

$$MO \rightarrow M + \tfrac{1}{2}\,O_2 \tag{13.3}$$

Equation 13.4 represents the electrochemical incineration of the organic pollutants, which are converted to CO_2 and water when physisorbed or free $M(^{\bullet}OH)$ reacts directly with them.

$$a\,M(^{\bullet}OH) + R \rightarrow M + mCO_2 + nH_2O + xH^+ + e^- \tag{13.4}$$

where R is an organic molecule that contains m carbon atoms but no heteroatoms, requiring $a = (2m + n)$ oxygen atoms to completely mineralize to CO_2. The efficiency of the oxidation reaction (Equation 13.4) may be completely impacted by the O_2 evolution reaction (Equation 13.5) [46].

$$M(^{\bullet}OH) \rightarrow M + \tfrac{1}{2}O_2 + H^+ + e^- \tag{13.5}$$

It is important to remark that the possible anodic reactions (organic oxidation and discharge of water) at non-active or active anodes occur within the reaction cage or near the Nernst layer. In the former case, there are no notable interactions between the reactants, products, and the electrode surface. However, in the latter case, the electrode surface is covered by electrolyzable H_2O molecules, forming at least one adsorbed layer. The diverse hydroxyl radicals generated through water discharge, which participate in the oxidation of organic pollutants, promote electrochemical incineration or combustion in the meantime [46, 48].

13.5 BDD

BDD is a crucial member of the carbon electrode family due to its highly pure carbon sp^3 state, which results in remarkably fascinating electrochemical properties. The interval is represented by the closed interval notation. The electrocatalytic properties of BDD are heavily influenced by factors such as the amount of and type of doping agent (boron, nitrogen, fluor, sulfur, etc.) present on its surface, the characteristics of the substrate materials, the surface termination, the ratio of sp^3 to sp^2 carbon bonds, and the thickness of the BDD layer, as well as its morphological factors and crystallographic orientation. The distinct characteristics of BDD have garnered significant interest, prompting the exploration of its potential as a preferred electrode over conventional options. BDD is applied to support materials such as Si, Ti, Nb, W, etc. using the chemical vapor deposition (CVD) technique. The material should possess favorable characteristics, including a low level of electrical resistance, a high level of mechanical durability, and a lack of reactivity in electrochemical processes. Furthermore, the adaptability and scalability of the treatment process by using BDD films depends on the expense of the substrate material.

13.6 BDD FOR PHARMACEUTICAL DEGRADATION

BDD is a unique and versatile material that has found applications in various fields, including electrochemistry, sensing, and wastewater treatment. In the context of pharmaceuticals, BDD has been explored for its potential applications in electro-chemical sensing and analytical chemistry, as such electrochemical sensors, bio-sensors, and drug delivery systems. In the case of water environments, BDD has been used for water quality monitoring for determining the water quality and detecting trace amounts of pharmaceutical residues in water bodies. This is important for assessing environmental impact and ensuring water safety. On the other hand, BDD has been used as an anode in the EO-based technologies for the electrochemical degradation of pharmaceuticals. This has been a relevant application for addressing concerns about pharmaceutical residues in the environment. However, the most important goal is not only to degrade pharmaceuticals in the context of wastewater treatment but also to achieve mineralization, which involves the complete conversion of organic pollutants into inorganic by-products such as carbon dioxide and water.

The application of BDD in pharmaceutical degradation is an area of ongoing research, and factors such as electrode configuration, operating conditions, and a wide range of pharmaceuticals (at different concentrations and toxicities as well as with diverse environmental impacts), are being investigated. The efficacy of EO with BDD anodes has been demonstrated for a diverse range of pharmaceutical compounds (Table 13.1). Within this framework, some representative examples will be described and discussed.

Brillas and co-workers investigated the EO of paracetamol using a Si/BDD electrode [49] at a wide range of pH values (2.0–12.0), and they found that the rate of oxidation is not affected by the pH of the solution. However, they observed that the oxidation rate increases as the applied current and temperature are increased. In a similar manner, Brillas et al. examined the process of oxidizing diclofenac using BDD and Pt electrodes in a neutral buffer solution with a pH value of 6.5 [50]. BDD electrodes effectively eliminated diclofenac by generating a significant amount of the powerful oxidant BDD($\bullet$OH). The performance of Pt electrodes was hindered by the limited generation of reactive Pt($\bullet$OH) species. Brinzila et al. studied the electrochemical degradation of tetracycline (150 mg L^{-1}) using a BDD electrode in a recirculation reactor operating in an up-flow batch mode (flow rate of 100 L h^{-1}) [51], obtaining 93% of COD removal and 87% of TOC removal efficiencies in 4 h. The average mass transfer coefficients passed from 5.07×10^{-6} m s^{-1} to 8.33×10^{-6} m s^{-1} when an increase in the flow rate was attained from 37 to 100 L h^{-1}. This increase in the mass transfer conditions was attributed to a significant reduction in the width of the diffuse layer, favoring the participation of $\bullet$OH to degrade tetracycline. De Amorim et al. conducted EO of antibiotics with Nb/BDD, specifically sulfamethoxazole and trimethoprim, in a filter press electrochemical reactor [52]. According to the report, the decay rate of the first order was observed under conditions where mass transfer was limited and the applied current exceeded the limiting current during electrolysis. At the optimal pH conditions (~pH 5.0), 36 mA cm^{-2} during 3 h at a 5.0 L min^{-1} of flow rate, COD and TOC were removed by about 90%. The dual function of BDD, electrochemical detection and EO of tetracycline molecules in pharmaceutical wastewater in a specific electrochemical cell setup, was performed by Vasilie and co-workers [53]. The process of tetracycline mineralization was examined through indirect oxidation, and the detection of electrochemical activity was carried out through direct oxidation during the analysis of cyclic voltammetry.

TABLE 13.1

Selected Results Using BDD Electrodes for Treating Pharmaceuticals in Different Water Matrices

Pharmaceutical	System (Anode, Cathode)	Experimental Remarks	Remarks	Ref.
Sulfamethazine	Diachem® BDD (16 cm²)	[Sulfamethazine] = 50 mg L^{-1}; j = 20–120 mA cm^{-2}; pH conditions varying in the range of 3, 4.5, 6.0, 7.5, and 9.0 for 60 min	Comparison with EO, electro-Fenton (EF) and EO/BDD-persulfate (PS) to remove 50 mg L^{-1} of sulfamethazine in 15 min EO/BDD-PS was superior to other technologies using the same reactor.	[54]
Carbamazepine, bisphenol S, propranolol, and sulfamethoxazole	BDD, stainless steel (8 cm²)	[Pollutant] = 5 µM; j = 1–20 mA m^{-2}; T = 25 °C using 5.0 mM NaClO$_4$ as the supporting electrolyte for 30 min	Over 80% of all the contaminants were removed within 30 min. It exhibited a good synergistic effect between persulfate and electrolysis on the degradation of pharmaceuticals.	[55]
Caffeine	BDD, Ti (63.5 cm²)	[Caffeine] = 0.01 M; Na$_2$SO$_4$ 0.05 M; j = 20–60 mA cm^{-2}, 240 min Sodium dodecyl sulfate (SDS) was also used as an additive in the supporting electrolyte.	Synergic behavior between SDS and the sulfate from the supporting electrolyte to produce PS at BDD	[56]
Diclofenac and acetaminophen	BDD, stainless steel (5 cm²)	[Diclofenac] = 170 mg L^{-1} and [acetaminophen] = 190 mg L^{-1}; 0.05 M Na$_2$SO$_4$ at pH 6.5; j = 100, 200, and 300 mA cm^{-2} at 35 °C for 210 min	AO-BDD mineralized diclofenac and acetaminophen. Intensification by coupling EO with photocatalysis	[57]
Florfenicol	Nb/BDD, stainless steel (25 cm²)	[Florfenicol] = 30–50 mg L^{-1}; pH 3, 6, and 9; i = 60–250 mA; different electrolytes (SO$_4^{2-}$, NO$_3^-$, and ClO$_3^-$); 150 min	A 91% TOC removal was observed at 360 min of electrolysis. Radical species (•OH, and SO$_4^{•-}$) played an important role in the drug degradation at pH 3–9, particularly at acidic pH.	[58]
Ampicillin	BDD, platinum (8 cm²)	[Ampicillin] = 1.1 mg L^{-1} at 25 °C for 30 min; 0.1 M Na$_2$SO$_4$ and Na$_2$S$_2$O$_8$ (100 to 500 mg L^{-1}); j = 5–110 mA cm^{-2}	Sodium persulfate seems to be activated electrochemically, thus acting as a source of reactive radicals and significantly enhancing the removal of ampicillin.	[59]

Dexamethasone	BDD, stainless steel (8 cm^2)	[Dexamethasone] = 0.25–2 mg L^{-1}; j = 0.02–0.2 A m^{-2}; and [sodium persulfate] = 50–250 mg L^{-1}	BDD EO completely degrades dexamethasone. The rate of degradation depends on [dexamethasone], j, the cathodic material, and the water matrix.	[60]
4-Aminophenazone	BDD, carbon felt (24 cm^2)	[4-Aminophenazone] = 44.4 mg L^{-1}; i = 30, 60, 120, and 250 mA; 0.05 M Na$_2$SO$_4$ with or without 0.2 mM Fe^{2+}	In all trials, 4-aminophenazone was completely destroyed, following pseudo-first-order kinetics with the rate constant values increasing with j. BDD was superior to the Ti$_4$O$_7$ anode under similar conditions.	[61]
Enrofloxacin	BDD, Ti (2.5 cm^2)	[Enrofloxacin] = 0.44 and 0.05 mmol dm^{-3} in Na$_2$SO$_4$ at pH = 3.0; j = 400 A m^{-2} at 35 °C	Properties of the diamond layer, particularly the ratio diamond/graphite carbon, strongly influence the results of the electrochemical oxidation of enrofloxacin.	[62]
Ciprofloxacin, salbutamol, and sulfamethoxazole	BDD, zirconium (69 cm^2)	[Ciprofloxacin] = 0.0695 mmol L^{-1}; [salbutamol] = 0.0671 mmol L^{-1}; [sulfamethoxazole] = 0.0596 mmol L^{-1} in K$_2$SO$_4$ 0.02 mol L^{-1}; j = 7.25 and 1.45 mA cm^{-2}. The flow rate was 360 L h^{-1} at 30 °C.	Higher removal rates, complete COD and TOC removals by the participation of strong oxidants electrogenerated from salts precursors in the solution	[63]
Tetracycline	Diachem® BDD, Stainless steel (20 cm^2)	[Tetracycline] = 100 and 150 mg L^{-1}; flow rates = 37, 75, and 100 L h^{-1}; supporting electrolyte anhydrous sodium sulfate (5 g L^{-1}) at 25 °C), 300 A m^{-2}, 200 mL of solution	Tetracycline was eliminated after 4 h. No tetracycline detected by HPLC for all experiments. Flow = 100 L h^{-1}; [tetracycline] = 150 mg L^{-1}, 4 h; COD and TOC removals were 93% and 87%, respectively. The organic nitrogen is mainly converted to ammonium, nitrate, and nitrite.	[51]
Tetracycline	BDD, stainless steel (280 cm^2)	[Tetracycline] = 25–100 mg L^{-1}; 0.1 M Na$_2$SO$_4$ in 120 min; pH = 3, 5, and 8; j = 25 to 100 A m^{-2}	Higher j, accelerating the degradation process TC = 25–100 mg L^{-1}, j = 50 A m^{-2} at pH 5 led to a degree of mineralization of 80% in 120 min.	[53]

(Continued)

TABLE 13.1 (*Continued*)

Selected Results Using BDD Electrodes for Treating Pharmaceuticals in Different Water Matrices

Pharmaceutical	System (Anode, Cathode)	Experimental Remarks	Remarks	Ref.
Piroxicam	BDD, Stainless steel (8 cm^2)	[Piroxicam] = 245–975 µg L^{-1}; presence of humic acids (5 and 10 mg L^{-1}) in 0.1 M Na$_2$SO$_4$ j = 2.7 to 80 mA cm^{-2}; pH = 3, 6, and 9; [inorganic ions] = 10 to 200 mg L^{-1} like bicarbonates, chlorides, and nitrates Other experimental variables were all examined.	[Piroxicam] = 245 µg L^{-1} at 26.7 mA cm^{-2} on the BDD anode in 15 min The presence of organic compounds like humic acid reduced the process's efficiency.	[64]
17α-ethynylestradiol	BDD, zirconium (19 cm^2)	[stock estrogen solution] = 1, 100, 800 mg L^{-1}; j = 0.9 to 2.6 mA cm^{-2}; adding 0.1 mol L^{-1} NaCl and Na$_2$SO$_4$ Process applicability was also tested for 17α-estradiol and bisphenol A.	The complete removal can be achieved within 5–7 min at 2.6–2.1 mA cm^{-2}.	[65]
Carboplatin	BDD, stainless steel (25 cm^2)	[Carboplatin] = 0.5 mg L^{-1}; 350 mL of 0.1 M Na$_2$SO$_4$; pH 4–9; j = 15, 30, and 45 mA cm^{-2}	Carboplatin was completely degraded by the BDD in 10 min. Higher degradation rate with an increase in the concentration of the supporting electrolyte (Na$_2$SO$_4$)	[66]
Norfloxacin	BDD, stainless steel (23.75 cm^2)	[Norfloxacin] = 100 mg L^{-1}; 0.1 mol L^{-1} Na$_2$SO$_4$; pH = 3, 7, 10; j = 10, 20, and 30 mA cm^{-2}; temperature = 10, 25, and 40 °C Boron content to sp^2/sp^3 carbon ratio of the BDD anode (100 ppm, 215; 500 ppm, 325; and 2500 ppm, 284).	The oxidation power of the tested BDD anodes is not affected by the values of their boron content to sp^2/sp^3 carbon ratio. Furthermore, in BDD anodes, the intermediate compounds (aromatic or carboxylic acids) that were identified were similar.	[67]
Paracetamol and diclofenac	BDD, stainless steel (64 cm^2)	[Paracetamol and diclofenac] = 50 and 100 mg L^{-1} at pH 3.0 for 3 h Current-controlled electrolysis = 1.56, 3.12, 4.68, and 6.25 mA cm^{-2}	The mineralization process for diclofenac and paracetamol drugs is accompanied by the release of NH$_4^+$ and NO$_3^-$ ions.	[68]

Pharmaceutical	Electrode system	Conditions	Observations	References
Acetaminophen	BDD, Ti/RuO$_2$–TiO$_2$–SnO$_2$ (1 cm^2)	[Acetaminophen] = 100 mg L^{-1} in 0.5 M Na$_2$SO$_4$; j = 20, 30, and 40 mA cm^{-2}	Total mineralization of acetaminophen was achieved on planar and porous BDD electrodes under all employing current densities.	[69]
Aspirin	BDD, Ti/RuO$_2$–TiO$_2$–SnO$_2$	[Aspirin] = 100 mg L^{-1} in 0.1 M Na$_2$SO$_4$; j = 20, 30, and 40 mA cm^{-2}	Aspirin was mineralized completely on BDD electrodes by direct and indirect EO mechanisms.	[70]
Norfloxacin	BDD, stainless steel (12 cm^2)	[Norfloxacin] = 100 mg L^{-1} in 2 g L^{-1} of Na$_2$SO$_4$ with a volume of 250 cm^3; j = 33, 50, and 83 mA cm^{-2}	A 92% removal of TOC but not complete mineralization was achieved by BDD. Residual carboxylic acids in the solution	[71]
Ketoprofen	BDD, Pt (11.25 cm^2)	[Ketoprofen] = 5 µM; j = 4.4, 8.9, and 13.3 mA cm^{-2}; pH = 3.0, 6.0, and 9.0; supporting electrolytes NaCl, Na$_2$SO$_4$ and NaNO$_3$ (0.1 M)	An 80% removal of TOC Formation of refractory chlorinated compounds in the presence of NaCl	[72]
Atenolol	BDD, Pt (11.25 cm^2)	[Atenolol] = 2.25 µM; j = 4.44, 8.88, and 13.3 mA cm^{-2}; pH = 3.0, 6.0, and 9.0 and supporting electrolytes NaCl, Na$_2$SO$_4$, and NaNO$_3$ (0.1 M)	The anodic oxidation of atenolol is significantly influenced by the electrolyte's composition. Atenolol mineralization by BDD with SO$_4^{2-}$ medium is enhanced by applying 13.3 mA cm^{-2}, consuming 57.6 Wh L^{-1}.	[73]
5-fluorouracil	BDD, stainless steel	[5-fluorouracil] = 5 and 50 mg L^{-1}; volumetric flow rate q_v = 4 and 13 L h^{-1}; j = 10, 50, and 150 A m^{-2}; pH = 7.6 and 2 Undivided and divided cells	Under the same operating conditions, the performance of the 5-fluorouracil solution at 50 mg L^{-1} becomes independent of the electrochemical system utilized.	[74]
Losartan	BDD, stainless steel (5 cm^2)	[Losartan] = 0.189 and 0.754 mM; j = 10–80 mA cm^{-2}; pH = 3, 7, and 10; electrolyte = Na$_2$SO$_4$ or KCl at 0.05 M	A 71% removal of TOC under optimal conditions (10 mA cm^{-2}, 0.05 M Na$_2$SO$_4$, pH 7.0, 25 °C, and 360 min of electrolysis) Carboxylic acid intermediates (oxalic, succinic, and oxamic)	[75]
Nitrofurazone	BDD (2 cm^2)	[Nitrafuranzone] = 10–100 mg L^{-1}; j = 0.1 A cm^{-2}; 1.33 A L^{-1} for 30 min; [Cl$^-$] = 0.001–0.010 M	Complete electrochemical degradation of nitrofurazone at BDD anode, leading to a higher rate of active species generation	[76]

(Continued)

TABLE 13.1 (*Continued*)

Selected Results Using BDD Electrodes for Treating Pharmaceuticals in Different Water Matrices

Pharmaceutical	System (Anode, Cathode)	Experimental Remarks	Remarks	Ref.
Losartan and irbesartan	BDD, (un)modified reticulated vitreous carbon	[Losartan and irbesartan] = 10 mg L^{-1} and 500 ng L^{-1}; pH = 3.0; [Na_2SO_4] = 0.05 M; i = 50 mA, 1 h	The in situ produced H_2O_2 significantly enhanced the electro-oxidation rate of losartan and irbesartan. Pseudo-first-order kinetics achieved the complete removal of 500 ng L^{-1} of both compounds in 30 min.	[77]
Ofloxacin	BDD, electrochemically exfoliated graphene-based cathode (24 cm^2)	[Ofloxacin] = 0.1 mM in 0.05 M Na_2SO_4; pH 3; j = 8.3 mA cm^{-2}; flow rate of 0.75 L min^{-1} to maintain an O_2-saturated level	High current densities were found to be beneficial to the generation of BDD($^{\bullet}$OH) because of more converging decay kinetics rates of ofloxacin in the AO process.	[78]
Metformin	BDD, Stainless steel (6 cm^2)	[Metformin] = 358 mg L^{-1}; 2 g L^{-1} Na_2SO_4; j = 5–40 mA cm^{-2} for 4 h; pH adjusted by adding either sulfuric acid or sodium hydroxide solutions (1 mol L^{-1})	COD completely removed under optimal conditions (j = 30 mA cm^{-2}, Na_2SO_4 = 2 g L^{-1}, NaCl = 0.8 g L^{-1}, pH = 2 at 25 °C)	[79]
Benzodiazepine and carbamazepine	BDD, stainless steel (20 cm^2)	[Benzodiazepines] (alprazolam, clonazepam, diazepam, lorazepam) and carbamazepine = 100 µg L^{-1} each; j = 25 to 1000 A m^{-2}; 25 °C; pH = 7.1–7.4; and 1.5 g NaCl L^{-1}	EO was able to completely remove all pharmaceuticals from ultrapure water (after 20 min) and surface water (after 90 min). For municipal wastewater, only 50% was removed. TOC (71%), COD (80%), and total nitrogen (82%) at 1000 A m^{-2}	[80]
Rifampicin	BDD, different cathodes (stainless steel, carbon felt, graphite, and Ti) (16 cm^2)	[Rifampicin] = 0.25 mM in 150 mL; 50 or 100 mM of Na_2SO_4, Na_2CO_3, $NaClO_4$, $NaNO_3$, or NaCl as a supporting electrolyte	Carbon felt was the best cathode for generating H_2O_2. Elevated concentrations of chloride ions in surface and ground water improved the COD removal. The use of Na_2SO_4, Na_2CO_3, or NaCl favors the production of strong secondary oxidant species. Carboxylic acids were the final organic by-products of rifampicin degradation.	[81]

The electrochemical decomposition of diclofenac was recently accomplished through the utilization of BDD and Ta/PbO_2 anodes. Diclofenac was efficiently oxidized by $\cdot OH$ on BDD and Ta/PbO_2, which are considered non-active anodes [82]. The degradation efficiency of diclofenac on the anode composed of Ta/PbO_2 exhibited an upward trend as both the current density and temperature increased (97% of removal efficiency). Also, a decrease in the effectiveness of diclofenac degradation (approximately from 97 to 60%) was observed when the initial diclofenac concentration and pH conditions increased. Mass transport constituted the limiting step of the EO process under optimal conditions, and pseudo-first-order kinetics adequately described the COD removal for both anodes. The diclofenac degradation on the Ta/PbO_2 anode exhibited a higher oxidation rate and current efficiency in comparison to the BDD anode. These results suggest that anodic oxidation using the Ta/PbO_2 electrode holds promise as a viable approach to address pharmaceutical effluents.

Meanwhile, the efficacy of electrolysis employing mixed metal oxide (MMO) and BDD anodes to eliminate penicillin G from urine media was assessed [83]. Diamond anodes were initially used to electrolyze various water matrices (sulfate, chloride, and urine) in order to establish the role of mediated oxidizing species and their mechanisms. Also, it was determined that the water matrix had a substantial impact on the removal of penicillin G at the lowest current density, as follows: chloride > sulfate > urine media. The total mineralization of synthetic effluents was obtained by using BDD anodes in urine media at the lowest current density and with a mineralization efficiency of 68.8 mg_{TOC} A^{-1} h^{-1}, even when initial TOC concentration was 15-fold that of sulfate or chloride media. Electrolysis employing MMO anodes in urine media enabled the comprehensive elimination of penicillin G, though at a comparatively lower efficiency than that achieved with BDD anodes. Likewise, the mineralization process attained about 7.0% at the lower current density because of variations in the electrocatalytic properties. Electrolysis utilizing BDD anodes yields chlorates and perchlorates as by-products, whereas MMO anodes yield solely chlorates for recovery. Nevertheless, the existence of these chlorine species becomes virtually inconsequential at the applied electric charge levels at which penicillin G is entirely eliminated (5 Ah dm^{-3}, irrespective of the water matrix or applied current density ranging from 10 to 100 mA cm^{-2}). Penicillin G is subject to a significantly more pronounced oxidation process than the other organic compounds that are detected in urine. By applying only the electric charge required to deplete the antibiotic, it is possible to generate an effluent with an extremely high concentration of the other organic compounds and a negligible concentration of chlorate, especially when operating at low current densities.

The findings of this research demonstrated that for the degradation of acetaminophen in synthetic solution [84], an interelectrode distance of 500 μm and an optimal current density of 4 mA cm^{-2} were specified for BDD microfluidic electrochemical reactors. In order to optimize the reaction conditions for BDD, which involves heterogeneous oxidation, it is necessary to maximize the contact between the solution and the BDD surface. This should be the case when the interelectrode gap is reduced until an optimal gap is identified in this study, thereby maximizing the "BDD surface/cell volume" ratio. The reduction in ideal mass transfer (12.6×10^{-3} m s^{-1}) caused by the formation of gas bubbles at the cathode and anode at short electrode intervals

(50 to 250 μm) accounts for this phenomenon. Nevertheless, the mass transfer exhibited a magnitude two to ten times that of a traditional EAOP reactor. Under these circumstances, the specific energy required to decompose acetaminophen by 84% could be decreased to 0.18 kWh g^{-1}. Moreover, an increase in solution conductivity from 0.23 to 2.0 mS cm^{-1} resulted in a five-fold decrease in energy demand at 85% pollutant degradation. Nevertheless, the results indicated that electrolysis could be executed utilizing a solution containing a low concentration of supporting electrolyte (e.g., a municipal WWTP outlet with 1 mS cm^{-1}) conductivity, without significant disappearance loss and with greater energy efficiency than with a solution containing a high concentration of conductivity. Hence, the necessity for electrolyte addition could be eliminated, specifically in the pragmatic wastewater experiments that showcased similar energy efficiency (0.15 kWh g^{-1}) for the degradation of pollutants by as much as 60%. This bodes well for the development of microfluidic electrochemical reactors in such fields. Despite this, specific constraints must be overcome in order to maximize the application of microfluidic reactors. These encompass their comparatively diminished capacity for treatment when compared to traditional reactors and their elevated vulnerability to clogging. Current investigation is focused on this impending challenge.

The accuracy of a user-friendly model that predicts the variation in concentration of a target organic during EO with a BDD anode has been validated through experimental results [85]. The contribution of the subsequent types of transfers to the oxidation process has been emphasized: oxygen atoms via reaction with hydroxyl radicals, electrons (via direct oxidation), and strong electrogenerated oxidants. The model that was suggested by Serrano and co-workers [85] demonstrated that a robust correlation among the empirical data collected under different operational conditions, namely sulfate concentrations and current densities, is well fitted. Due to the fact that actual solutions invariably comprise additional organic compounds, an investigation was conducted while considering the existence of a prevalent organic compound (urea) or another pharmaceutical compound. Consistently, experimental findings and theoretical models indicated that the presence of these compounds did not significantly affect the degradation of the target compound, ciprofloxacin. Furthermore, the influence of salts was integrated during the execution phase of this model. In fact, the acceleration of target molecule degradation can occur due to the generation of potent oxidants induced by the polarization of specific anions. For instance, research has demonstrated that the addition of sulfate to electrolysis results in a 50% increase in time required to completely remove the CIP antibiotic. To summarize, the model considers the following input parameters concerning the target molecule: the number of electrons exchanged for direct oxidation, represented as n; the electroactivity on BDD; and the rate of reaction involving the electrogenerated oxidants from the salts, when present (as described and estimated using a pseudo-first-order kinetic model in this investigation). The model's ability to predict the oxidation characteristics of electroactive and non-electroactive organics in single and mixed solutions at various applied current densities is supported by the strong correlation between experimental results and predicted values obtained from model simulations under all conditions.

The EO of 30 pharmaceuticals (including hormones, antibiotics, antihistamines, anti-inflammatory agents, depressants, hypertensives, and ulcer agents) was analyzed in solutions containing 0.05 M Na_2SO_4, 0.05 M NaCl, and 0.05 M Na_2SO_4 + 0.05 M NaCl as supporting electrolyte media at 25 °C. [86]. By applying a current of 40 mA cm^{-2} in the presence of 0.05 M Na_2SO_4 + 0.05 M NaCl, a TOC removal efficiency surpassing 95% was accomplished. The source of this was the hydroxyl radicals and active chlorine species produced at the BDD anode. Twenty-five intermediates that are produced as a result of EO have been identified. Distinct variations in the compounds formed were observed depending on whether the supporting electrolyte was NaCl or Na_2SO_4. Ions of NH_4^+, NO_3^-, and SO_4^{-2} were generated in each electrolytic medium that was utilized. In total, 88% of TOC was removed from a sample of secondary effluent water containing 30 pharmaceuticals using EO at a neutral pH and 6 mA cm^{-2}, without the addition of a supporting electrolyte. The outcomes of this study demonstrated that EO is a viable method for removing emerging concerns regarding organic contaminants, including pharmaceuticals, due to the absence of pH alterations and chemical additions to the wastewater. Moreover, the removal of pharmaceuticals required an energy consumption of 18.95 kW h m^{-3}, or 2.90 USD m^{-3}. This indicates that EO is a cost-effective and efficient method for treating emergent contaminants, including pharmaceuticals, in wastewater and other types of water [86].

Terbuthylazine (TBA) has emerged as a prevalent pesticide in natural waters, having supplanted atrazine in numerous European member states [87]. TBA is an emerging concern compound on account of its persistent nature, toxicity, and established ability to disrupt the endocrine systems of both wildlife and humans. The effectiveness of water treatment plant methods in eliminating this pesticide is only limited, and the production and disposal of by-products, some of which have demonstrated estrogenic properties similar to atrazine, are not adequately addressed. A study conducted by Tasca et al. [87] assessed the effectiveness of an innovative electrochemical cell designed to degrade organic pollutants. The cell utilized a solid polymer electrolyte (SPE) positioned in between a Ti/RuO_2 cathode and a BDD anode. A constant current drives the operation of the cell, which processes an aqueous solution of TBA. A significant increase in the applied current from 100 to 500 mA results in a notable surge in herbicide removal from 42 to 92% within the first 30 minutes of treatment. The maximum degradation occurs when the rate of degradation at 500 mA falls by 97% between 30 and 60 minutes. Conversely, at 100 mA, an augmentation in the initial concentration of TBA from 0.1 to 4 mg L^{-1} leads to a removal efficiency of 89%. Furthermore, the conversion of the herbicide into minor metabolites and desethylterbuthylazine amounts to less than 1%. No sludge is generated, and no chemicals are required. Additional research is warranted due to the potential of this technology to ensure the discharge-free elimination of various emerging pollutants, including those found in personal care products, pharmaceuticals, and pesticides.

Conversely, in order to ascertain the EO of the nonsteroidal anti-inflammatory drug meloxicam (MLX) across a wide pH range using a BDD electrode, two irreversible voltammetric steps are necessary [88]. The accuracy of the proposed mechanism of EO was established through the utilization of HPLC/MS to analyze MLX solutions electrolyzed on a carbon fiber brush electrode. The primary by-products of the initial oxidation step involving two electrons were determined to be benzoisothiazole and thiazole. The monoamide of oxalic acid, which is generated through the oxidation of the aldehyde group of thiazole, is presumed to be an element of the second two energy levels. By integrating BDD with DPV, a voltammetric technique was developed to determine MLX. The scan rate, pulse height, and pulse width, which are fundamental differential pulse voltammetric parameters, were optimized at a pH of 3 in accordance with the first anodic signal in BRB. It was decided that this signal was ideal for analytical purposes. The appropriate BDDE pretreatment method was also suggested. The LOD of 5.9×10^{-8} mol L^{-1} is remarkably low and exhibits a broad range of values (2.5×10^{-7}–8.5×10^{-5} mol L^{-1}) in comparison to the LODs documented for various modified electrodes. Furthermore, it is significantly lower than the LODs achieved with bulk CPE. Recovery rates for this method varied between 97.2 and 101.1% when model solutions were utilized. Pharmaceutical samples were effectively analyzed to validate the applicability of the proposed method to real-world environments.

The research published by a significative cooperation group from Czech Republic and Slovakia presented the outcomes of utilizing thin films composed of polycrystalline BDD in order to eradicate resistant bacteria and water micropollutants [89]. The results of the study indicated that BDD anodes effectively decreased the concentration of more than 51% of the pharmaceuticals and drugs that were analyzed. With regard to coliforms and staphylococci, the level of efficiency approached 100%. The objective of this study was to determine the frequency with which 29 distinct psychoactive substances and their metabolites were detected in 18 wastewater treatment facilities (WWTPs) located in Slovakia and the Czech Republic [89]. It is noteworthy to mention that the geographical region examined accommodates a population exceeding 1.5 million individuals. Cotinine, tramadol, methamphetamine, and venlafaxine were identified as the substances with the highest concentrations in the effluent and input analyses performed with in-line SPE-LC-MS/MS. The degradation efficiency of compounds such as tramadol, venlafaxine, oxazepam, and citalopram was limited to 30% at all the WWTPs that were observed. Additionally, it was discovered that the wastewaters from Slovakia contained antibiotic-resistant staphylococci and coliforms, while the effluents from the Czech Republic did not contain any strains of *E. coli* and *S. aureus* resistant to chloramphenicol and tetracycline [89].

At present, researchers are examining the effect of supporting electrolytes on the EO of aqueous solutions containing carbamazepine (CBZ), sulfamethoxazole (SMX), and propranolol (PRO) [90]. The by-products that were identified and were utilized to propose degradation mechanisms, and the acute toxicity of each electrolyte was assessed in that study. The experiments were carried out in batch mode using a 2 L undivided reactor and an anode composed of Nb/BDD mesh, with an applied current of 2.5 A. Various supporting electrolytes (Na_2SO_4, NaCl, or NaBr) were used to understand the presence of these ionic species at an identical concentration of 7 mM. The findings indicated that the rates of degradation were considerably accelerated in assays that employed NaCl and NaBr. Through the utilization of GC–MS, the reaction by-products were ascertained. Specifically, ten by-products were identified when Na_2SO_4 was employed as the electrolyte. In contrast, 19 by-products (12 non-halogenated and seven halogenated) and 20 by-products (ten non-halogenated and ten halogenated) were identified when NaCl and NaBr were utilized, respectively [90]. It was deduced, on the basis of the by-products produced, that the primary mechanism is the indirect oxidation of electrogenerated reactive halogen species (RHS) using halide ions as electrolytes. The proposed degradation pathways consist of bond rupture and transformation (hydroxylation, deamination, desulfonation, and halogenation), which result in the formation of compounds with a decreased molecular weight. These compounds then undergo further transformations until they are completely degraded. By-products of bromination and chlorination validate halogenation processes. Conversely, the electrogenerated RHS markedly impeded the growth of *Vibrio fischeri*. Nevertheless, the application of Na_2SO_4 as the electrolyte at a pharmaceutical concentration of 5 g L^{-1} did not result in acute toxicity. The supporting electrolyte is of utmost importance in the EO process due to its substantial influence on the rate of degradation, by-product formation, and acute toxicity.

13.7 REMARKS

The use of BDD anodes has allowed researchers to determine that this type of electrode is efficient to eliminate pharmaceuticals from different water matrices. The price of BDD anodes should be not a limitation in designing, constructing, and using diamond electrochemical devices. Conversely, the translation of lab diamond technologies to real applications will allow researchers to commercialize different kinds of electrochemical devices to attend to different environmental problems. Nevertheless, it is evident that the use of diamond anodes efficiently decreases the pharmaceuticals concentrations by about 90%, attending the environmental limits of these compounds in waters. Also, the noxious and toxic effects related to the concentrations of pharmaceuticals dissolved in water can be significantly decreased because the water matrices are successfully decontaminated by the EO approach. More studies are needed in order to determine the possible re-uses of the treated waters as well as the combination of renewable energies to supply electricity to the EO-based technologies.

13.8 ACKNOWLEDGMENTS

Financial support from Conselho Nacional de Desenvolvimento Científico e Tecnológico (CNPq, Brazil) (306323/2018–4, 312595/2019–0, 439344/2018–2, 315879/2021–1, 409196/2022–3, 408110/2022–8), and from Fundação de Amparo à Pesquisa do Estado de São Paulo (Brazil), FAPESP 2014/50945–4 and 2019/13113–4, is gratefully acknowledged.

REFERENCES

[1] Mezzelani M, Gorbi S, Regoli F. Pharmaceuticals in the aquatic environments: Evidence of emerged threat and future challenges for marine organisms. Mar Environ Res. 2018;140:41–60.

[2] Monisha RS, Mani RL, Sivaprakash B, et al. Green remediation of pharmaceutical wastes using biochar: A review. Environ Chem Lett. 2021 20:1 [Internet]. 2021 [cited 2023 Dec 9];20:681–704. Available from: https://link.springer.com/article/10.1007/s10311-021-01348-y.

[3] Yang Y, Ok YS, Kim KH, et al. Occurrences and removal of pharmaceuticals and personal care products (PPCPs) in drinking water and water/sewage treatment plants: A review. Sci Total Environ. 2017;596–597:303–320.

[4] Shi X, Leong KY, Ng HY. Anaerobic treatment of pharmaceutical wastewater: A critical review. Bioresour Technol. 2017;245:1238–1244.

[5] aus der Beek T, Weber FA, Bergmann A, et al. Pharmaceuticals in the environment—Global occurrences and perspectives. Environ Toxicol Chem [Internet]. 2016 [cited 2023 Dec 9];35:823–835. Available from: https://onlinelibrary.wiley.com/doi/full/10.1002/etc.3339.

[6] De Andrade JR, Oliveira MF, Da Silva MGC, et al. Adsorption of pharmaceuticals from water and wastewater using nonconventional low-cost materials: A review. Ind Eng Chem Res [Internet]. 2018 [cited 2023 Dec 9];57:3103–3127. Available from: https://pubs.acs.org/doi/abs/10.1021/acs.iecr.7b05137.

[7] Wang XT, Deng X, Zhang T Di, et al. Biocompatible self-healing hydrogels based on boronic acid-functionalized polymer and laponite nanocomposite for water pollutant removal. Environ Chem Lett [Internet]. 2022 [cited 2023 Dec 9];20:81–90. Available from: https://link.springer.com/article/10.1007/s10311-021-01350-4.

[8] Abdel Maksoud MIA, Fahim RA, Bedir AG, et al. Engineered magnetic oxides nanoparticles as efficient sorbents for wastewater remediation: A review. Environ Chem Lett. 2021 20:1 [Internet]. 2021 [cited 2023 Dec 9];20:519–562. Available from: https://link.springer.com/article/10.1007/s10311-021-01351-3.

[9] Zare EN, Mudhoo A, Khan MA, et al. Water decontamination using bio-based, chemically functionalized, doped, and ionic liquid-enhanced adsorbents: Review. Environ Chem Lett. 2021;19:4 [Internet]. 2021 [cited 2023 Dec 9];19:3075–3114. Available from: https://link.springer.com/article/10.1007/s10311-021-01207-w.

[10] Li Y, Zhang L, Ding J, et al. Prioritization of pharmaceuticals in water environment in China based on environmental criteria and risk analysis of top-priority pharmaceuticals. J Environ Manage. 2020;253:109732.

[11] Rathi BS, Kumar PS, Show PL. A review on effective removal of emerging contaminants from aquatic systems: Current trends and scope for further research. J Hazard Mater. 2021;409:124413.

[12] Pereira A, Silva L, Laranjeiro C, et al. Selected pharmaceuticals in different aquatic compartments: Part I—source, fate and occurrence. Molecules. 2020;25:1026 [Internet]. 2020 [cited 2023 Dec 9];25:1026. Available from: www.mdpi.com/1420-3049/25/5/1026/htm.

[13] Mackuľak T, Černanský S, Fehér M, et al. Pharmaceuticals, drugs, and resistant microorganisms—environmental impact on population health. Curr Opin Environ Sci Health. 2019;9:40–48.

[14] Khan HK, Rehman MYA, Malik RN. Fate and toxicity of pharmaceuticals in water environment: An insight on their occurrence in South Asia. J Environ Manage. 2020;271:111030.

[15] Godoy AA, Kummrow F. What do we know about the ecotoxicology of pharmaceutical and personal care product mixtures? A critical review. Crit Rev Environ Sci Technol [Internet]. 2017 [cited 2023 Dec 9];47:1453–1496. Available from: www.tandfonline.com/doi/abs/10.1080/10643389.2017.1370991.

[16] Pleskov Y V. Electrochemistry of diamond: A review. Russ J Electrochemistry [Internet]. 2002 [cited 2023 Dec 9];38:1275–1291. Available from: https://link.springer.com/article/10.1023/A:1021651920042.

[17] Alfaro MAQ, Ferro S, et al. Boron doped diamond electrode for the wastewater treatment. J Braz Chem Soc. 2006;17:227–236.

[18] Luong JHT, Male KB, Glennon JD. Boron-doped diamond electrode: Synthesis, characterization, functionalization and analytical applications. Analyst [Internet]. 2009 [cited 2023 Dec 9];134:1965–1979. Available from: https://pubs.rsc.org/en/content/articlehtml/2009/an/b910206j.

[19] Kraft A. Doped diamond: A compact review on a new, versatile electrode material. Int J Electrochem Sci. 2007;2:355–385.

[20] Macpherson JV. A practical guide to using boron doped diamond in electrochemical research. Phys Chem Chem Phys [Internet]. 2015 [cited 2023 Dec 9];17:2935–2949. Available from: https://pubs.rsc.org/en/content/articlehtml/2015/cp/c4cp04022h.

[21] Chaplin BP. Critical review of electrochemical advanced oxidation processes for water treatment applications. Environ Sci Process Impacts [Internet]. 2014 [cited 2023 Dec 9];16:1182–1203. Available from: https://pubs.rsc.org/en/content/articlehtml/2014/em/c3em00679d.

[22] McCreery RL. Advanced carbon electrode materials for molecular electrochemistry. Chem Rev [Internet]. 2008 [cited 2023 Dec 9];108:2646–2687. Available from: https://pubs.acs.org/doi/abs/10.1021/cr068076m.

[23] Pelskov Y V., Sakharova AY, Krotova MD, et al. Photoelectrochemical properties of semiconductor diamond. J Electroanal Chem Interfacial Electrochem. 1987;228:19–27.

[24] Haenni W, Rychen P, Fryda M, et al. Chapter 5 Industrial applications of diamond electrodes. Semiconduct Semimet. 2004;77:149–196.

[25] Bundy FP, Hall HT, Strong HM, et al. Man-made diamonds. Nature. 1955 176:4471 [Internet]. 1955 [cited 2023 Dec 9];176:51–55. Available from: www.nature.com/articles/176051a0.

[26] Cobb SJ, Ayres ZJ, Macpherson J V. Boron doped diamond: A designer electrode material for the twenty-first century. https://doi.org/101146/annurev-anchem-061417-010107 [Internet]. 2018 [cited 2023 Dec 9];11:463–484. Available from: www.annualreviews.org/doi/abs/10.1146/annurev-anchem-061417-010107.

[27] Eversole WG. US patents 3,030,187; 3,030,188 [Internet]. 1962 [cited 2023 Dec 10]. Available from: https://scholar.google.com/scholar_lookup?title=US%20patents%203%2C030%2C187%2C%203%2C030%2C188%2C%20filed%20July%2023%2C%201958%2C%20issued%20April%2017%2C%201962%3B%20Canadian%20patent%20628%2C567&publication_year=1961&author=W.G.%20Eversole.

[28] Zavázalová J, Barek J, Pecková K. Boron doped diamond electrodes in voltammetry: New designs and applications: An overview. 2013.

[29] Srikanth VVSS, Sampath Kumar P, Kumar VB. A brief review on the in situ synthesis of boron-doped diamond thin films. Int J Electrochem. 2012;2012:1–7.

[30] Hutton LA, Iacobini JG, Bitziou E, et al. Examination of the factors affecting the electrochemical performance of oxygen-terminated polycrystalline boron-doped diamond electrodes. Anal Chem [Internet]. 2013 [cited 2023 Dec 9];85:7230–7240. Available from: https://pubs.acs.org/doi/full/10.1021/ac401042t.

[31] Ramesham R, Rose MF. Kinetic studies of hydroquinone/quinone at the boron-doped diamond electrode by cyclic voltammetry. J Mater Sci Lett. 1997;16:799–805.

[32] Panizza M, Brillas E, Comninellis C. Application of boron-doped diamond electrodes for wastewater treatment. J Environ Eng Manage [Internet]. 2008;18:19–153. Available from: www.researchgate.net/publication/228801323.

[33] Martin HB, Argoitia A, Landau U, et al. Hydrogen and oxygen evolution on boron-doped diamond electrodes. J Electrochem Soc. 1996;143:L133–L136.

[34] Chen X, Chen G, Yue PL. Anodic oxidation of dyes at novel Ti/B-diamond electrodes. Chem Eng Sci. 2003;58:995–1001.

[35] Yu X, Zhou M, Hu Y, et al. Recent updates on electrochemical degradation of bio-refractory organic pollutants using BDD anode: A mini review. Environ Sci Pollut Res Int [Internet]. 2014 [cited 2023 Dec 9];21:8417–8431. Available from: https://pubmed.ncbi.nlm.nih.gov/24777320/.

[36] Fryda M, Herrmann D, Schafer L, et al. Properties of diamond electrodes for wastewater treatment. New Diam Front C Tec [Internet]. 1999 [cited 2023 Dec 9];9:229–240. Available from: https://infoscience.epfl.ch/record/79837.

[37] Chaplin BP, Wyle I, Zeng H, et al. Characterization of the performance and failure mechanisms of boron-doped ultrananocrystalline diamond electrodes. J Appl Electrochem [Internet]. 2011 [cited 2023 Dec 9];41:1329–1340. Available from: https://link.springer.com/article/10.1007/s10800-011-0351-7.

[38] Garcia-Segura S, Keller J, Brillas E, et al. Removal of organic contaminants from secondary effluent by anodic oxidation with a boron-doped diamond anode as tertiary treatment. J Hazard Mater. 2015;283:551–557.

[39] Cañizares P, Sáez C, Sánchez-Carretero A, et al. Influence of the characteristics of p-Si BDD anodes on the efficiency of peroxodiphosphate electrosynthesis process. Electrochem Commun. 2008;10:602–606.

[40] Martínez-Huitle CA, Ferro S. Electrochemical oxidation of organic pollutants for the wastewater treatment: Direct and indirect processes. Chem Soc Rev [Internet]. 2006 [cited 2023 Dec 9];35:1324–1340. Available from: https://pubmed.ncbi.nlm.nih.gov/17225891/.

[41] Nidheesh PV., Zhou M, Oturan MA. An overview on the removal of synthetic dyes from water by electrochemical advanced oxidation processes. Chemosphere [Internet]. 2018 [cited 2023 Dec 9];197:210–227. Available from: https://pubmed.ncbi.nlm.nih.gov/29366952/.

[42] Panizza M, Cerisola G. Direct and mediated anodic oxidation of organic pollutants. Chem Rev [Internet]. 2009 [cited 2023 Dec 9];109:6541–6569. Available from: https://pubmed.ncbi.nlm.nih.gov/19658401/.

[43] Comninellis C. Electrocatalysis in the electrochemical conversion/combustion of organic pollutants for waste water treatment. Electrochim Acta. 1994;39:1857–1862.

[44] Marselli B, Garcia-Gomez J, Michaud PA, et al. Electrogeneration of hydroxyl radicals on boron-doped diamond electrodes. J Electrochem Soc. 2003;150:D79–D83.

[45] Rajkumar D, Palanivelu K. Electrochemical treatment of industrial wastewater. J Hazard Mater [Internet]. 2004 [cited 2023 Dec 9];113:123–129. Available from: https://pubmed.ncbi.nlm.nih.gov/15363521/.

[46] Fawell J, Ong CN. Emerging contaminants and the implications for drinking water. In: Tortajada C, editor. Water Quality Policy and Management in Asia [Internet]. 1st ed. London: Routledge; 2014 [cited 2023 Dec 9]. pp. 61–77. Available from: www.taylorfrancis.com/chapters/edit/10.4324/9781315829593-5/emerging-contaminants-implications-drinking-water-john-fawell-choon-nam-ong.

[47] Larsson DGJ, de Pedro C, Paxeus N. Effluent from drug manufactures contains extremely high levels of pharmaceuticals. J Hazard Mater [Internet]. 2007 [cited 2023 Dec 9];148:751–755. Available from: https://pubmed.ncbi.nlm.nih.gov/17706342/.

[48] Migliorini FL, Braga NA, Alves SA, et al. Anodic oxidation of wastewater containing the Reactive Orange 16 Dye using heavily boron-doped diamond electrodes. J Hazard Mater [Internet]. 2011 [cited 2023 Dec 9];192:1683–1689. Available from: https://pubmed.ncbi.nlm.nih.gov/21803493/.

[49] Brillas E, Sirés I, Arias C, et al. Mineralization of paracetamol in aqueous medium by anodic oxidation with a boron-doped diamond electrode. Chemosphere. 2005;58:399–406.

[50] Brillas E, Garcia-Segura S, Skoumal M, et al. Electrochemical incineration of diclofenac in neutral aqueous medium by anodic oxidation using Pt and boron-doped diamond anodes. Chemosphere. 2010;79:605–612.

[51] Brinzila CI, Pacheco MJ, Ciríaco L, et al. Electrodegradation of tetracycline on BDD anode. Chem Eng J. 2012;209:54–61.

[52] De Amorim KP, Romualdo LL, Andrade LS. Electrochemical degradation of sulfamethoxazole and trimethoprim at boron-doped diamond electrode: Performance, kinetics and reaction pathway. Sep Purif Technol. 2013;120:319–327.

[53] Vasilie S, Manea F, Baciu A, et al. Dual use of boron-doped diamond electrode in antibiotics-containing water treatment and process control. Process Saf Environ Prot. 2018;117:446–453.

[54] Nashat M, Mossad M, El-Etriby HK, et al. Optimization of electrochemical activation of persulfate by BDD electrodes for rapid removal of sulfamethazine. Chemosphere. 2022;286:131579.

[55] Yao J, Zhang Y, Dong Z. Enhanced degradation of contaminants of emerging concern by electrochemically activated peroxymonosulfate: Performance, mechanism, and influencing factors. Chem Eng J. 2021;415:128938.

[56] Escalona-Durán F, Ribeiro da Silva D, Martínez-Huitle CA, et al. The synergic persulfate-sodium dodecyl sulfate effect during the electro-oxidation of caffeine using active and non-active anodes. Chemosphere. 2020;253.

[57] Peralta-Hernández JM, de la Rosa-Juárez C, Buzo-Muñoz V, et al. Synergism between anodic oxidation with diamond anodes and heterogeneous catalytic photolysis for the treatment of pharmaceutical pollutants. Sustain Environ Res. 2016;26:70–75.

[58] Periyasamy S, Lin X, Ganiyu SO, et al. Insight into BDD electrochemical oxidation of florfenicol in water: Kinetics, reaction mechanism, and toxicity. Diam Relat Mater. 2016;69:152–159.

[59] Frontistis Z, Mantzavinos D, Meriç S. Degradation of antibiotic ampicillin on boron-doped diamond anode using the combined electrochemical oxidation—Sodium persulfate process. J Environ Manage. 2018;223:878–887.

[60] Grilla E, Taheri ME, Miserli K, et al. Degradation of dexamethasone in water using BDD anodic oxidation and persulfate: Reaction kinetics and pathways. J Chem Tech Biotechnol. 2021;96:2451–2460.

[61] Ganiyu SO, Oturan N, Trellu C, et al. Abatement of analgesic antipyretic 4-aminophenazone using conductive boron-doped diamond and sub-stoichiometric titanium oxide anodes: Kinetics, mineralization and toxicity assessment. Chem Electro Chem. 2019;6:1808–1817.

[62] Guinea E, Centellas F, Brillas E, et al. Electrocatalytic properties of diamond in the oxidation of a persistant pollutant. Appl Catal B. 2009;89:645–650.

[63] Lan Y, Coetsier C, Causserand C, et al. On the role of salts for the treatment of wastewaters containing pharmaceuticals by electrochemical oxidation using a boron doped diamond anode. Electrochim Acta. 2017;231:309–318.

[64] Kouskouki A, Chatzisymeon E, Mantzavinos D, et al. Electrochemical degradation of piroxicam on a boron-doped diamond anode: Investigation of operating parameters and ultrasound synergy. Chem Electro Chem [Internet]. 2019 [cited 2023 Dec 11];6:841–847. Available from: https://onlinelibrary.wiley.com/doi/full/10.1002/celc.201800971.

[65] Frontistis Z, Brebou C, Venieri D, et al. BDD anodic oxidation as tertiary wastewater treatment for the removal of emerging micro-pollutants, pathogens and organic matter. J Chem Tech Biotechnol [Internet]. 2011 [cited 2023 Dec 9];86:1233–1236. Available from: https://onlinelibrary.wiley.com/doi/full/10.1002/jctb.2669.

[66] Barışçı S, Turkay O, Ulusoy E, et al. Electrochemical treatment of anti-cancer drug carboplatin on mixed-metal oxides and boron doped diamond electrodes: Density functional theory modelling and toxicity evaluation. J Hazard Mater. 2018;344:316–321.

[67] Coledam DAC, Aquino JM, Silva BF, et al. Electrochemical mineralization of norfloxacin using distinct boron-doped diamond anodes in a filter-press reactor, with investigations of toxicity and oxidation by-products. Electrochim Acta. 2016;213:856–864.

[68] García-Montoya MF, Gutiérrez-Granados S, Alatorre-Ordaz A, et al. Application of electrochemical/BDD process for the treatment wastewater effluents containing pharmaceutical compounds. J Ind Eng Chem. 2015;31:238–243.

[69] He Y, Dong Y, Huang W, et al. Investigation of boron-doped diamond on porous Ti for electrochemical oxidation of acetaminophen pharmaceutical drug. J Electroanal Chem. 2015;759:167–173.

[70] He Y, Huang W, Chen R, et al. Anodic oxidation of aspirin on PbO2, BDD and porous Ti/BDD electrodes: Mechanism, kinetics and utilization rate. Sep Purif Technol. 2015;156:124–131.

[71] Mora-Gomez J, Ortega E, Mestre S, et al. Electrochemical degradation of norfloxacin using BDD and new Sb-doped SnO2 ceramic anodes in an electrochemical reactor in the presence and absence of a cation-exchange membrane. Sep Purif Technol. 2019;208:68–75.

[72] Murugananthan M, Latha SS, Bhaskar Raju G, et al. Anodic oxidation of ketoprofen—An anti-inflammatory drug using boron doped diamond and platinum electrodes. J Hazard Mater. 2010;180:753–758.

[73] Murugananthan M, Latha SS, Bhaskar Raju G, et al. Role of electrolyte on anodic mineralization of atenolol at boron doped diamond and Pt electrodes. Sep Purif Technol. 2011;79:56–62.

[74] Ochoa-Chavez AS, Pieczyńska A, Fiszka Borzyszkowska A, et al. Electrochemical degradation of 5-FU using a flow reactor with BDD electrode: Comparison of two electrochemical systems. Chemosphere. 2018;201:816–825.

[75] Salazar C, Contreras N, Mansilla HD, et al. Electrochemical degradation of the antihypertensive losartan in aqueous medium by electro-oxidation with boron-doped diamond electrode. J Hazard Mater. 2016;319:84–92.

[76] Kuznetsov VV, Ivantsova NA, Kuzin EN, et al. Study of the process of electrochemical oxidation of active pharmaceutical substances on the example of Nitrofurazone ((2E)-2-[(5-Nitro-2-furyl)methylene]hydrazine Carboxamide). Water (Switzerland) [Internet]. 2023 [cited 2023 Dec 11];15:3370. Available from: www.mdpi.com/2073-4441/15/19/3370/htm.

[77] Ali I, Barros de Souza A, Liu Z, et al. Improving the removal of losartan, irbesartan and their transformation products through in situ produced hydrogen peroxide in electrochemical oxidation processes. J Water Proc Engineering. 2023;55:104133.

[78] Yang W, Oturan N, Liang J, et al. Synergistic mineralization of ofloxacin in electro-Fenton process with BDD anode: Reactivity and mechanism. Sep Purif Technol. 2023;319:124039.

[79] Chaabene R, Khannous L, Samet Y. Electrochemical degradation of aqueous metformin at boron-doped diamond electrode: Kinetic study and phytotoxicity tests. Int J Environ Sci Technol [Internet]. 2023 [cited 2023 Dec 11];20:5169–5182. Available from: https://link.springer.com/article/10.1007/s13762-022-04325-2.

[80] Souza-Chaves BM de, Bosio M, Dezotti M, et al. Advanced electrochemical oxidation applied to benzodiazepine and carbamazepine removal: Aqueous matrix effects and neurotoxicity assessments employing rat hippocampus neuronal activity. J Water Proc Engineering. 2022;49:102990.

[81] Brito LRD, Ganiyu SO, dos Santos E V., et al. Removal of antibiotic rifampicin from aqueous media by advanced electrochemical oxidation: Role of electrode materials, electrolytes and real water matrices. Electrochim Acta. 2021;396:139254.

[82] Ji Z, Liu T, Tian H. Electrochemical degradation of diclofenac for pharmaceutical wastewater treatment. Int J Electrochem Sci. 2017;12:7807–7816.

[83] Cotillas S, Lacasa E, Herraiz M, et al. The role of the anode material in selective Penicillin G Oxidation in urine. Chem Electro Chem [Internet]. 2019 [cited 2024 Jan 30];6:1376–1384. Available from: https://onlinelibrary.wiley.com/doi/full/10.1002/celc.201801747.

[84] Mousset E, Puce M, Pons MN. Advanced electro-oxidation with boron-doped diamond for acetaminophen removal from real wastewater in a microfluidic reactor: Kinetics and mass-transfer studies. Chem Electro Chem [Internet]. 2019 [cited 2024 Jan 30];6:2908–2916. Available from: https://onlinelibrary.wiley.com/doi/full/10.1002/celc.201900182.

[85] Lan Y, Coetsier C, Causserand C, et al. An experimental and modelling study of the electrochemical oxidation of pharmaceuticals using a boron-doped diamond anode. Chem Eng J. 2018;333:486–494.

[86] Calzadilla W, Espinoza LC, Diaz-Cruz MS, et al. Simultaneous degradation of 30 pharmaceuticals by anodic oxidation: Main intermediaries and by-products. Chemosphere. 2021;269:128753.

[87] Tasca AL, Puccini M, Clematis D, et al. Electrochemical removal of terbuthylazine:Boron-doped diamond anode coupled with solid polymer electrolyte. Environmental Pollution. 2019;251:285–291.

[88] Šelešovská R, Hlobeňová F, Skopalová J, et al. Electrochemical oxidation of anti-inflammatory drug meloxicam and its determination using boron doped diamond electrode. J Electroanal Chem. 2020;858:113758.

[89] Mackuľak T, Medvecká E, Vojs Staňová A, et al. Boron doped diamond electrode—The elimination of psychoactive drugs and resistant bacteria from wastewater. Vacuum. 2020;171:108957.

[90] García-Espinoza JD, Mijaylova Nacheva P. Effect of electrolytes on the simultaneous electrochemical oxidation of sulfamethoxazole, propranolol and carbamazepine: Behaviors, by-products and acute toxicity. Environ Sci Pollut Res [Internet]. 2019 [cited 2024 Jan 30];26:6855–6867. Available from: https://link.springer.com/article/10.1007/s11356-018-4020-9.

Index